Adhesion of Cells, Viruses and Nanoparticles

Kevin Kendall · Michaela Kendall
Florian Rehfeldt

Adhesion of Cells, Viruses and Nanoparticles

Springer

Professor Kevin Kendall
Chemical Engineering
University of Birmingham
Edgbaston B15 2TT
UK
K.Kendall@bham.ac.uk

Dr Michaela Kendall
Peninsula College of Medicine
and Dentistry
The Knowledge Spa
Truro
Cornwall TR1 3HD
UK
m.kendall@ex.ac.uk

Dr Florian Rehfeldt
Georg-August-Universität Göttingen
3rd Institute of Physics - Biophysics
Friedrich-Hund-Platz 1
37077 Göttingen
Germany
rehfeldt@physik3.gwdg.de

ISBN 978-90-481-2584-5 e-ISBN 978-90-481-2585-2
DOI 10.1007/978-90-481-2585-2
Springer Dordrecht Heidelberg London New York

Library of Congress Control Number: 2010937429

Printed on acid-free paper

Springer is part of Springer Science+Business Media (www.springer.com)

Preface

It is now 100 years since adhesion of cells was shown to be vital for their growth and reproduction. Ross Granville Harrison, Fig. 1, invented the technique for culturing cells, a technique which is now of massive importance for studying genetics, cancer, tissue engineering and disease processes.

Harrison was a 37 year old lecturer at Johns Hopkins in the USA, observing the growth of nerve fibres in embryos, when he found that he could insert solid blood clot material into the animal and the cells would continue to propagate along the

Fig. 1 Ross Granville Harrison 1870–1959[1] (with permission of Royal Society)

foreign material. Subsequently in 1907 he found that the nerve cells would also grow on the blood clot in a dish outside the embryo. In his 1914 paper,[2] he then showed that the shape of the cells depended on the solid substrate by testing the cells on clotted plasma, spider web fibres and glass cover slips. This was the first indication that adhesion was essential for shape and differentiation of cells.

It is interesting that 100 years have also elapsed since the discovery that viruses can cause animal disease. Ellerman and Bang[3] in 1908 showed that leukaemia could be transmitted to chickens by injecting cell free material. A few years later, Rous[4] in 1910 and 1911 showed that solid tumours could be transferred from chicken to chicken to spread the disease and also isolated the infective agent in a cell free filtrate. It later became clear that the virus particles, which at that time could not be imaged by microscopy, were adhering to the cells to cause infection. Rous received the Nobel Prize for this work in 1966, more than 50 years after his observations. There was no model at that time to describe the mechanism by which the virus particle attached to and entered the cell.

The idea that fine particles in smoke caused damage to humans goes back much further: it is said that the city of London imposed smoke control in the thirteenth century because coal fires were 'prejudicial to health'.[5] However, the understanding of the mechanisms of toxicity has only recently emerged. The particles in smoke, which are approximately the same size as cells or viruses, somehow adhere to the lung surfaces and cause organ failure, even heart disease. How does this adhesion process occur? Similarly, it was observed[5] in the nineteenth century that chimney sweeps suffered disease from the soot which contacted them; 'I have known 8 or 9 sweeps lose their lives by the soot cancer. The parts which it seizes are entirely eaten off'. We examine some of the processes which contribute to nanoparticle toxicity in Chapter 11.

Ever since Robert Hooke[6] viewed a slice of cork using his early microscope (Fig. 2), showing for the first time 'Cells distinct from one another', but clearly adhering very strongly to form the strong lightweight porous wood material, we have been fascinated by the adhesion forces which hold large multicellular organisms together. The purpose of this book is to address the description, definition and understanding of these adhesion forces in relation to three systems.

- Inanimate fine particles
- Virus particles
- Cells

In recent times, the theoretical idea which has dominated the field is that of the adhesion molecule, a complex protein like fibronectin for example, as described by Hynes.[7] Such molecules have been thought to control the adhesion of cells. Indeed, an enormous amount of work has been done by thousands of scientists to define various adhesion molecules, whose range, variety, complexity and nomenclature have expanded substantially over the past decades.[8–19] Unfortunately, the 'lock and key' model on which this science has been based, also around a century old,[20] is unacceptable. While there is no doubt that a coating of fibronectin on a surface definitely helps cell adhesion, we aim to show in this book that the adhesion

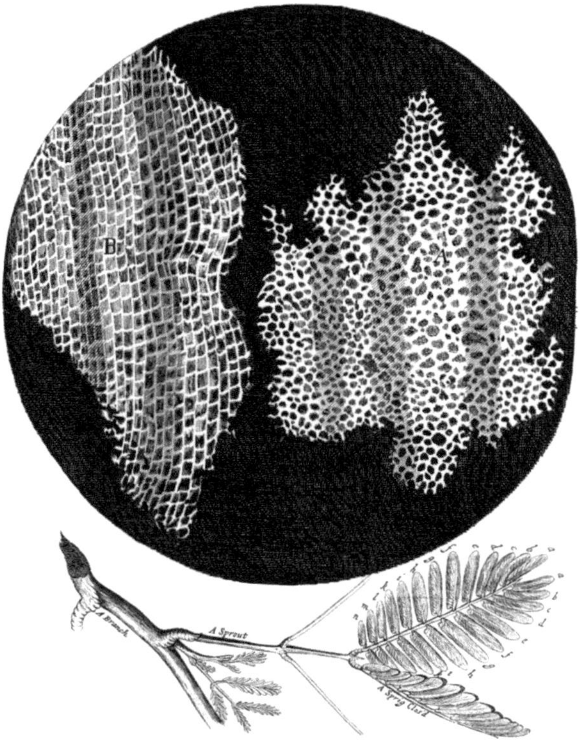

Fig. 2 Illustration from Robert Hooke's book which first showed cells and the adherence between them[6] (with permission of Royal Society)

molecule is only one factor in the equation. Van der Waals forces are the key cause of the adhesion. Substrate elasticity and geometry are also important. In addition we aim to show that there are complex mechanisms such as Brownian motion and surfactant molecules in solution which display major effects.

Originally, the theoretical ideas used in this book were defined[21,22] in 1970–1971. By considering the contact of elastic bodies, it became evident that three parameters generally entered the equation for adhesive force F, as indicated below.

$$F = K\ [WEd^3/(1-\nu^2)]^{1/2} \tag{1}$$

where K was a constant, W the work of adhesion in $\mathrm{Jm^{-2}}$, E the elastic modulus in Pa, ν the Poisson's ratio and d the dimension in metres. From this model, it is clear that the adhesion molecules have an effect on W, but elasticity E,ν is equally influential and the geometry d is much more important. The most surprising thing about this new theory was that adhesion force was strongest when the surfaces were absolutely smooth and clean, with no adhesion molecules present. In other words the effect of adhesion molecules was to reduce the adhesion force, not to cause the adhesion force as intimated in references.[7–19] The purpose of this book is to show that this new theory gives a much more satisfactory account of the results than the simplistic adhesion molecule lock and key models which have dominated during the last century. Part 1 deals with the fundamentals, the phenomenology and the theory, Part II goes on to describe the mechanisms and measurements at both macroscopic and nanoscopic levels, then Part III looks at detailed research in adhesion of nanoparticles, viruses and cells.

In this endeavour, we have been assisted by many colleagues.

In particular KK and MK express thanks to Patricia Kendall for constant support over 40 years.

KK thanks the late David Tabor who set him on the path to studying this field and to Ken Johnson, Alan Roberts and Alan Gent who have been partners since 1966.

MK wishes to thank Morton Lippmann, Bob Maynard, Teresa Tetley and Howard Clark for timely advice, to Uludag University for solid support, and KK for making it all fun.

FR wishes to thank his mentors Erich Sackmann, Motomu Tanaka and Dennis E Discher for continuous support and inspiration and Andre EX Brown and Peter Zwiauer for critical reading and fruitful discussions.

If you have any comments on the ideas expressed here, please email us on k.kendall@bham.ac.uk, m.kendall@ex.ac.uk and rehfeldt@physik3.gwdg.de

February 2010 Kevin Kendall, Michaela Kendall and Florian Rehfeldt

References

1. Abercrombie, M, Biographical Memoirs of fellows of the Royal Society, 7 (1961) 110–126.
2. Harrison, R.G., The reaction of embryonic cells to solid structures, J Expt Zool 17 (1914) 521–44.
3. Ellerman, C. and O. Bang. Centralbl. Bakteriol. 46 (1908) 595–609.
4. Rous, P. J. Exp. Med. 12 (1910) 696–705; Rous, P. J. Exp. Med. 13 (1911) 397–411
5. www.ace.mmu.ac.uk; British Parliamentary Papers, Session No.13, 1863: Report from the Committee of the House of Commons.
6. Hooke, R., Micrographia, Royal Society London 1665, available on www.gutenberg.net.
7. Hynes, R.O., Fibronectins, Springer Verlag, New York 1990.
8. Springer, T.A., Adhesion receptors of the immune system, Nature 346 (1990) 425–34.
9. Hortsch, M., Nott, P., New Cell Adhesion research, Nova Science 2009.
10. Umemori, H., The sticky synapse: Cell adhesion molecules, Springer Berlin 2009.
11. Garrod, D. R., Structure and function in cell adhesion, Portland Press 2008.
12. Cress, A.E., Nagle, R.B., (eds.), Cell adhesion and cytoskeletal molecules in metastasis, Springer, Dordrecht 2006.
13. Beckerle, M.C., (ed.), Cell Adhesion, Oxford University Press 2002.
14. Ley, K., (ed.), Adhesion molecules: function and inhibition, Birkhauser, Basel 2007.
15. Reutter, W., Schuppan, D., Tauber, R., Zeitz, M., Falk symposium, Cell adhesion molecules in health and disease, Kluwer 2003.
16. Coutts, A.S., Adhesion protein protocols, Humana Press London 2nd Ed 2007.
17. Behrens, J., Nelson, W.J., Cell adhesion, Springer Berlin 2004.

18. Collins, T., Leukocyte recruitment, endothelial cell adhesion molecules..., Kluwer Dordrecht 2001.
19. Barker, J., McGrath, J., (eds.), Cell adhesion and migration..., Harwood Academic Amsterdam 2001.
20. Kaufmann, S.H.E., Elie Metchnikoff and Paul Ehrlich's impact on infection biology, Microbes and Infection 10 (2008) 1417–9.
21. Kendall, K., The adhesion and surface energy of elastic solids, J PhysD: Appl Phys 4 (1971) 1186–95.
22. Kendall, K., Molecular adhesion and its applications; the sticky universe, Kluwer, New York 2001, chapter 12.

Contents

Part I
Fundamentals

Chapter 1
Background to Adhesion of Cells, Viruses and Nanoparticles

three principal factors ... the tissue ... the fluid ... the solid support

Harrison (1914)[1]

Adhesion of particles in biological systems is of paramount importance. It allows particles to attach onto cells which may then ingest them; it causes virus parasites to adhere to a cell surface before infecting and taking over the nucleus; it is responsible for cell communication by making contact and exchanging molecules; consequently it is needed for mating of cells; it is related to infection where a cell can invade another colony; it allows amazingly large aggregates of cells to build up, so that about 10^{13} cells are adhering in the human body. Without biological adhesion there would be no multicellular organisms, no complex plants and animals, and therefore no human race.

Typically, a list of adhesion functions is

- Toxicity of particles
- Specificity of virus infection
- Sexual interaction of cells
- Signalling between contacting cells
- Embryonic development
- Formation of nervous system
- Holding tissues together in adults
- Inflammation and wound healing
- Metastasis of tumours

This importance is reflected in the large numbers of papers written in technical journals by specialists in biological adhesion. Around 12,000 papers are written each year containing adhesion as a key word, mainly about adhesion molecules, the large (i.e. 1–10 nm) proteins or glycoproteins which seem to regulate the adhesive interactions between cells. A single academic institution may typically contain ten academic staff working on the adhesion between particles, viruses or cells. Such endeavour builds up across the planet to a total of about 50,000 individuals who are attempting to understand the complexity of adhesion phenomena in biological studies.

K. Kendall et al., *Adhesion of Cells, Viruses and Nanoparticles*,
DOI 10.1007/978-90-481-2585-2_1,

1.1 The Problem of Understanding Adhesion

The problem is that this large number of practitioners would find it difficult to agree on the basic principles of the adhesion which they are studying. A number of errors and fallacies are being propagated in the literature, causing confusion among students and researchers alike. The purpose of this book is an attempt to focus on the basic physical principles of adhesion which this large number of students and researchers can then use to build their arguments.

A key stimulus for writing this monograph was the chapter on geckos by Kellar Autumn in the recent book on biological adhesives.[2] Surprisingly, the gecko (Fig. 1.1) sticks to surfaces without any adhesive. It does not need adhesion molecules. The question is 'How can materials stick together without adhesive?' Usually, when trying to explain the adhesion of particles, viruses or cells, our first inclination is to find the adhesion molecules which cause and control the adhesion force. This book shows that this is a simplistic argument which describes only one part of the story. Adhesion is often a much more complex process where two more parameters are vital: geometry and elasticity. We must change our philosophy from the simplistic statement 'Adhesion molecules cause cells to stick', to 'van der Waals forces cause cell adhesion'.

Fig. 1.1 The lizard Gekko gecko with one foot adhering to a glass plate (*foreground*) and the SEM picture of the hairy nanostructures (setae) which adhere to the glass (*background*)[3] (Copyright National Academy of Sciences USA, with permission)

As Kellar Autumn explains, the controversies about biological adhesion stretch back over centuries. In his review of the seven postulated mechanisms (at least) of gecko adhesion, Autumn quickly dismisses the idea that an adhesive is involved. Dewitz in 1882[4] had ruled this out because geckos have no glandular tissue on their toes and so cannot secrete sticky material. Dewitz also considered and rejected suction which is another perennial fallacy. Dellit[5] in 1934 carried out experiments in a vacuum to show that suction was not involved. He then employed x-rays to ionise air to show that electrostatic forces were not activated when geckos walked on metal surfaces. But Dellit fell into the trap of thinking the gecko used tiny hooks to grip roughnesses on the surfaces. This false theory of keying or interlocking cannot be supported because geckos stick best to ultra-smooth silica surfaces.[6] The only logical conclusion is that van der Waals forces are the cause of gecko adhesion, as first proposed by Haase in 1900.[7]

1.2 van der Waals Forces Cause Adhesion

Van der Waals produced his theory of atomic attractions to explain the behaviour of real gas molecules compared to perfect gas kinetic theory. Because real gas molecules do attract each other slightly, the perfect gas laws must be modified to account for this adhesion force. It is a well-understood intermolecular force which does not require any extra mechanisms like adhesion molecules, suction, tiny hooks or adhesives to predict adhesion. These ideas were fully recounted in the previous book 'Molecular adhesion and its applications'.[8]

A key feature of van der Waals force is its short range of action, much less than electrostatic, magnetic or gravitational bonds. This means that the mechanism of failure is brittle cracking which is best described by an energy balance theory, outlined in Chapter 2. The key parameters which influence such adhesion are work of adhesion, geometry and elastic modulus. Of course, adhesion molecules play a part in this theory because they have a large influence on the magnitude of the work of adhesion. The message in this book is that adhesion molecules cannot explain the whole picture, which must also contain geometrical and elastic parameters of great importance.

The present book emerged from a discussion between the authors after a paper at the Adhesion Society Conference in 2007 where it was demonstrated that cells can sense the elastic properties of the underlying substrate. Consequently, the cell adhesion area, cell shape and other parameters are strongly dependent on the elastic modulus irrespective of the adhesion molecules present.[9] Additionally, a study by Engler et al. showed that the matrix elasticity can guide the differentiation of stem cells in otherwise identical biochemical conditions.[10] We therefore got together to summarise the fundamental principles of cell, virus and nanoparticle adhesion, based on the idea that work of adhesion, geometry and elasticity govern the force. Cells and viruses should therefore behave like inanimate polymer nanoparticles in adhesion terms. Adhesion molecules at the surface will then reduce the adhesion force.

1.3 Adhesion Molecules: Learning from the Gecko

Although our premise is that van der Waals forces dictate adhesion, complex molecules, structures and mechanisms are present. Indeed, biological adhesive systems present fabulous arrays of features which will occupy experts for centuries to come. But these do not cause adhesion; they are merely variants of mechanisms which modulate van der Waals adhesion in captivating ways.

Consider the example of water influencing the adhesion of a single gecko foot-hair to glass as shown in Fig. 1.2.[3,12] It is clear that as the humidity was increased, the adhesion force also increased. Our first inclination is to suggest that the water is acting as an adhesive. We could even suggest that water is an adhesion molecule which causes molecular bonding between the hair spatula and the surface. But this reasoning is too simplistic and it is necessary to study the mechanisms more carefully.

The reality is more complex because it is well-known that water molecules reduce the van der Waals adhesion of polymers on glass. Johnson, Kendall and Roberts[11] demonstrated that dry, smooth polymer sticks best. Water reduces the van der Waals

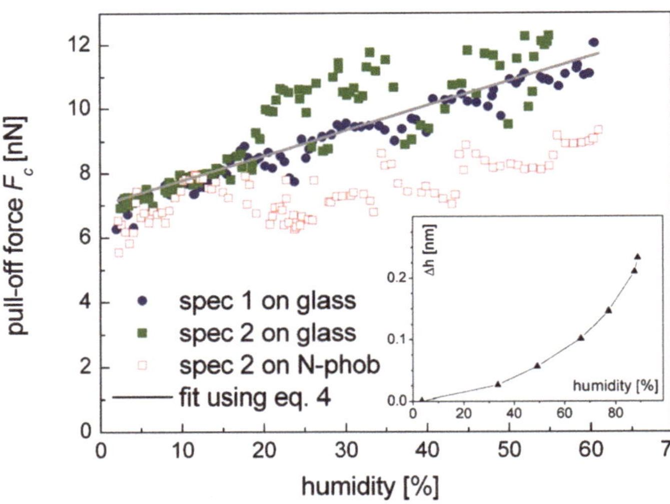

Fig. 1.2 Atomic force microscope measurements of adhesion force for a single Gecko seta on both glass and hydrophobed glass showing how humidity influenced the results. One water monolayer would be at 88% humidity[3,12] (Copyright National Academy of Sciences USA, reprinted with permission)

attraction by an order of magnitude and the adhesive force should fall, if other factors are maintained equal. In the case of the spatula experiment above, Kellar Autumn has argued that other things are not equal; first the water softens the keratin of the spatula to allow it to make improved geometrical contact with the glass; second, the water increases the loss modulus of the keratin to dissipate more energy in the peeling (Autumn, private communication). These two effects increase the adhesion sufficiently to overcome the fall in the van der Waals work of adhesion resulting from the presence of water. In addition, large water bridges can build up around 70% humidity, increasing the contact area by a large factor.[12] The conclusion is that apparently obvious adhesion results can only be correctly interpreted by considering all the relevant parameters; the work of adhesion, the geometry and the elastic properties.

Thus, the key point which we learn from the gecko is that the weak van der Waals forces which have been known for more than a century are the true cause of adhesion and that these can be linked with and modulated by other features such as geometry, material elastic properties, test conditions and surface molecules to give the solution to biological adhesion questions.

1.4 False Hypotheses

This new theory is important because a number of false statements recur in the literature. For example it was claimed[13] that 'It seems obvious that extracellular structures are essential for adhesion to a surface'. This is an apparently self-evident argument that was destroyed by Isaac Newton in the seventeenth century when he showed that clean, smooth surfaces stick best.[14] You do not need hooks, Velcro fastenings, or adhesion molecules. The fact is that all small particles stick naturally as a result of van der Waals forces as shown later in this book.

Another fallacious statement is that 'Conidia of most fungal species must be alive in order for adhesion to occur'.[15] This suggests that adhesion is not simply inanimate van der Waals forces, but requires living material; another false concept. In fact the opposite may be true: a living material is required to prevent adhesion in some cells which keep themselves separate using special mechanisms.

A further recent statement was 'A large number of microbes, fungal and algal spores, microscopic invertebrates and invertebrate larvae use adhesive polymers to stick to whatever surface they encounter.[16] This clearly emphasises the concept of the adhesive, i.e. the material which many researchers think is needed between two surfaces to make them stick. However, it has been clarified in a number of papers[17–19] that surfaces stick without any adhesive providing they are clean and smooth. Indeed, adhesives which wet the surfaces reduce the adhesion between the smooth surfaces. 'Adhesives' should really be called 'sealants' because they are essentially gap fillers which improve the extent of contact while weakening the molecular attachments. The truth about biological adhesion is nearer to what Federle[20] proposes: many separate nano-contacts each acting with van der Waals force, depending on elastic and geometrical parameters. If you consider cells crawling across a surface, a permanent adhesive bond would prohibit such motion. Therefore, adhesion is more a dynamic equilibrium between attractive and repulsive forces whose complex interplay can lead to adhesion or release across a number of contact spots.

Another key issue is the variety of cells, viruses and nanoparticles to be investigated. Shapes and sizes vary enormously, from simple sphere-like bodies like yeast or adenovirus, to cylinders like E-coli or tobacco mosaic virus, to complex shapes which can vary or flatten against surfaces like human cells or flu virus, or to spiky balls like puff-ball fungal spores. In this book we consider a range of shapes, recognising that geometry is one of the most important features of the adhesion process. It is obvious that animal cells that flatten onto a surface must stick better than spherical yeast cells which merely deform at a small contact spot, and yeast will stick better than spiky spores which only touch at very small areas.

Although several books are available discussing bio-adhesion[21–28] none gives the broad picture described above. This is the first book to attempt this overview, starting from the idea of van der Waals force.

1.5 van der Waals Force

The most important step in understanding adhesion forces arose from the kinetic theory of matter which was developed towards the end of the nineteenth century.[29] This theory was first developed on the assumption that no adhesion existed between the atoms or molecules of a gas. Then van der Waals showed that this assumption was not quite true.

In 1827, the botanist Brown had been observing pollen from the American plant Clarkia through his microscope and he saw tiny grains inside which were jiggling in the water. The water prevented the grains from sticking together, giving a close approximation to zero adhesion. Brown could see that each particle, around 1 μm in diameter, was dancing in the watery suspension as though bombarded by invisible impacts, randomly hitting the particle from all directions. Brown tested a number of other particles, including dead pollen, and showed this movement was a general phenomenon, not due to fluid flow or living matter, unlike previous observers. The conclusion was that the liquid is composed of very small atomic or molecular particles, too tiny to be visible in his microscope, in constant motion. The collisions of the invisible particles on the pollen grains were causing the dancing movements, as shown in Fig. 1.3.

The immediate effect of Brown's observation was to stimulate theoretical argument about the properties of gases which to a first approximation behaved as though there was no adhesion between their constituent atoms. On this assumption, Clausius, Maxwell, Boltzmann and their co-workers generated the mathematical theory describing the behaviour of perfect gases.[29,30] This was verified experimentally by Perrin[31] between 1908 and 1913 by studying the behaviour of non-adhering nanoparticles suspended in water following Einstein's analysis of the random walk of Brownian particles in 1905[32] For this breakthrough, Perrin received the 1926

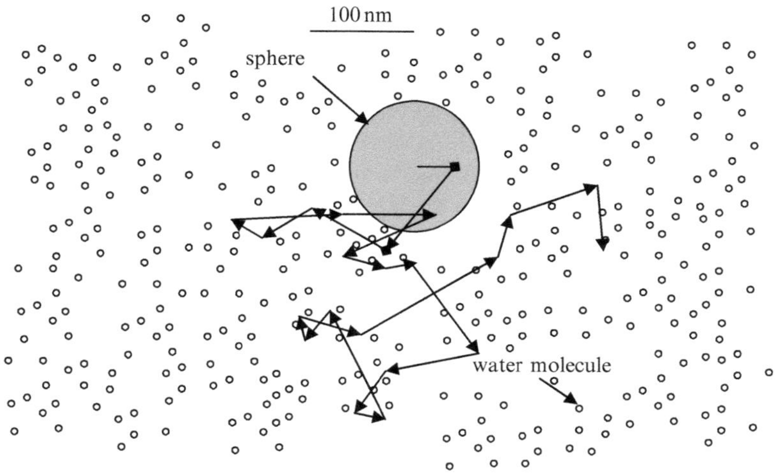

Fig. 1.3 Schematic showing a sphere immersed in moving water molecules which impact the sphere incessantly to move it along a random walk, as described in Perrin's book[31]

Nobel prize in Physics when he gave credit to Brown and his predecessors in stating that the effect 'predicted by Lucretius, suspected by Buffon, and established with certainty by Brown, constitutes the *Brownian movement*'.

Van der Waals in 1873 recognised that gases were not perfect and that deviation from the ideal gas laws could be explained by a universal attraction between all molecules.[33] This idea, that particles must attract each other with a considerable adhesive force because of the individual atomic attractive forces, was the beginning of a logical theory of adhesion.

1.6 Difference Between van der Waals and Electrostatic Forces: Yeast Adhesion

It is important to distinguish van der Waals force, often called dispersion or London forces, from simple electrostatic attractions. Adhesion of bodies as a result of electrical charging was known to the Greeks who had rubbed glass or amber with cloth to attract small pieces of litter. The effect is best illustrated by rubbing a balloon on cloth. The rubber material picks up an electrical charge, which can cause sparking electrical discharges, allowing the balloon to stick to a window. It is evident that such adhesion is different from van der Waals force because moisture, or nuclear radiation, allows the charge to leak away and the balloon drops off, whereas van der Waals forces, being induced dipole attractions, are not so much affected. Also, electrostatic forces can be both attractive and repulsive whilst van der Waals force is always attractive. These differences have been well described in a number of excellent texts[34,35] so it is not necessary to repeat them here.

Let us consider a demonstration of cell adhesion which shows that electrostatic forces are not involved, yet considerable adhesion between a glassy surface and a sheet of cells is observed. Figure 1.4 shows a peeling experiment in which a layer of yeast cells is detaching from a very smooth plastic plate by wedging with a blade. The adhesion of yeast cells is industrially important for preparing spray dried powder which can be easily dispersed to make bread formulations, as described originally in 1968 by Fantozzi and Trevelyan.[36] A 16% dispersion of yeast cells in water was sprayed through a centrifugal nozzle to produce fine droplets which were dried for 10 s in a 90°C air stream, producing rounded 100 μm diameter particles of 33% solids. Further drying in air at 37°C gave spheres of dried yeast at 97% solids, for long life and excellent activity (Fig. 1.4b, c).

The sheet of cells was made by mixing dried yeast (Saccharomyces Cerevisiae) with water into thick paste, then pressing a film of this material onto a smooth acrylic sheet using a thin sheet of paper as a backing to absorb the moisture. When held upside down, the sheet of cells adhered to the glass and a perfect black contact could be seen through the transparent material, which was then partly dried to a relative humidity of 25%. By prising the edge of the cell film with a sharp blade, the film of cells could be made to start peeling and a crack was readily observed moving along the interface.

No free electrostatic charge can exist in this experiment because the water vapour leaks it away. Also, there was no capillary force at this low relative humidity.

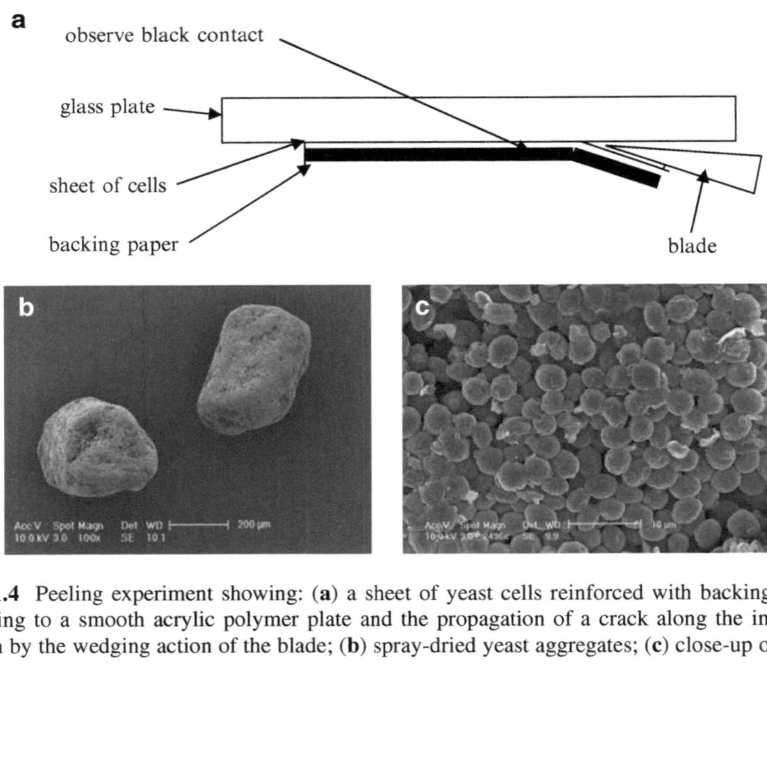

Fig. 1.4 Peeling experiment showing: (**a**) a sheet of yeast cells reinforced with backing paper adhering to a smooth acrylic polymer plate and the propagation of a crack along the interface driven by the wedging action of the blade; (**b**) spray-dried yeast aggregates; (**c**) close-up of yeast cells

The cells appeared to form a dense film as a result of the compaction under pressure. Consequently, the experiment appeared similar to the peeling of a soft elastomer described later in 2.5. We conclude that van der Waals attraction is holding the cell film to the acrylic polymer. The attraction clearly does not act over a large gap. This was tested by pushing a detached piece of cell film back towards the polymer surface. No adhesion was seen until the film touched and jumped into contact with the plastic. Then the film spread out along the surface, appearing to wet the acrylic material. So the adhesion was reversible but the force was low. Typically, a force of 0.2 g was required to detach a 10 mm wide film in peeling. This is the correct magnitude for van der Waals force between yeast cells in air, giving a work of adhesion around 200 mJm^{-2}, similar to rubber.

The effects of geometry, elasticity and contamination were also readily demonstrated with this experiment. A simple geometric change, pulling the film along the surface rather than wedging it, showed that the force had to increase by an order of magnitude to detach the cells. Similarly, increasing the bending modulus by increasing thickness of the cells made it more difficult to wedge off the layer, apparently increasing the adhesion. Finally, flooding the film with liquid water made the cells detach much more readily, showing that water diminishes the adhesion.

1.7 Fall in Adhesion Under Water

Similar experiments have been carried out on sheets of animal cells cultured under water on standard plastic plates.[37] Quail myoblast cells were grown in special culture media on glass plates which had been coated first with aminosilane, then strips of collagen on top, as shown in Fig. 1.5. The silane inhibited cell growth on the glass so that strips of fused cells ultimately grew on the collagen and could be peeled off by sucking on a micropipette.

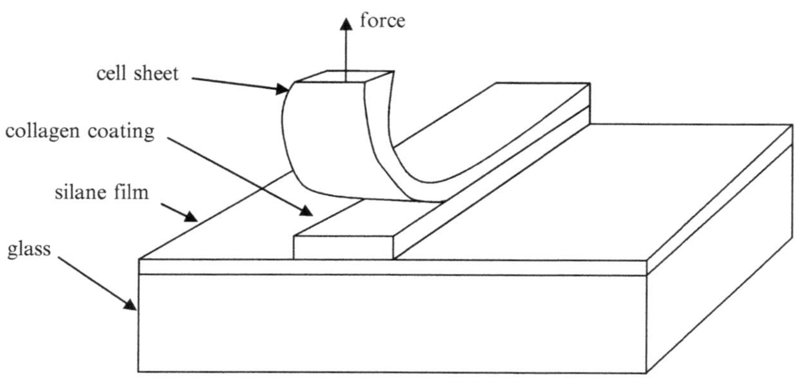

Fig. 1.5 Diagram of muscle cell strips being peeled from collagen pattern laid on silanated glass[37]

The width of the muscle cell strip was about 10 μm and the peeling force at low velocity was 1 nN. Later in Chapter 2.5 we will see that this corresponds to a work of adhesion around 0.1 mJm^{-2}, two orders of magnitude less than the peeling of a yeast or elastomer strip from dry glass. It is clear from this experiment that the presence of water and contaminant molecules has had a huge effect on reducing the adhesion of the cells. You do not need the molecular models of contact similar to those of Dembo and Bell,[38,39] describing specific lock and key contact to explain these results, because the contamination has reduced adhesion, not caused it. It is now necessary to consider the critical distinctions between capillary forces, electrostatic attractions and van der Waals adhesion.

1.8 The Short Range of van der Waals Force

A simple way to distinguish van der Waals adhesion from electrostatic attraction is to observe the range of action of the force between two spherical particles as they are pulled apart. To give a reference force, it is useful to compare these two examples with the force necessary to separate two particles with a droplet of liquid in between.

When a liquid droplet acts to glue two balls together; the force of adhesion is almost constant as the balls separate[40] as shown schematically in Fig. 1.6. Thus this 'liquid bridge' type of adhesion is 'tough'; the force stays high with distance and so a great deal of energy is required to pull the bodies apart. The energy, i.e. the integration of force times distance increment, is the area under the line. Electrostatic adhesion is not so tough because now the force falls off with the square of distance from the centres. But the force is still of long range and can be measured with distance on a large scale, for example with a meter ruler.

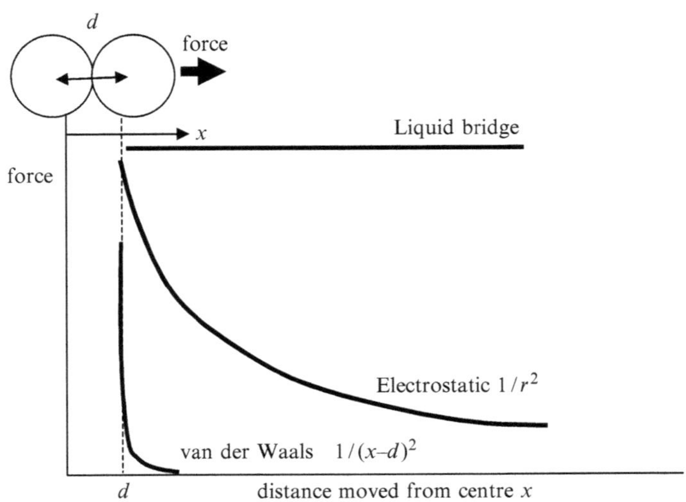

Fig. 1.6 Comparison of three types of adhesion force as two spheres are pulled apart

However, the van der Waals adhesion is very different.[41] This falls off in a very short distance of separation. In consequence, these molecular adhesion forces cannot be measured with a metre ruler, but need a nanometre scale. The adhesion force may be high when the molecules are touching, but even a separation of one nanometre causes the force to drop almost to nothing. Thus the surfaces snap apart in a brittle fashion, totally different from the other types of adhesion force. The area under the curve is very small. In other words, the energy of van der Waals adhesion may be negligible.

The other feature of van der Waals force which is evident from this comparison is seen when the balls are brought back together. The spheres with liquid between make contact again very easily. Also the electrostatic bond is readily renewed as the balls touch again, to give the same strength as before. But the molecular adhesion is not easily regained. The smallest speck of dust, or the contamination by a single layer of foreign molecules, can prevent the van der Waals bonding. In other words, the surfaces cannot be replaced in exactly the same position to reinstate the original bond. Van der Waals adhesion is not reliable or repeatable, because molecules cannot easily be put back in exactly the same position.

1.9 Measurement of van der Waals Force on Polymer Spheres

Viruses and cells are made up of polymer materials and so are best treated as soft, almost elastomeric bodies. Experiments to measure van der Waals forces on polymer spheres were pioneered by Johnson, Kendall and Roberts in 1971.[11] The experiments used rubber spheres because they were elastic, transparent and adhered easily to each other. Roberts had developed a way of moulding rubber in concave glass lenses to produce remarkably smooth elastomeric spherical surfaces as shown in Fig. 1.7. The rubber composition was mixed and then pressed hot into the glass lens. After cooling, the rubber lens could be peeled out of the glass mould. Two such rubber spherical surfaces were then moved towards each other to make contact (Fig. 1.7b).

As the two smooth spherical surfaces approached each other, within a few micrometres of contact, the familiar Newton's ring pattern could be seen in the narrow gap between the smooth surfaces. Then, as the rubber lenses were moved still nearer, a sudden jumping together of the rubber was observed and the black contact spot grew rapidly to a large size as the rubber deformed and spread under the influence of the van der Waals adhesion (Fig. 1.7c). The large black spot was an indication of large adhesion.

Consider pulling the spheres apart to measure adhesion. At zero load the black spot is about 1 mm diameter (Fig. 1.8a).When a small tensile force is applied to pull the spheres apart (Fig. 1.8b) the black contact spot shrinks to a new equilibrium circle of smaller radius as a crack runs through the contact. But when a larger tensile force is applied, the circle shrinks continuously until fracture of the adhesive bond occurs (Fig. 1.8c). This pull-off force is another measure of the molecular adhesion between the spheres.

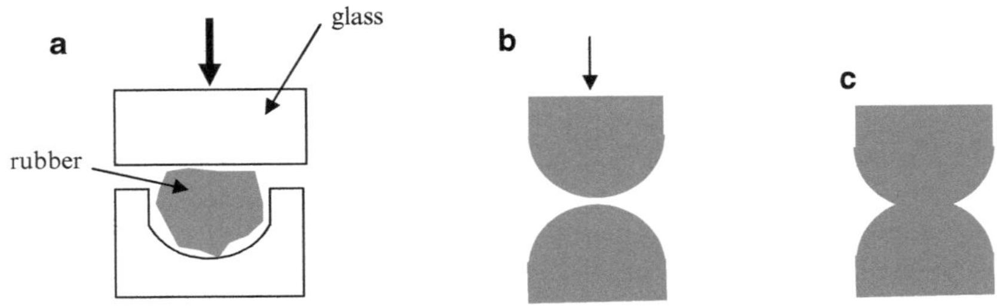

Fig. 1.7 (**a**) Moulding rubber lenses in glass formers; (**b**) bringing two rubber lenses together; (**c**) spreading of rubber to form a large contact spot

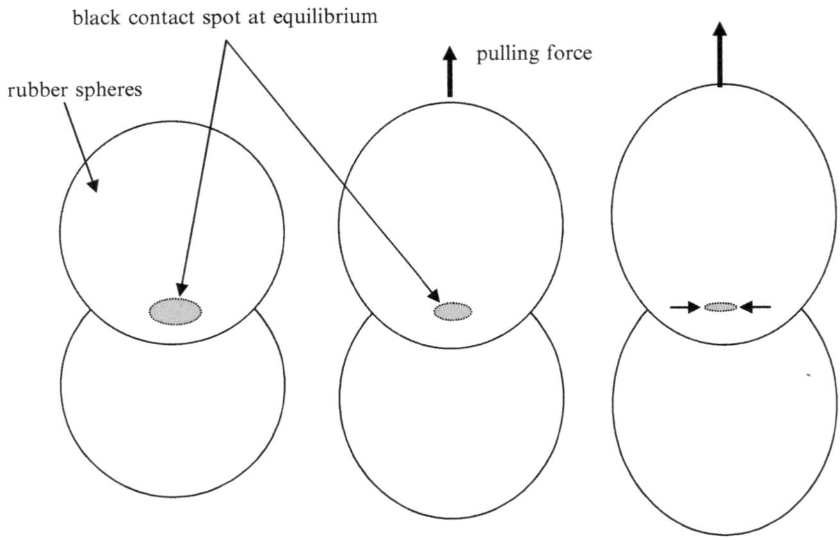

Fig. 1.8 Two transparent rubber spheres in molecular contact over the black contact spot region; (**a**) zero applied force; (**b**) Small tensile force gives smaller equilibrium circle; (**c**) a large tensile force causes the crack to run through the contact

After examining the way in which two spheres stick together, as above, it becomes apparent that adhesion is not a single process, but one which we can separate into three different but related actions; jumping into contact, achieving a certain black spot size, then cracking apart as a tensile force is applied.

The first adhesion phenomenon is the most convincing; all particles leap spontaneously together as a result of van der Waals attractions. No adhesion molecules, adhesive materials nor keying structures are required. Thus adhesion can be measured by looking at the distance covered by the leap. A long jump means strong adhesion.

The second phenomenon, the achievement of a black spot resulting from adhesion forces, was first mentioned by Newton as a measure of true molecular contact. This black spot at equilibrium balances the van der Waals adhesion forces trying to enlarge the contact, in competition with the elastic forces in the rubber trying to push the particles apart. This balance defines the size of adhesion. A large spot means large adhesion. Clearly, this is a dynamic equilibrium at the molecular level, even though the black spot seems to be static when viewed with an optical microscope.

The third phenomenon, that of detachment of the particles from each other by applying a pull-off force, is the test of adhesion which is most familiar to us. We increase the tension force applied to the particles until they just come apart, and define that force as the adhesion force. A large force means large adhesion. However, this is a very difficult experiment to carry out because the final detachment is an instability which is hard to reproduce exactly each time. Thus, many different values of adhesion can be found for the same samples in such tests, depending on the rate of loading, the precise moment of detachment etc., leading to considerable unreliability in such measurements.

1.10 Effect of Contaminant Molecules on the Surfaces

These three adhesion measures in the contact make-and-break process also allow us to test the effect of contamination, showing that contaminant molecules on the surface generally decrease the attraction between bodies. In other words, adhesion molecules reduce adhesion. Consider immersing the rubber spheres in water as in Fig. 1.9 and repeating the adhesion experiment. The results show that all three indicators of molecular adhesion; the jumping into contact, the size of the contact spot, and the pull-off force; are diminished by the presence of the water molecules. Adding surfactant molecules like proteins or adhesion molecules reduces adhesion even further. Contaminants in general reduce van der Waals adhesion because the dielectric properties of the molecules shield the electromagnetic van der Waals forces.

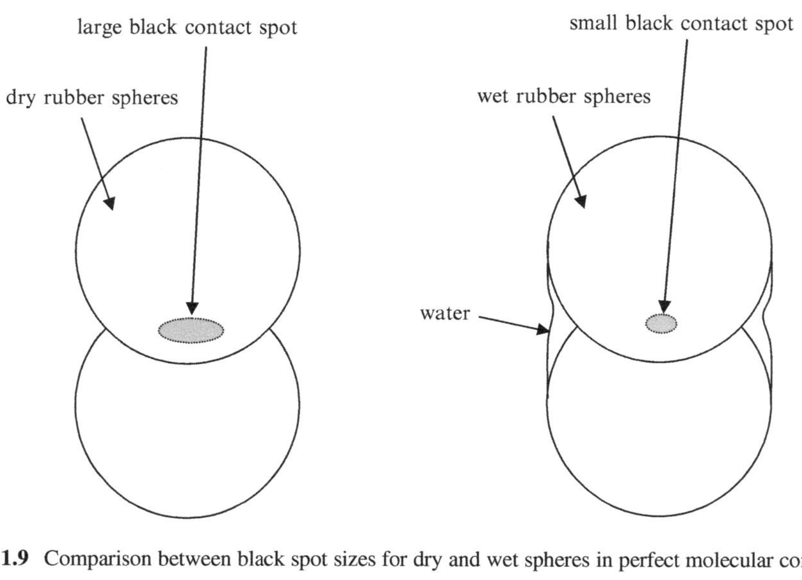

Fig. 1.9 Comparison between black spot sizes for dry and wet spheres in perfect molecular contact

When the rubber spheres are brought together under water, they can approach much closer before the jumping occurs. This suggests that the attractive force pulling the spheres together is reduced. Once the spheres have jumped into contact, the contact spot can be seen expanding to its equilibrium value. But now in contaminated conditions, the contact spot size is much smaller, indicating that the adhesion is less. Similarly, when the spheres are pulled apart under water, the force required is about ten times less than in clean conditions, showing a much reduced adhesion. Thus it is evident that the presence of contaminant molecules on the rubber surfaces has diminished the van der Waals adhesion.

The surprising conclusion from these observations is that contaminant molecules on the surface of the polymer spheres (e.g. adhesion molecules) reduce adhesion and do not cause adhesion.

1.11 Effect of Roughness

Roughness has a similar effect to contamination in reducing the adhesion by causing separation between the attracting molecules. The study of the effect of surface roughness on van de Waals adhesion was carried out systematically by Fuller and Tabor in 1975.[41] They used silicone rubber which was moulded into smooth glass concave lenses to produce spherical bodies which could be contacted with an acrylic plastic flat of varying roughness, prepared by bead blasting, as shown in Fig. 1.10.

The pull-off force was measured as the roughness was increased. The other variables investigated were the curvature of the rubber surfaces and their elastic modulus. Curvature was found to make little difference to the results, but the elasticity was found to be very important. When the rubber was made stiffer, by cross-linking it more strenuously to give a high elastic modulus, the adhesion decreased significantly. The results are shown in Fig. 1.11 for three different stiffness of rubber. On the left-hand axis, the adhesion force relative to that of smooth surfaces is plotted against centre-line-average roughness on the bottom axis. This roughness was measured using a stylus profilometer instrument and ranged from 0.1 to 2 μm in value. It was clear that adhesion fell off systematically with roughness, especially for the stiffer rubber which gave almost no adhesion at 1 μm roughness.

The random roughness of surfaces can be modelled by a statistical distribution, as first shown by Johnson[42] and later much expanded by others.[43] Using such a statistical theory, Fuller and Tabor defined an adhesion parameter which was the

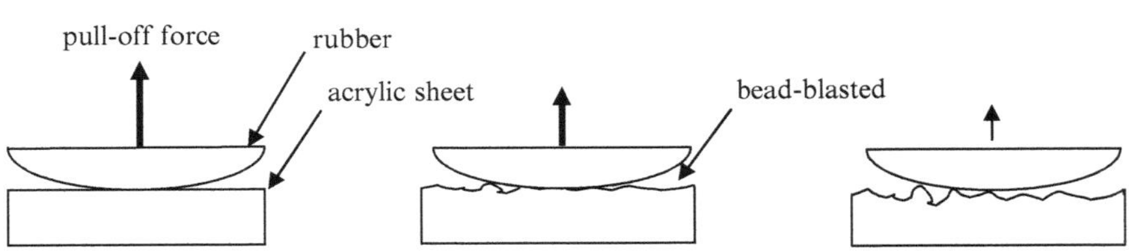

Fig. 1.10 Smooth rubber adhering to surfaces of varying roughness[41]

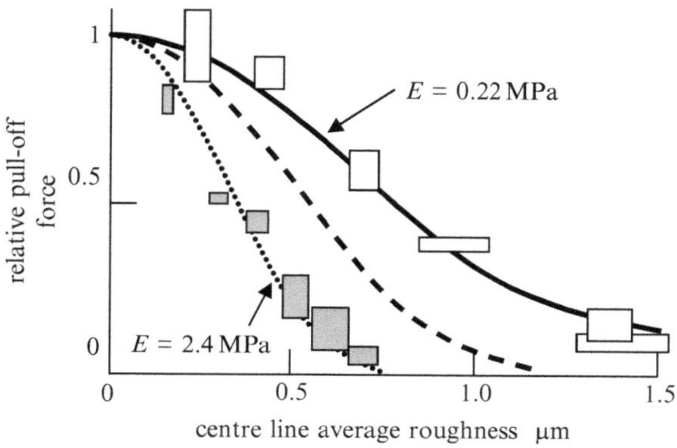

Fig. 1.11 Results of Fuller and Tabor showing adhesion falling with surface roughness

asperity height divided by the maximum extension an asperity could withstand before adhesive fracture. This adhesion parameter increased with roughness and elastic modulus but decreased with work of adhesion and asperity radius. Thus the increase in modulus was shown to be equivalent to an increase in roughness, thereby explaining why compliant materials stick best.

A surprising conclusion follows from this: structures on the surface (e.g. large adhesion molecules which cause separation of the materials) reduce adhesion of smooth surfaces.

1.12 Effect of Elasticity on Adhesion

Normally adhesion testing is strongly influenced by elasticity. The reason is that elasticity allows movement, and this movement is essential for breaking adhesive bonds. In terms of the energy balance theory above, as the material stretches, more energy (i.e. force times stretch distance) is pumped into the crack, which then converts the energy into new surface. Thus the lower the elastic modulus, the lower the adhesion force. Indeed, if there is sufficient elastic energy stored in the elastomer, then no external force is required to break the bond and the surfaces separate spontaneously.

An example of this elastic movement mechanism is shown in Fig. 1.12 which shows how an elastic layer, e.g. of cells, can be scraped or split from a substrate as described earlier in Fig. 1.4. As the scraper is pushed under the film, the elastic film material bends elastically and consequently, the elastic modulus E of the film material must be taken into account to obtain the adhesion force F. The equation describing the relationship between force F and work of adhesion W, derived by the energy balance method, in this case is,[44]

$$F = b\left(WEh^3 / 6c^2\right)^{1/2} \tag{1.1}$$

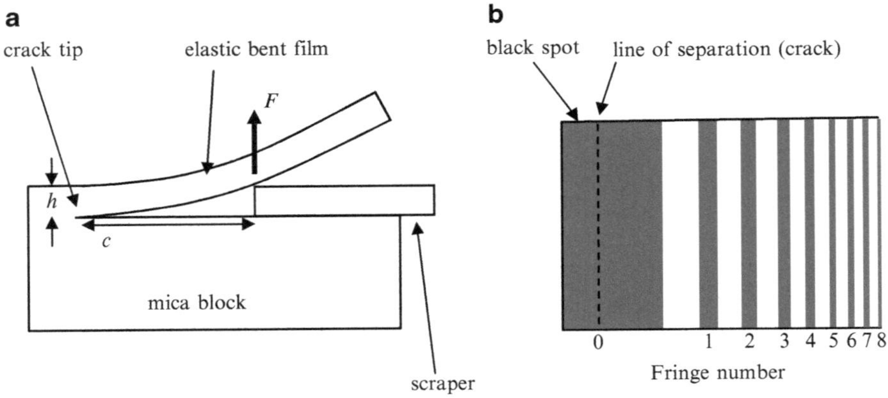

Fig. 1.12 (**a**) Scraping or wedging an elastic film from a substrate. (**b**) View through the transparent film showing the interference pattern

where h is the thickness of the film. Three points emerge from this equation: the first is that the force is lower for lower elastic modulus; secondly, the adhesion force is substantial because E is such a large number; third, it is clear that the adhesion force can be different depending on the adhesion test configuration, in this case changing with crack length c, even though the van der Waals work of adhesion W is the same.

This adhesion test is very interesting because it was first used in 1930 by Obreimoff[44] to prove the energy balance theory of adhesive fracture. His paper was most significant because it identified for the first time the three processes involved in adhesion; the jump to contact, the equilibriation of the joint, and the pulling apart of the mica sheets. In addition, Obreimoff saw that evacuating the apparatus improved the adhesion, and also found electrical discharges which proved that adhesion was essentially an electromagnetic phenomenon.

After splitting the mica, Obreimoff found that the surfaces would spontaneously jump back together when he removed the glass wedge. He measured the energy of this spontaneous adhesion and it was about 0.8 Jm^{-2}, substantially less than the original adhesion energy. Then he pushed the wedge back in to measure the adhesion formed between the foil and the block by the jumping process and found the adhesion energy was around 1.2 Jm^{-2}, so it was taking more energy to split the adhering mica than was recovered on the jumping together. Thus he found some energy loss or adhesive hysteresis in this process. He also realised that it took some time for the splitting to reach equilibrium; the fringes moved for quite a time after the wedge was fixed, around 15 s. This was the first observation of 'adhesive drag', the rate effect on adhesion.

Perhaps of most significance was the result obtained when the air was evacuated from the vessel around the mica. Adhesion was increased as the air was pumped out, showing that removal of contaminant molecules from surfaces gives increased adhesion. The surprise is that adhesion molecules reduce adhesion between clean smooth surfaces. As these contaminant molecules were removed by the evacuation, the energy of adhesion was then increased to 20 Jm^{-2}, and impressive electrical discharges were seen around the mica samples at 1 nanobar pressure. This proved that the adhesion was connected with electromagnetic forces between the atoms in the mica crystal, not with lock and key, Velcro, adhesion molecule or other mechanisms. The conclusion is that adhesion of cells, viruses and nanoparticles depends on the van der Waals force, on geometry and on elasticity.

References

1. Harrison, R.G., The reaction of embryonic cells to solid structures, J Expt Zool 17 (1914) 521–44.
2. Autumn, K., Properties, principles and parameters of the gecko adhesive system, in Biological Adhesives, (AM Smith and JA Callow eds) Springer, Berlin 2006, p. 225–256.
3. Huber, G., Mantz, H., Spolenak, R., Mecke, K., Gorb, S.W., Arzt, E., Evidence for capillary contributions to Gecko adhesion from single spatula nanomechanical measurements, PNAS 8 (2005) 16293–6.
4. Dewitz, H. Wie ist es den Stubenfliegen und vielen anderen Insekten möglich an senkrechten Glaswänden emporzulaufen? Sitz Ges Naturf Freunde, (1882) 5–7; West T., The foot of the fly…, Trans Linn Soc London 23 (1862) 393–421
5. Dellit, W.-D. Zur Anatomie und Physiologie der Geckozehe. *Jena. Z. Naturw*, 68 (1934) 613–656.
6. Autumn, K., Liang, Y.A., Hsieh, S.T., Zesch, W., Chan, W.P., Kenny, W.T., Fearing, R. and Full, R.J., Nature, 405 (2000) 681–5.
7. Haase, A., Arch Naturgesch 66 (1900) 321–45.
8. Kendall, K., Molecular adhesion and its applications, Kluwer, New York 2001, ch. 3.
9. Rehfeldt, F., Engler, A.J. and Discher, D.E., Stem cell adhesion and microsensing, The Adhesion Society 30th Annual Meeting, Feb 18–21, 2007, Tampa Bay Fl USA.
10. Engler, A.J. et al., Matrix elasticity directs stem cell lineage specification, Cell 126 (2006) 677–689.
11. Kendall, K., The adhesion and surface energy of elastic solids, J PhysD: Appl Phys 4 (1971) 1186–95; Johnson, K.L., Kendall, K. and Roberts, A.D., Surface energy and the contact of elastic solids, Proc R Soc Lond **A324** (1971) 301–313.
12. Nguyen, T., Chen, L., Giu, X., Effects of surface free energy and relative humidity, Proc Adhesion Society, 33 Annual Meeting Feb 21, 2010, Daytona Beach Fl.
13. Epstein, L. and Nicholson, R.L., Adhesion and Adhesives of Fungi and Oomycetes, in Biological Adhesives, (AM Smith and JA Callow eds) Springer, Berlin 2006, p 47.
14. Newton, I., Opticks, Smith and Walford, London 1704, reprinted Dover, New York, 1952, p. 201.
15. Landini, P., Jubelin, G. and Dorel-Flamant, C., The molecular genetics of bioadhesion and biofilm formation, in Biological Adhesives, (AM Smith and JA Callow eds) Springer, Berlin 2006, p. 22.
16. Smith, A.M. and Callow, J.A. (eds), Biological adhesives, Springer, Berlin 2006.
17. Kendall, K., Sticky solids, Contemporary Physics **21** (1980) 277–97.
18. Kendall, K., Theoretical aspects of solid-solid adhesion, Sci Prog Oxf, **72** (1988) 155–71
19. Kendall, K., Adhesion: Molecules and mechanics, Science **263** (1994) 1720–25
20. Federle, W., Why are so many adhesive pads hairy?, J Experimental Biology 209 (2006) 2611–21
21. Adams, J.C., (ed.) Methods in cell-matrix adhesion, Academic Press, New York 2002.
22. King, M.R., Principles of cellular engineering, Academic Press, New York 2006.

23. Cress, A.E., Nagle, R.B., Cell Adhesion and Cytoskeletal Molecules in Metastasis (Series: Cancer Metastasis - Biology and Treatment, Vol. 9), Springer 2006.
24. Beckerle, M.C., Cell adhesion, Oxford UP, Oxford 2001.
25. Roberts, J., Gonzales-Carranza, Z., Plant cell separation and adhesion, (series Annual Plant Reviews) Wiley Blackwell, 2007.
26. Symons, M., Rho GTPases, Springer 2004.
27. Denker, H.W., Molecular Approaches to cell-cell adhesion, Karger, Basel 2002.
28. Garrod, D., North, A.J., Chidgey, M.A.J., The adhesive interactions of cells, (series, Advances in molecular and cell biology 28) Elsevier, 1999.
29. Maxwell, J.H., Theory of Heat, Longman Green, London, 1871; reprinted Dover, New York, 9th ed 2001.
30. Brush, S.G., Statistical Physics and the atomic theory of matter, Princeton University Press, Princeton, New Jersey, 1983.
31. Perrin, J.B., Atoms, Ox Bow Press, New York, 1990 (French original, 1913).
32. Einstein, A. Über die von der molekularkinetischen Theorie der Wärme gefordete Bewegung von in ruhenden Flüssigkeiten suspendierten Teilchen (On the Motion—Required by the Molecular Kinetic Theory of Heat—of Small Particles Suspended in Stationary Liquid), Annalen der Physik **17** (1905) 549–560
33. van der Waals, J.D. PhD Thesis, Over de Continuïteit van den Gas- en Vloeistoftoestand (On the continuity of the gas and liquid state). Leiden 1873.
34. London, F., The general theory of molecular forces, Trans Faraday Soc. **33** (1937) 8–26.
35. Rowlinson, J.S., Cohesion: A scientific history of intermolecular forces, Cambridge University Press, 2002.
36. Fantozzi, E., Trevelyan, W., Patent Application Number 04/777160 (1971)
37. Ra, H.J., Picart, C., Feng, H., Sweeney, H.L., Discher, D.E., Muscle cell peeling from micro-patterned collagen: direct probing of focal and molecular properties of matrix adhesion, J Cell Science 112 (1999) 1425–36.
38. Dembo, M., Torney, D.C., Saxman, K., Hammer, D., The reaction-limited kinetics of membrane to surface adhesion and detachment, Proc R Soc Lond B234 (1988) 55–83.
39. Bell, G.I., Models for the specific adhesion of cells to cells, Science 200 (1978) 618–627.
40. Murase, K., Mochida, T. and Sugama, H., Experimental and numerical studies on liquid bridge formed among three spheres, Granular Matter **6** (2004) 111–119.
41. Fuller, K.N.G. and Tabor, D., The effect of surface roughness on the adhesion of elastic solids, Proc R Soc Lond **A345** (1975) 327–42.
42. Johnson, K.L., Contact mechanics, Cambridge University Press, 1985, ch.13.
43. Hills, D.A., Nowell, D. and Sackfield, A., Mechanics of elastic contacts, Butterworth-Heinemann, Oxford, 1993, ch.14.
44. Obreimoff, J.W., The splitting strength of mica, Proc R Soc Lond **A127** (1930) 290–297.

Chapter 2
Phenomenology of Adhesion: From Macro- to Nano-Systems

the solid support influences the form and arrangement assumed by the moving cells

Harrison (1914)

Adhesion of macroscopic polymer particles and films has been explained in the previous chapter. The difficulty is to understand adhesion of cells and viruses which are much smaller; from μm to nm in diameter. To overcome this problem of moving from macro- to nano-systems, we have to prepare and measure surfaces with nano-scale perfection.

Two hundred years before van der Waals, Isaac Newton was the first person to suggest that two particles brought into close contact should adhere strongly when he wrote 'two polish'd marbles, … by immediate contact stick together'.[1] However, the experimental demonstration of this phenomenon for larger spheres was not so readily achieved. Newton reasoned that the imperfection of surfaces was the most important factor inhibiting contact, either through surface roughness or through contamination as described in the previous chapter. Therefore he developed better methods for polishing glass lenses to upgrade their quality. This was the work that brought him recognition in 1671 when he first displayed at the Royal Society his marvellous new reflecting telescope lenses for improved observation of the stars and planets. But Newton only observed adhesion between the smooth glass lenses sporadically. There seemed to be asperities and dust preventing perfect contact.

It turns out that Newton would have been much more successful in proving his ideas on adhesion if he had used the rubbery material which Columbus had brought back from the new world. Such soft material sticks far better than glass. Also, it has become evident that large bodies, like Newton's glass lenses, are less likely to show adhesion than small ones. If Newton had done his experiments on fine polymer beads or yeast cells, then he would have seen the adhesion he was expecting. The conclusion is that van der Waals adhesion is observed best on small, smooth, soft objects like polymer spheres or biological cells.

K. Kendall et al., *Adhesion of Cells, Viruses and Nanoparticles*,
DOI 10.1007/978-90-481-2585-2_2,

2.1 Showing the van der Waals Adhesion Force

Our argument is that van der Waals forces attract cells towards each other. Of course, the key question is how cells self-assemble. In other words, how do they forge a suitable contact and then produce close-packed aggregates of cells, as in the slime mould shown in Fig. 2.1a–g, Dictyostelium Discoideum,[2] which can exist as a single cell that aggregates into a multicellular organism when food is scarce. The 'Sultan of Slime' Prof John Bonner has a movie showing this on his web-site and a book describing his work. The detailed molecular effects at the cell contacts have been described in recent papers but the overall mechanics depending on geometry and elasticity are not normally considered.[2]

It is easy to show that, even when cells do not spontaneously self-assemble, they can adhere strongly. For example, a dispersion of yeast cells (Saccharomyces cerevisia) in water when spray-dried forms small spherical aggregates with good

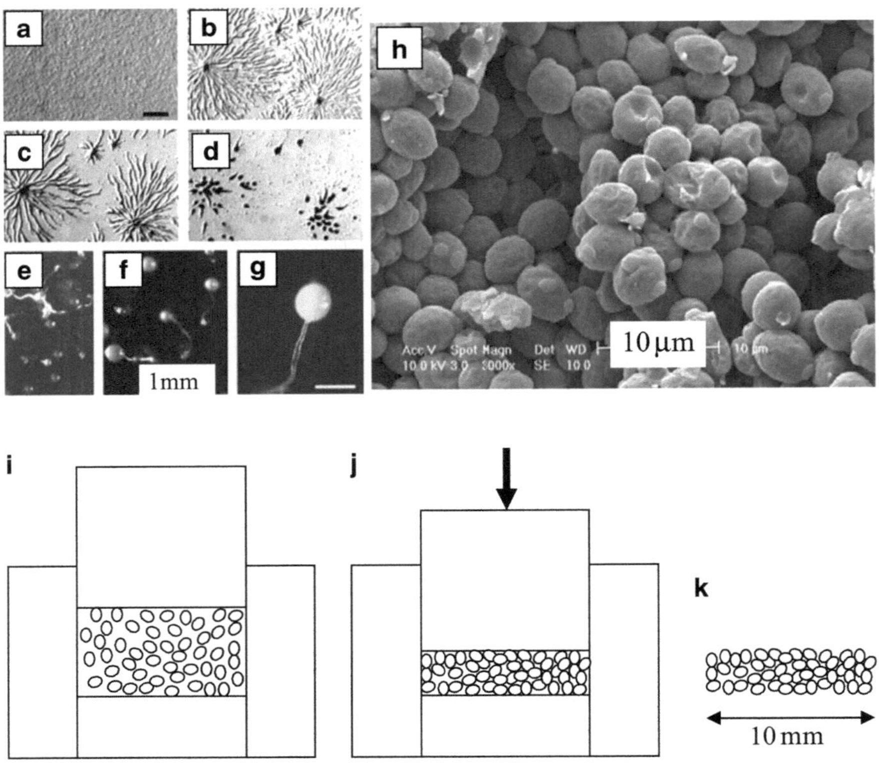

Fig. 2.1 (**a**) Slime mould cells growing on no-nutrient agar plate; (**b**)–(**d**) starved cells aggregating; (**e**)–(**g**) fruiting bodies [2]; (**h**) yeast cells before compaction; (**i**) loose powder loaded into steel die; (**j**) powder compressed by die; (**k**) pellet after release from die (with permission of the Royal Society)

strength (Fig. 2.1h). The individual 4.5 μm diameter cells can be seen to form extended contact spots with each other about 1 μm in size, indicating a work of adhesion around 200 mJm^{-2} from JKR theory.[10]

One gram of yeast cells in the form of spray dried, almost spherical, aggregates was poured into a hard steel pelleting die, as shown in Fig. 2.1i. Electrostatic forces were prevented by the presence of 0.2 g moisture which leaked away any stray electrons.

After squeezing the powder with a pressure of 10 MPa, as in Fig. 2.1j, the pellet was ejected from the die in one piece (Fig. 2.1k). The yeast cell particles then stuck together with considerable strength. Of course, the tablet was porous because the yeast cells did not pack together perfectly to exclude all the pores. But where the yeast particles touched each other, the molecules of the cell membranes were in close proximity as a result of the large force which had urged them into molecular contact. Then, the short range van der Waals adhesion forces discussed above pulled the particles strongly together, thereby resisting external stresses and giving substantial adhesion.

By testing the pellet in bending or tension, it is easy to find that the compacted pellet has the properties of a gel. It is viscoelastic and brittle, but not so strong as a dense elastomer which has no pores to weaken it. The adhesion energy was measured by cutting a radial slit into the pellet with an abrasive file and pulling the disc apart with an edge force. From the 1 mN force applied across the 2 mm width of the pellet, the adhesion fracture energy would be 200 mJm^{-2}, about that expected for van der Waals forces.

2.2 The First Demonstration of Nanoparticle Adhesion

William Hyde Wollaston[3] first described this type of adhesion experiment in 1829. He was interested in making dense and strong wires from platinum and other rare metals such as palladium and osmium which he had just discovered. Platinum is so hard and refractory that it is extremely difficult to work by ordinary melting and casting techniques. Wollaston prepared the platinum in fine particle form by precipitating the metal from an acid solution which had been used to remove impurities. This produced a mixture of water and particles which were cleaned by washing, then dispersed by milling in a wooden mortar and pestle. Wollaston called the mixture ‘mud’ whereas we would now describe this as a nanoparticle dispersion.

The difficulty was converting the fine metal powder into a dense block. To achieve this, Wollaston needed to get the particles sticking together in a dense packing. First he constructed a mechanical press, shown in Fig. 2.2, consisting of a brass cylinder into which fitted an iron piston around 25 mm diameter. The barrel of the cylinder was tapered so that the pellet could be ejected after compaction. After placing the platinum mud in the barrel, then covering it with blotting paper to allow the water to soak out, and greasing the piston with lard, Wollaston pressed on the ram to apply a force of about 30 tonnes weight to the powder mass. This force was sufficient to increase the packing of the fine particles from a loose state of 20%

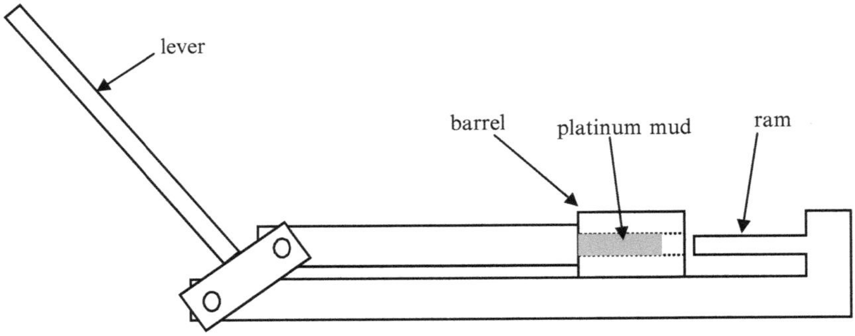

Fig. 2.2 Wollaston's press for compacting platinum powder together into an adhering pellet

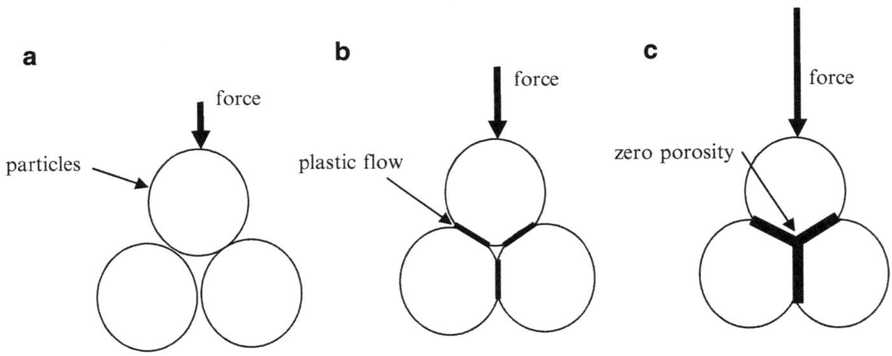

Fig. 2.3 Three stages in compaction of plastic particles; (**a**) low load; (**b**) higher load causes plastic deformation; (**c**) higher load removes all pores

packing to a porous pellet near 50% dense. This pellet was ejected in one piece and was 'hard and firm' suggesting that each platinum particle was adhering to its neighbours by van der Waals forces.

Heating of the pellet with burning charcoal was sufficient to remove moisture and organic lubricant. Then the pellet was raised to white heat in a Staffordshire coke furnace. This caused the pellet to contract as the particles sintered together. Pounding the hot pellet with a hammer produced a material which was 99% dense and which produced platinum wire of the 'highest tenacity'.

The adhesion between plastic particles during compaction is even more striking than that between elastic platinum grains. This is readily demonstrated by compressing potassium bromide powder in a steel die, as routinely done for infra red analysis. During the compression of the grains, the pressure at the contact points becomes larger than the yield pressure and consequently the contact spots enlarge until all the porosity has been excluded, as indicated schematically in Fig. 2.3.

On removal from the die, the compacted pellet is seen to be fully transparent and completely dense. The pellet is also very strong, elastic and brittle, comparable to a piece of solid potassium bromide made by other methods. Thus the conclusion is that the compaction force has brought the grains into molecular contact, generating adhesion. Further force sheared the material close to the contacts, allowing plastic flow to produce more intimate molecular contact until all the particle surfaces adhered strongly.

The lessons we learn from such demonstrations are fourfold; all bodies can be made to adhere together by van der Waals attractions but finer, softer particles stick more easily; force is usually required to overcome friction or repulsive forces and to push the bodies into molecular contact; and deformation, especially plastic or diffusive flow, allows more extensive contact to give maximal adhesion.

2.3 Arguments Against van der Waals Adhesion

Sceptics say that the adhesion developed in the above experiments can be explained by other well-known ideas. Four sorts of argument levelled against molecular adhesion phenomena can be listed. There is first the suction argument, which says that the particles are acting as rubber suction pads, and merely sealing around the edges; this is easily shown to be false because the pellets are just as strong in vacuum, where suction pads fall apart. Secondly, there is the mechanical keying argument, which suggests that the particles behave like Velcro, with little hooks and eyes to cause adhesion. This is easily answered because when you look at the surfaces by electron microscopy, the particles are often extremely smooth and shiny. There are no hooks and eyes. Moreover, the smoother particles show stronger adhesion than rough particles. A third argument is that there is some adhesive material, e.g. adhesion molecules, on the particle surfaces acting like a glue to bind the particles together. This is simply disproved by cleaning up the surfaces; the cleaner the particles, the better they stick. In fact adhesives are known to reduce adhesion between particles as we showed in Chapter 1. Finally, a fourth argument is that the particles are oppositely charged, to give electrostatic attractions. This is readily disproved by doing the experiment in the presence of moisture or ionising radiation to leak away the electrons: Adhesion is not affected.

Having dismissed all the plausible yet simplistic theories of adhesion between particles, we can now conclude; adhesion is dominated by van der Waals forces.

2.4 Definition of van der Waals Adhesion

Consider a definition of adhesion which allows it to be distinguished from all the other known forms of attractions between bodies: van der Waals adhesion is the force experienced when bodies make contact at the molecular level, with gaps around 1nm, near molecular dimensions, allowing van der Waals forces to dominate.

This definition raises a number of questions which will be addressed in the following chapters. The obvious question relates to the origins and laws of van der Waals adhesion. How can one measure and interpret such phenomena? Clearly molecular adhesion forces have the same origins as the forces of cohesion which hold the molecules of solids and liquids together. These can be understood in terms of the heats of melting or evaporation, the elastic stiffness or the chemical reactivity of materials, as described later.

These van der Waals forces had been measured for the first time in the 1920s as de Boer, London and others were formulating the basic equations of the instantaneous dipole attractions. Tomlinson in 1928 heated and drew fresh fibres of fused silica to perform adhesion experiments, bringing the crossed fibres together to observe the contact point, which he estimated to have a black spot size less than 1 μm, and measuring the force of adhesion by elastic deflection of the fibre, as shown in Fig. 2.4. He also formed spherical blobs on the ends of the fibres and tested these in the same way.[4]

Tomlinson was careful to release electric charges by ionising the air, and also made sure that the silica was dry to avoid questions about surface moisture. The best adhesion was observed immediately on cooling from red heat because there was then no roughness or contamination. Also marked damage was seen after adhesion of the surfaces, proving that the forces of adhesion were large enough around the black spot to crack the glass material.

A typical fibre was 60 μm in diameter, about the same as a human hair, and the adhesion force measured was 30 μN. This force was readily determined by the 5 mm deflection of the fibre just before detachment. The interesting feature of the experiment reported by Tomlinson was the sudden attachment of the fibres when contact was approached. The surfaces jumped into contact! He took trouble to apply no force pushing the surfaces together, and concluded that 'the molecular attractions acting at the instant of geometrical contact are sufficient to draw the fibres together'. The adhesion force seemed to be in proportion to the diameter of the bodies and was related to the energy of adhesion per square metre of interface, the work of adhesion W.

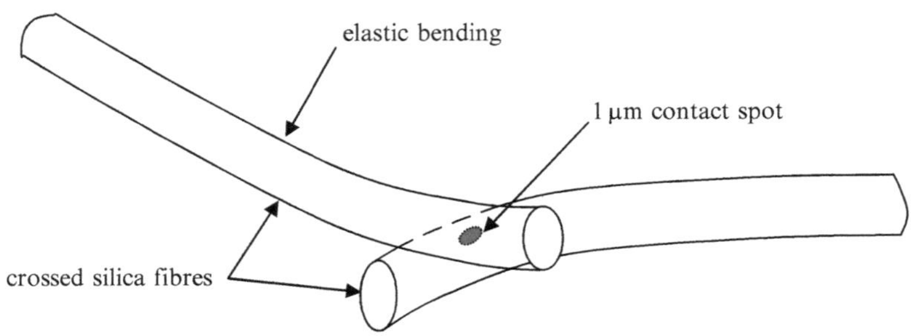

Fig. 2.4 Tomlinson's experiment on adhesion of silica fibres

2.5 Definition of *W*, the Work of Adhesion in Peeling

In order to quantify van der Waals forces, it is necessary to take into account the area of contact of the bodies in contact. With spheres this is difficult because the contact spot is often very small and difficult to measure. Therefore it is best to start with an extended uniform contact such as that between a sheet of cellular tissue and a smooth flat acrylic surface. Also van der Waals forces vary rapidly with separation distance, so it is best to deal with the energy of the adhesion process than with the force/distance curves. Therefore we define a work of adhesion W as the energy to break reversibly 1 m^2 of contact. W can then readily be measured by finding the energy needed to separate a known area of contact. The simplest test is that first defined by Rivlin in 1944 to describe paint peeling from a smooth surface.[5] He analysed the test using the energy balance approach described below.

Imagine an elastic film, made of cells for example, peeling under a force F from a rigid smooth substrate as in Fig. 2.5. A crack can be observed moving at steady speed along the interface by looking through the transparent substrate with reflected light. After a while, the crack has moved a distance c. The area of interface broken by this crack movement is bc where b is the width of the peeling film. Therefore the energy expended to create new surfaces by breaking the van der Waals bonds is Wbc where W is the thermodynamic work of adhesion (i.e. the reversible energy required to break one square metre of van der Waals bonds at the interface). The work done by the force is force times distance i.e. Fc which is all presumed to go into the surface energy Wbc, because energy must be conserved. Therefore the peel equation is (see box)

$$F = Wb \tag{2.1}$$

Of course there is elastic deformation energy in the bent elastic film, from the time when the force was first hung on the film. But this remains constant during peeling and so does not supply any energy to the surfaces. It is merely a constant energy term which moves along with the crack. Consequently it does not change during

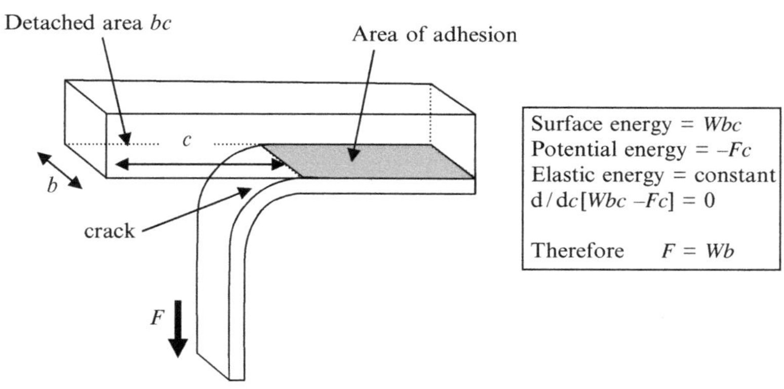

Fig. 2.5 Energy conservation theory applied to peeling of an elastic film from a transparent substrate

the energy balance. We also assume that there are no stretching or dissipation energy terms as the film is detached.

This theory presumes that the crack can also heal at the same force. In practice, the force has to be slightly reduced for healing to be seen. For the most perfect elastic system, there is a force which can be suspended on the film whereby the crack does not know whether to peel or heal. The crack is essentially in thermodynamic equilibrium in which a slight increase in force will cause separation, and a slight decrease will cause healing. This is the situation to which the peel equation $F = Wb$ applies.

The peel equation derived by Rivlin above is most interesting because it seems to have no connection with the strength of the interface, that is the stress required to pull the interface apart. The idea that solid materials require a stress or pressure to tear them apart goes back to Galileo and his treatise on the two sciences of mechanics and strength of materials.[6] This has been a remarkably persistent idea which has not been justified by work on many materials. For example, it has been known for many years that glass can fail at a whole range of different stresses depending on the chemistry at the surface. Indeed, that was the whole point of Griffith's original work,[7] to understand why the failure stress of glass can vary so much. Here we reject the concept of a constant fracture stress because it has no place in the equilibrium theory of van der Waals bond breakage adopted in this chapter. Equation 2.1 shows quite clearly that the work of adhesion W (i.e. the energy of the van der Waals bonds) is the main material property resisting cracks. Only the width of the strip is then relevant to the adhesion force, not the area of contact.

2.6 Nature of Bonds in the Equilibrium Theory of Adhesion

The model of adhesion described above is a macroscopic thermodynamic model which satisfies the conservation of energy principle, averaging the behaviour of all the bonds in the system over the large scale sample. The quantity W is the measurable parameter which assumes that the bodies behave as elastic continua.

But such a continuum theory must fail as we approach molecular dimensions where adhesive failure is actually occurring. In order to reconcile the macroscopic picture with our knowledge of the molecular reality, consider the situation depicted in Fig. 2.6 which shows a film of polymer peeling from a glass surface. This shows that, at large scales (2.6a) we should treat the materials by continuum mechanics, but at the molecular scale (2.6b) we must consider molecular models where adhesion is governed by an interaction potential and by statistical mechanics of a large number of molecular bonds.

Magnifying the crack tip where the separation of molecules is occurring as in Fig. 2.6b shows the polymer molecules making and breaking contact continuously as Brownian motion arises at the molecular level. At macroscopic resolution, the crack seems to be in equilibrium at a particular loading (Fig. 2.6a). There does not seem to be any motion at the crack tip. However, when viewed at the atomistic level the crack tip is seen to be in rapid thermal Brownian motion. The attracted

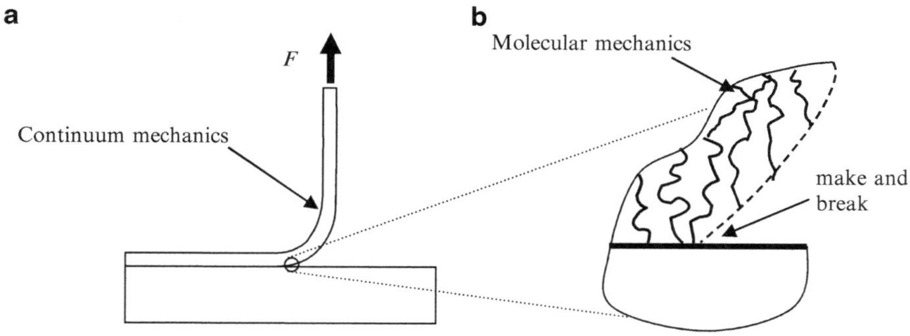

Fig. 2.6 Schematic picture showing how continuum mechanics applies at large scale (**a**) but molecular modelling takes over at nanometre scales (**b**)

molecules form the adhered region to the left of the crack tip, whereas the unattracted molecules lie to the right at the open crack surface. The crack tip is not a static point in this model. It is wandering kinetically from right to left as the molecules spontaneously break and then rebond. Cracking is thus viewed as a reversible van der Waals interaction between molecules at the crack tip. The force applied to open or close the crack is not the cause of interaction, i.e. peeling or healing, at the crack tip. The interaction is happening spontaneously and equally in both directions, causing the crack to open and close spontaneously at the molecular scale. Applying the crack driving force merely shifts the van der Waals equilibrium in one particular direction, either opening or closing the crack.

Work of adhesion must be distinguished from surface energy which is often used to describe liquid surface attractions. Surface energy γ (i.e. surface tension) is extremely useful for liquids because many liquid surfaces have no resistance to flow at low rates, and so reach an equilibrium shape which is dominated by surface energy. Thus, surface energy of liquids is readily measured by the deformation of liquid surfaces in surface tension measurements such as Wilhelmy plate or sessile drop.[8] Unfortunately, the surfaces of solids are elastic under ordinary conditions. These elastic forces are so much larger than surface tension that the measurement of solid surface energy has proved very difficult. Certainly, the adhesion of bodies depends on the work of adhesion W and if the bodies are identical and smooth, we can take $W=2\gamma$ where γ is the surface energy of the solid as shown by Johnson et al.[10] However, surface energy of solids is seldom enough to overcome elasticity, which is the next major variable to be discussed.

2.7 Bradley's Adhesion Rule

Bradley[4] had read Tomlinson's paper and developed an improved method of measuring the adhesion together with a better theory based on London's wave mechanics theory of the van der Waals forces between molecules. By adding up the

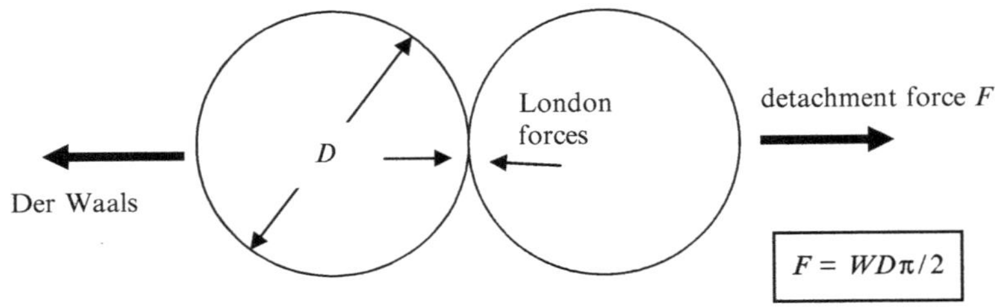

Fig. 2.7 Bradley's theory of attraction between rigid spheres based on London van der Waals forces

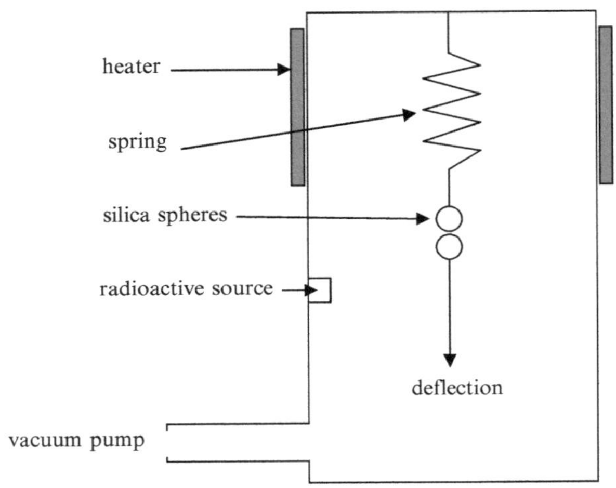

Fig. 2.8 Schematic of Bradley's apparatus for measuring the adhesion of silica spheres

forces for all the molecules in two rigid spheres, Bradley came to the conclusion that the adhesive force required to separate them should be proportional to the sphere diameter, as shown in Fig. 2.7. He also showed that the force should be proportional to the work of adhesion W of the spheres, that is the energy required to separate 1 m^2 of interface reversibly. Thus he produced his famous equation for adhesion of spheres.

Bradley then constructed an apparatus (Fig. 2.8) for measuring the force required to separate two silica spheres from adhesive contact. The rig could be evacuated to remove moisture and other contamination. Heaters were used to bake the glass. A radioactive source ionised the gas to leak away any stray charges. The silica balls were heated to incandescence immediately before the measurements. A deflection was applied to the bottom sphere until it detached from the upper sphere, giving the results shown in Fig. 2.9.

Bradley found that the adhesion remained constant as gas was evacuated, so water was obviously not the cause of the adhesive force. However, the spheres were probably too large to give the molecular contact which Tomlinson had

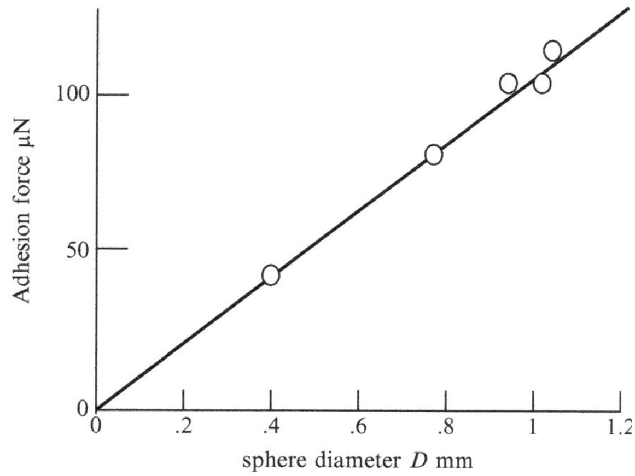

Fig. 2.9 Bradley's results for adhesion of silica spheres

observed on his ten times smaller fibres. The force was proportional to diameter which fitted the theory, but the force was several times smaller than expected, and smaller than Tomlinson observed. This problem became evident when Bradley attempted to use sodium borate spheres instead of fused silica. Although adhesion was seen, it was variable and Bradley suggested the surfaces were rough as a result of reaction with water. The conclusion was that these spheres were still too large and insufficiently smooth to obtain reliable van der Waals adhesion.

Another significant issue was the deformation and flattening of the spheres at the point of contact. Newton had stated that this occurred as the spheres were pulled together by adhesion forces. Tomlinson worked out from the well-known Hertz equations of elasticity, described below in 2.9, that the contact spot diameter should be around 1 µm for the contact of his silica fibres. Bradley took no account of this idea and his theory was based on the assumption of rigid spheres, obviously inapplicable to elastic particles. Derjaguin[9] attempted a solution of this question in 1934 by combining the Hertz and Bradley ideas, but his answer was not quite right. The final solution was obtained in 1971, showing that Bradley's equation requires only slight numerical modification, by about a factor 2, to $F = WD\pi 3/8$ when elastic deformation is taken into account[10] So elastic stiffness hardly affects adhesion of spheres, and therefore it should not matter whether we stick stiff diamond or compliant bacterial spheres together, we should get the same adhesion force result for the same work of adhesion. This is a surprise because the contact areas are obviously much bigger for the cells than the diamonds. Contact area and adhesion force are not proportional. In fact, the surprising conclusion from these arguments is that the adhesion of particles is in proportion to the particle diameters, not the contact diameters. This result, Bradley's rule, is of outstanding importance because it means that adhesion dominates all other forces at dimensions below 1 µm particle size.

2.8 The Significance of Bradley's Rule

The discovery that solid bodies jump into adhesive contact under the influence of the van der Waals attractions was enormously significant. This experiment was certainly known to Tomlinson[4] in 1928 and was studied both by Obreimoff in 1930 as described in Chapter 1, then by Derjaguin and Abrikossova[15] sometime later. Once you see this phenomenon, you become convinced that molecular adhesion due to van der Waals force exists. When you observe it in different situations; on mica, on glass, on metals, on polymers, then you realise it is a universal observation that applies to all bodies when there are no contaminant molecules or surface roughness to stop the adhesive electromagnetic interaction of the particles.

In a sense, although it came 100 years after Brown, this observation stands with Brownian motion as a critical break-through. Before Brownian motion was seen, engineers thought that matter was continuous and static. Suddenly they were aware that it was molecular and moving, though this molecularity and movement could often be ignored at the macroscopic level because the atoms and the energies were so small. In a similar vein, engineers normally treat objects as non-adhering. Wheels roll and particles flow macroscopically without sticking. But it turns out that this is an illusion brought about by the small values of the work of adhesion, and by Bradley's rule. Bodies should always stick but this depends greatly on size.

Consider the wheel of a truck, which is 1 m in diameter. This does not adhere significantly to the road surface, and can be lifted up without significant sticking. However, applying Bradley's rule that the adhesion force should be the product of the diameter and the work of adhesion, which we know to be about 0.1 Jm^{-2}, then the adhesion force should be 0.1 Newton. This is the theoretical prediction shown in Fig. 2.10 on the Bradley's rule line. It is a small force, compared to weight, as can be seen by the logarithmic scales in the diagram.

Compared to the gravity force acting on the truck tyre, this theoretical adhesion force is a million times smaller. But we also know that the tyre is so rough that it does not make perfect molecular contact with the road surface, so the measured adhesion is another million times less, as shown on the dotted curve. Thus it is clear that the adhesion at the engineer's level is much smaller than the weight, and can normally be neglected. Racing car tyres are an exception to this because they are made much smoother to obtain better grip. In this case the adhesion is small but measurable.

The interesting thing about Fig. 2.10 is the way the forces change as the size of the body becomes smaller. The force of gravity falls with the cube of diameter and so drops rapidly for smaller bodies, whereas the adhesion falls more slowly, proportional to diameter. Thus for smooth particles, there is a transition around 1mm diameter, where gravity and adhesion are equal. There is a problem here because this suggests that 1 mm dust particles should cling to the ceiling if they are smooth enough to obey Bradley's rule. Fortunately, the surfaces are rough and so the bottom curve applies and the transition is near 20 μm. Thus, when we are dusting, we tend to find particles which are not 1 mm large but small enough to be just about visible. However, geckoes can use Bradley's rule, and can stick to ceilings by increasing the number of their contact spots as described in Chapter 6.

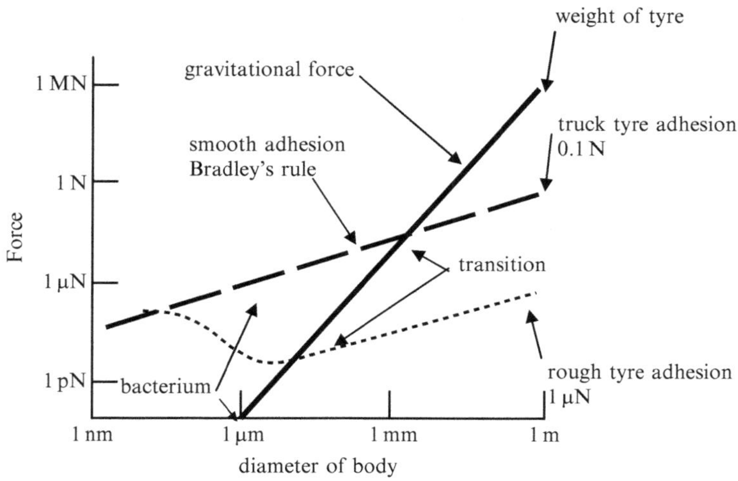

Fig. 2.10 Adhesion between bodies of different size, compared to gravity, showing Bradley's rule and the influence of roughness

Consider now a bacterium cell which is 1 μm in diameter. This now behaves in a completely different way to the truck tyre, as seen in Fig. 2.10. Because of the rapid decline in gravitational force with diameter, the weight of a bacterium is now extremely tiny, below 1 pN, less than a single weak chemical bond force. But adhesion according to Bradley's rule has not declined so fast and is around 1 μN for a smooth bacterium or 1 nN for a rough surface contact about the same as a chemical bond. Thus a bacterium in dry conditions will always stick to a surface and cannot behave like a truck tyre which exhibits near zero adhesion. A virus particle is 10–50 times smaller than the bacterium and consequently from Fig. 2.10 will be even more adherent compared to gravitational and fluid forces. Therefore it will be even more subject to van der Waals forces. Of course, the presence of water reduces adhesion considerably, and the addition of surface contamination (e.g. adhesion molecules) drops adhesion still further, even to zero or negative values.

The conclusion from this argument is that Bradley's rule can explain the transition from the clean macroscopic engineering world, where nothing sticks, to the nano-world of the virus where everything sticks.[11] The transition for smooth spheres in clean air is about 1 mm, depending on the density. Larger than 1 mm, ball bearings roll around and behave as we expect from our common experience. Below this they should stick, as smooth silicone rubber spheres are known to do. However, roughness introduces another dimension into this argument because true molecular contact is not achieved. This shifts the transition down to smaller particles, around 20 μm in size. Thus there is a grey area of transition where particles can behave in a schizoid way depending on surface roughness, sometimes sticking, sometimes not, between one micrometre and one millimetre in particle size. This is the size range occupied by cells which therefore should be expected to have complex adhesive behaviour. Contamination with water or adhesion molecules will reduce this still further. We would like to emphasise that this region, where the adhesion force of individual cells is on the same order of magnitude as their gravitational force, is essential for multicellular organisms and therefore the prerequisite for our very own existence. It allows for switching between an adherent and a free state due to small changes in topography or surface chemistry. One could say that no 'real' life is possible in the 'always sticky' and the 'never sticky' regime.

2.9 Adhesion of Spheres; Hertz Theory

Although Newton had measured the black spot at the contact of glass spheres, and also seen the black spot expand and contract reversibly as the spheres were pressed together, he did not pursue the relationship between the spot size and the load. Almost 200 years were to elapse before Hertz defined the connection published in two papers of 1881 and 1882.[12] Hertz was a 23 year old assistant to Helmholtz in Berlin when he was stimulated by Newton's rings and derived the elastic theory of sphere contact in his Christmas vacation in 1880.[13] He found that the spot diameter increased with the cube root of load F, showed that the elastic modulus E, Poisson's ratio ν and sphere diameter D were also important, and verified his equation

$$d^3 = 3\,(1-\nu^2)\;FD/E \qquad (2.2)$$

which applies to equal spheres, by measuring contact spots for glass and metal spheres, as shown in Fig. 2.11. Hertz was a prodigy; unfortunately he moved onto radio waves rather than studying the influence of adhesion on the sphere contacts. For spheres of different diameters D_1 and D_2 and materials of different elastic constants, the effective diameter $D = 2\,D_1\,D_2/(D_1 + D_2)$ and effective modulus $E/(1-\nu^2) = 2\,E_1E_2/(E_1 + E_2)$ could be substituted into Eq. 2.2, where $E_1 = E_1/(1-\nu_1^2)$ and $E_2 = E_2/(1-\nu_2^2)$.

The brilliant contribution of Hertz was to understand that the spheres press into each other to give a hemispheric pressure distribution, with maximum pressure P in the middle, falling as $P(1-z^2/d^2)^{1/2}$ to zero pressure at the edge of the contact spot, where z is the diameter of a circle within the contact spot of diameter d. He also knew that the centres of the two spheres approached each other because of such pressure by a distance $\delta = d^2/D$ given by

$$\delta^3 = 9\,(1-\nu^2)^2\;F^2/E^2D \qquad (2.3)$$

All these expressions assumed that no adhesion or friction existed at the contact between the spheres. This was an excellent assumption for large loads such as those experienced in ball bearings, under train wheels, and where car tyres meet a road surface. The experimental measurements then fitted the theory very well. However, when the load was zero, or even tensile, as in the experiments conducted by Bradley[4, 14] in the 1930s on adhering spheres, it was clear that the spheres were still

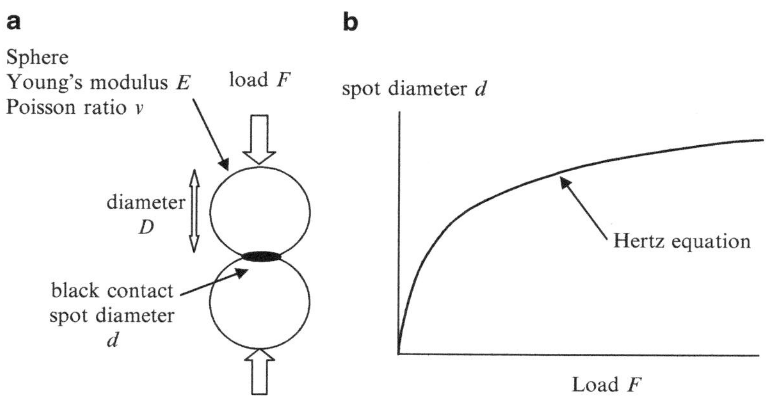

Fig. 2.11 (**a**) Two elastic spheres pressed into contact; (**b**) increase in spot diameter with load

deforming, but now the pressure was generated by the van der Waals attractions and not by the external load. The simplest way to take this into account was to assume that the van der Waals attractions were acting like an external load and to use Eq. 2.1 to describe the results, as Derjaguin did in 1934. His basic thermodynamic argument was correct because it equated the work done by the surface attractions against the work of deformation in the elastic spheres. However, the deformation Derjaguin used was not exactly correct because he did not take into account the effect of surface attractions on the pressure distribution at the contact.

2.10 The JKR Contribution

An improved solution to the problem of contact between elastic spheres with surface adhesion was obtained by Johnson, Kendall and Roberts[10] 37 years later. This came about because Roberts[16] and Kendall had both been supervised by Tabor while studying for doctorates in Cambridge, while Johnson had collaborated over many years with Tabor on the contact problems associated with friction and lubrication.[17]

Roberts,[18] while observing the contact of rubber windscreen wiper blades on glass, had noticed that the contact spot was much larger than he expected from Hertz theory under dry conditions, yet approached the Hertz predictions rather precisely when wetted with soapy water. Kendall[19] had been measuring the contact spot size between polymer, glass and metal surfaces using optical and ultrasonic methods, and became convinced that adhesion made the Hertz equation incorrect at low loads because the contact spot was larger than expected, as illustrated in Fig. 2.12.

The problem was to explain such increased values of contact size. Long before these experimental measurements, Johnson[20] had attempted to do this by showing that the pressure distribution within an adhesive contact could be described by adding two simple stress distributions together. However, he was puzzled by the

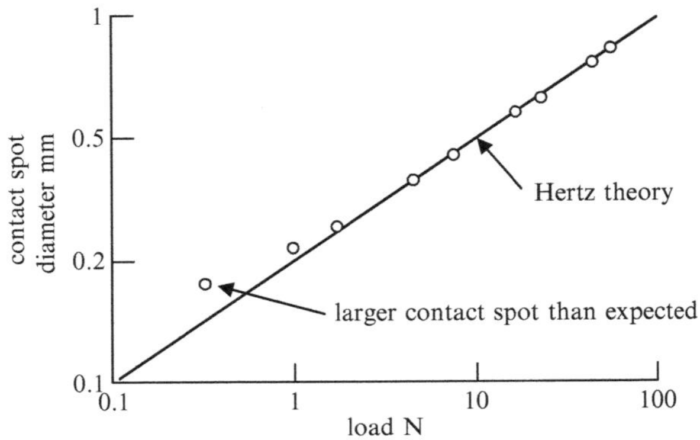

Fig. 2.12 Results for contacts between glass surfaces measured optically and by ultrasonics[19]

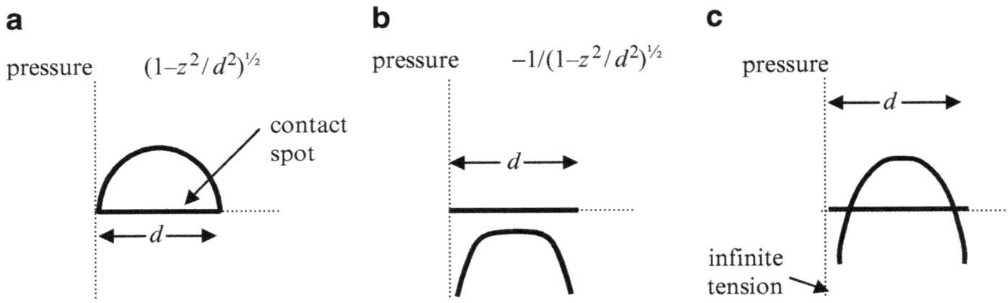

Fig. 2.13 (**a**) Hemispherical compressive stress at Hertzian contact spot; (**b**) rigid punch tensile stress distribution; (**c**) resultant pressure distribution obtained by adding (**a**) and (**b**)

resulting infinite stresses at the edge of the contact which he therefore expected to fail under the high tension. Figure 2.13 shows the way the component stresses add within the contact region.

It became clear in 1970 that the answer lay in applying the method of Eq. 2.5 to Johnson's stress distribution. Johnson did the mathematics to provide the correct equation, the so-called JKR equation, for the elastic contact spot diameter d of equal spheres, diameter D and elastic constants E and ν, with short-range work of adhesion W.

$$d^3 = 3\,(1-\nu^2)D\left\{F + 3\pi WD/4 + \left[3\pi WDF/2 + (3\pi WD/4)^2\right]^{1/2}\right\}/E \qquad (2.4)$$

Roberts and Kendall then did more experimental work on rubber/rubber and gelatin/poly(methyl methacrylate) contacts. These results fitted Eq. 2.4 extremely well, as shown in Fig. 2.14, and allowed the work of adhesion to be scaled to the observations. It later turned out that a similar mathematical argument had been produced by Sperling in 1964, but he had found no experimental evidence to support his theory.[21]

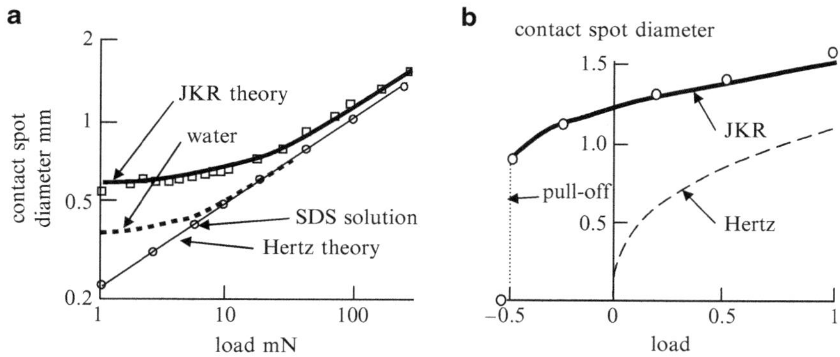

Fig. 2.14 (**a**) Results for contacts between two rubber spheres measured optically,[10] on log scale; (**b**) results for gelatin to glass contact plotted on linear scale showing tensile loads

From the contact spot size d_0 at zero load, where

$$d_0^{\,3} = 9\pi W D^2 (1 - \nu^2)/2E \tag{2.5}$$

the work of adhesion fitting the results for dry rubber contact was 71 mJm^{-2} and that for gelatin on poly(methyl methacrylate) was 105 mJm^{-2}. When water was present at the rubber contact, the work of adhesion dropped to 6.8 mJm^{-2}, and this was consistent with Young's equation for the contact angle of 66° measured for water droplets sitting on the smooth rubber. This was the first time that Young's equation had been verified by direct measurement. When 0.01 M sodium dodecyl sulphate (SDS) solution was the immersion medium, the rubber contact size fitted the Hertz equation down to the lowest loads obtainable, showing that the work of adhesion was less than 1 mJm^{-2}. This was consistent with the presence of the surfactant film preventing adhesive contact between the wetted rubber surfaces.

2.11 The Nature of Adhesive Contact Between Polymer Spheres

This theory and its supporting experimental evidence changed the conception of adhesive contact considerably (Fig. 2.15). Not only was it evident that van der Waals adhesion could have a considerable effect for small compliant particles and for small loads, where the particles were significantly attached by the adhesive forces, but also the elastic contact was acting as its own measuring device which sensed adhesion. In short, van der Waals adhesion between solids could be measured by observing the size of Newton's black spot, while knowing the elasticity and geometry of the particles. The black contact spot was an adhesion sensor.

The assumption of very short-range molecular force, which acts only within the contact spot, is reasonable for contact sizes larger than a few nanometres, though

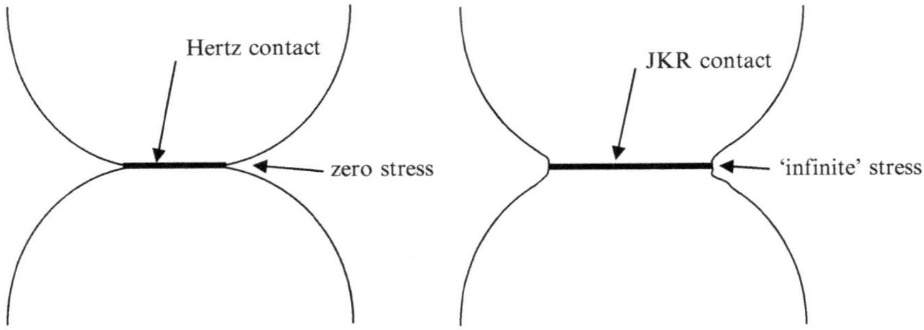

Fig. 2.15 Hertz contact on *left* becomes JKR contact on *right* as adhesion pulls the surfaces together to change the shape of the elastic surfaces

much discussion has continued on this.[22–24] There is then an 'infinite stress' at the contact edge, exactly the same kind of stress singularity found in cracking problems, as shown by Griffith. Obviously, such an 'infinite stress' cannot exist in reality. Near the crack tip, the molecules must be fluctuating rapidly with Brownian movement and the crack will be making and breaking many times per second. An applied compression force pushes the equilibrium towards more contact, as shown by considering a molecular dynamics model.[22]

On the other hand, when a tensile force is applied to pull the spheres apart, the black contact spot shrinks. An equilibrium contact spot size could be obtained as the load was reduced, but below a certain contact size, equilibrium could no longer be found and the surfaces then came apart rather quickly at a load given by the JKR analysis

$$F = -3\pi WD/8 \tag{2.6}$$

where D was the diameter of the equal spheres.

2.12 Application to Nanoparticles, Viruses and Cells

These ideas are relevant to both viruses and cells because we can treat living spherical organisms as complex polymer particles. In the same way that nanoparticles can be imaged and probed in contact with each other or adhering onto a surface, cells and viruses can be observed and prodded to understand their behaviour. But, because the cells and viruses are so small compared to the spheres described above, much more sensitive instruments are needed to measure the small forces. Typically, the Atomic Force Microscope (AFM) is required to detect the forces which range down from microNewtons to nanoNewtons.

Figure 2.16 shows an Atomic Force Microscope (AFM) experiment in which a chain aggregate of carbon nanoparticles prepared by evaporation was stretched to test the force of adhesion holding the individual grains together.[25] Clearly, this force depends on the contact spot size between particles and this can be estimated by JKR or Maugis-Dugdale models.[26] JKR analysis has also been applied to manipulation

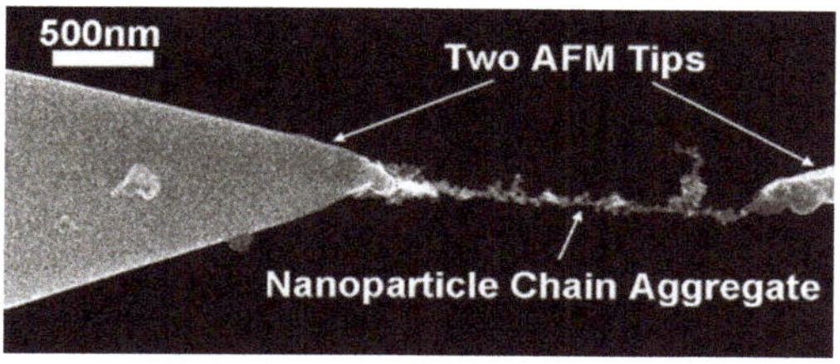

Fig. 2.16 A chain aggregate of carbon nanoparticles being stretched between AFM tips[25] with permission

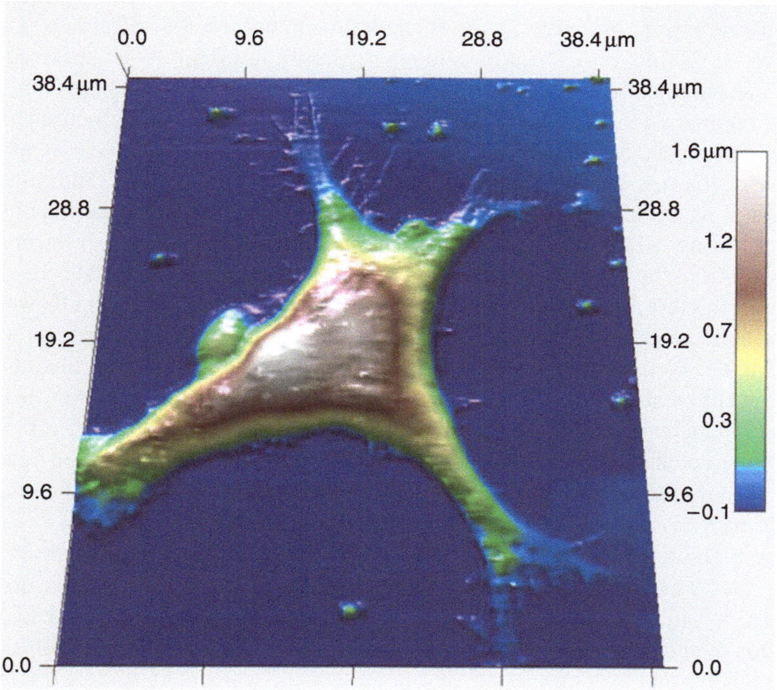

Fig. 2.17 Human lung fibroblast cell on glass coverslip surface as viewed by AFM (permission by Dr Murphy, Liverpool John Moores University)

of gold nanoparticles on silicon surfaces in order to understand the deformation of the particles, their sliding and pull-off behaviour.[27]

AFM can also be used to image cells using the standard silicon or Si3N4 probes as shown in Fig. 2.17. A human lung cell is shown, illustrating how the cell is pulled down almost flat onto the glass surface. If a virus particle is attached to the silicon tip, then the virus can be brought into contact with the cell and interaction forces measured. Such experiments have been carried out with virus particles, typically spherical virions of HIV.[28] The 100 nm diameter spheres were functionalised and attached to AFM tips, then brought into contact with a living GHOST parental cell, a well-defined cancer cell line.

After the virus particles were coated onto the AFM probe, the experiment was repeated and force curves were generated then interpreted in terms of single molecule interactions. The problem is that the deformations of the virion and the cell were not accounted for. There must be substantial contact between a sphere and a soft essentially flat cell surface and it seems unlikely that only one molecular bond of force 35 pN is formed as suggested by Wirtz et al. Figure 2.18 illustrates the model proposed by the authors. That model did not take into account simple elasticity considerations.

A more interpretable experiment was discussed by Roos et al. in 2007 on the compression of a virus particle against a plane solid surface using an AFM probe.[29] The mechanical properties of the virus particles could be measured by squeezing 40 separate virions and plotting the force deflexion curves, the stiffness was calculated to be 523 pN/nm. The adhesion energy could in principle be extracted from these measurements but this was not done.

By comparison, most force experiments on single cells have focused on the elastic and plastic deformations of the particles. For example, a typical experiment on yeast cells was carried out by Ren, Donald and Zhang in 2008.[30] The cells were imaged in an environmental scanning electron microscope and remained alive for 5 min, during which they could be compressed to understand their mechanical behaviour. Previously,[31] such cells had been probed between glass plates in air in a compression test machine to give the results shown in Fig. 2.18. The cells were not fully elastic and showed substantial time dependent deformation. However a computer model of the cells was found to fit the data. In this example, the adhesion forces were small, especially under water, and were neglected. However, from the contact diameter of 1 μm in air at zero load, and the cell diameter of 4.5 μm, taking the measured elastic modulus as 127 MPa, the work of adhesion between yeast and glass in humid air was calculated from Eq. 2.5 (modified for sphere/flat contact) to be 146 mJm^{-2}, comparable with that obtained by peeling.

These results were comparable with optical contact black spot measurements on swollen agarose beads[32] and on microcapsules.[33] Work of adhesion down to 0.01 mJm^{-2} in phosphate buffered saline solution could be measured by these methods to show the effects of osmotic pressure and temperature. It is clear from these results that contamination of the cells with water and surface molecules reduces the adhesion from about 100 mJm^{-2} in air to much lower values, less than 1 mJm^{-2}.

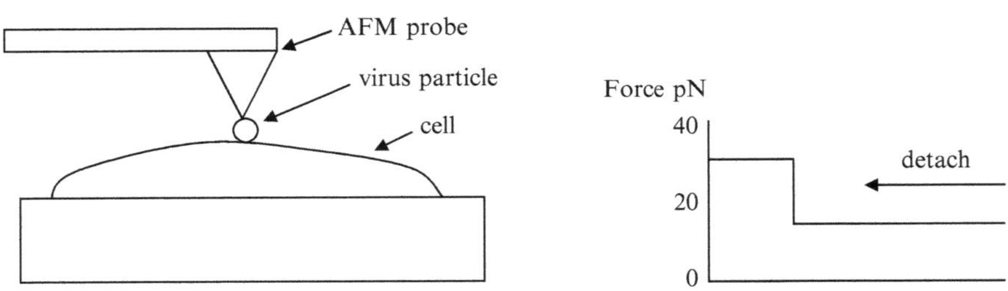

Fig. 2.18 *Left*: AFM probe with virion attached touching cell surface; *right*: result showing typical 30 pN jump on separation of the virion from the cell

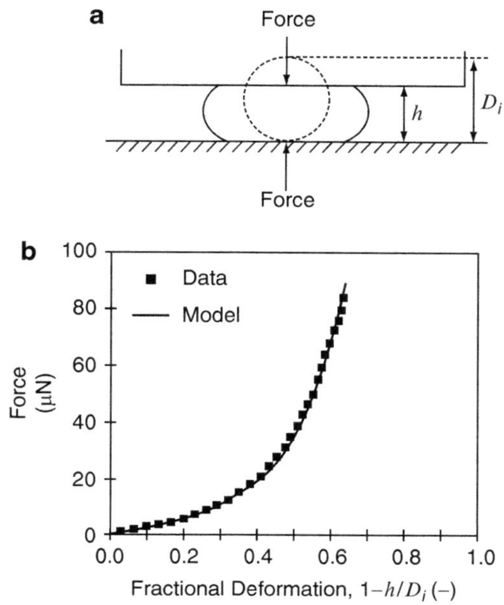

Fig. 2.19 (**a**) An optic fibre is positioned above a single cell and moved down to compress it, measuring the force. (**b**) Force versus deformation results fitted the finite element model[31] ignoring adhesion; reprinted with permission

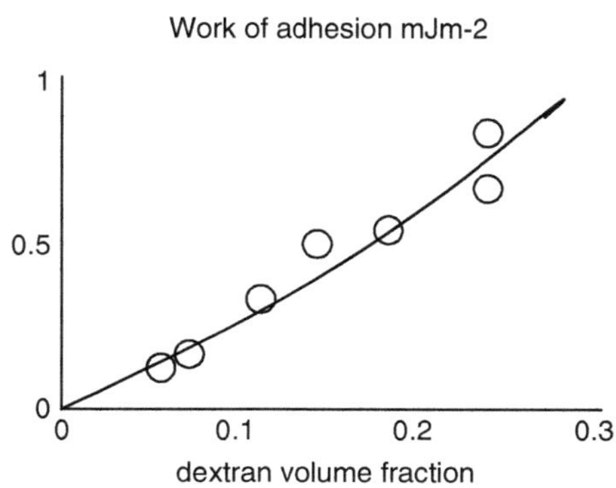

Fig. 2.20 Work of adhesion for S180 cells increasing with added dextran compared to the theoretical line from depletion theory[34]

Another interesting study of small adhesive forces was carried out by Chu et al. in 2005[34] where the measurement technique was advanced over the standard micropipette methods.[35] S180 cells, which usually do not adhere because of repulsive forces, were brought into contact using micropipettes and the work of adhesion was measured in solutions of dextran at increasing concentrations. Dextran is a non-adsorbing polymer which pushes particles together by the depletion effect, causing small adhesion which could be measured by the JKR method. The results are shown in Fig. 2.20, fitting the depletion theory of de Gennes.

2.13 Conclusions

It is evident from this chapter that particles in general adhere by van der Waals forces which can be defined in terms of W the work of adhesion, taking into account the geometry, elasticity and forces applied in the system. Viruses and cells can be viewed like polymer particles or capsules which can be subjected to the same measurement processes and analysed by conventional theories. Of course, cells are not perfectly smooth, nor spherical, nor elastic but valuable data can be collected in an effort to understand their adhesion behaviour as affected by surface topography and surface chemistry (e.g. adhesion molecules). Dry cells stick best and the effect of water is to reduce the adhesion considerably. Further addition of contaminant molecules gives additional changes in adhesion depending on the charge, the surface adsorption and the size of the molecules. Adhesion molecules reduce adhesion.

References

1. Newton, I., Opticks, Smith and Walford, London 1704, reprinted Dover, New York, 1952, p. 376.
2. Jang, W., Gomer, R.H., Combining experiments and modelling to understand size regulation in Dictyostelium Discoideum, J R Soc Interface 5 (2008) S45–48; http://dictybase.org/; Bonner, J.T., The social amoebae: Biology of Cellular slime molds, Princeton UP, 2008; Kendall, K., Molecular adhesion and its applications, Kluwer, New York 2001, p.297; b) Cavey, M., Lecuit, T., Molecular bases of cell-cell junctions stability and dynamics, CSH Perspectives in Biol 1 (2009) a002998;
3. Wollaston, W.H., On a method of rendering platina malleable, Phil Trans R Soc Lond 119 (1829) 1–8.
4. Tomlinson, G.A., Molecular cohesion, Phil Mag 6 (1928) 695–712; Bradley, R.S., The cohesive force between solid surfaces and the surface energy of solids, Phil Mag 13 (1932) 853–862.
5. Rivlin, R.S., The effective work of adhesion, Paint Technology 9 (1944) 215–18.
6. Galileo, Two Sciences (1638),translated by S. Drake, Wisconsin University Press.
7. Griffith, A.A., The phenomena of rupture and flow in solids, Phil Trans R Soc Lond. A221 (1920) 163–198.
8. Everett, D. H., Basic Principles of Colloid Science, Royal Society of Chemistry, Letchworth, 1988
9. Derjaguin, B.V., Kolloid Zeits 69 (1934) 155–64.
10. Johnson, K.L., Kendall, K. and Roberts, A.D., Surface energy and the contact of elastic solids, Proc R Soc Lond A324 (1971) 301–313.
11. Kendall, K., 'Adhesion: Molecules and Mechanics', Science, 263 (1994) 1720–25.
12. Hertz, H., Miscellaneous papers, (ed.) P. Lenard, Macmillan London, 1896, p. 146.
13. Johnson, K.L., Contact Mechanics, Cambridge University Press, Cambridge UK 1985; see also Johnson, K.L., and Greenwood, J.A., J Coll Int Sci, 192, 326–333 (1997); and Maugis, D. 'Contact, Adhesion and Rupture of Elastic Solids, Springer, Berlin (1999) 284–296.
14. Bradley, R.S., The cohesion between smoke particles, Trans Faraday Soc 32, 1088 (1936)
15. Derjaguin, B.V., Krotova, N.A. and Smilga, V.P., Adhesion of Solids, (translated by R.K. Johnson), Consultants Bureau, London 1978, p 423–442.
16. Roberts, A.D., Squeeze films between rubber and glass, J. Phys. D: Appl. Phys. 4 (1971) 423–432.

17. Bowden, F.P. and Tabor, D., Friction and Lubrication of Solids, Part1 (1950) & Part 2 (1964), Clarendon Press Oxford.
18. Roberts, A.D., Preparation of optically smooth rubber surfaces, Engng Mater Des 11, (1968) 579.
19. Kendall, K., The stiffness of surfaces in static and sliding contact, PhD Thesis, University of Cambridge, 1969.
20. Johnson, K.L., A note on the adhesion of elastic solids, Brit J Appl Phys 9 (1958) 199–200.
21. Sperling, G., Doktor-Ingenieurs Dissertation, Technischen Hochschule Karlsruhe, 1964, 75.
22. Gilabert, F.A., Krivtsov, A.M., Castellanos, A., A Molecular dynamics model for single adhesive contact, Meccanica 41 (2006) 341–349.
23. Johnson, K.L. and Greenwood, J.A., An adhesion map for contact of elastic spheres, J Coll Int Sci, 192 (1997) 326–333.
24. Maugis, D. 'Contact, Adhesion and Rupture of Elastic Solids, Springer, Berlin (1999) 284–296.
25. Rong, W., Ding, W., Mädler, L., Ruoff, R.S., Friedlander, S.K., Mechanical Properties of Nanoparticle Chain Aggregates by Combined AFM and SEM: Isolated Aggregates and Networks, 6 Nanoletters (2006) 2646–55.
26. Grobelny, J., Namboodiri Pradeep, N., Kim, D., Estimation of contact area of nanoparticles in chains using continuum elastic contact mechanics, J Nanoparticle Res. (2008) 10.1007/s11051-008-9434-8.
27. Korayem, M.H., Zakeri, M., Sensitivity analysis of nanoparticles pushing critical conditions in 2-D controlled nanomanipulation based on AFM, Int J Adv Manufacturing Technol (2008) 10.1007/s00170-008-1519-0.
28. Dobrowsky, T.M., Zhou, Y., Sun, S.X., Siliciano, R.F., and Wirtz, D., Monitoring Early Fusion Dynamics of Human Immunodeficiency Virus Type 1 at Single-Molecule Resolution, J Virology 82 (2008) 7022–33.
29. Roos, W.H., Ivanovska, I.L., Evilevitch, A. and Wuite, G.J.L., Viral capsids: mechanical characteristics, genome packaging and delivery mechanisms, Cell Mol Life Sci. 64 (2007) 1484–97.
30. Ren, Y., Donald, A.M., Zhang, Z., Investigation of the morphology, viability and mechanical properties of yeast cells in ESEM, Scanning (2008) 10.1002/sca.20126.
31. Smith, A.E., Zhang, Z., Thomas, C.R., Moxham, K.E., Middelberg, A.P.J., The mechanical properties of Saccharomyces cerevisiae, PNAS 97 (2000) 9871–4.
32. Moy, V.T., Jiao, Y., Hillmann, T, Lehmann, Sano, T, Adhesion energy of receptor mediated interaction measured by elastic deformations, Biophys J. 76 (1999) 1632–8.
33. Liu, K.K., Chan, V., Zhang, Z., Capsule-substrate contact deformation: determination of adhesion energy, Med biol eng comput 40 (2002) 491–495.
34. Chu, Y.S., Dufour, S., Thiery, J.P., Perez, E., Pincet, F., Johnson-Kendall-Roberts Theory Applied to Living Cells. Phys Rev Let, 94 (2005) 028102
35. Curtis, A.S.G., Lackie, J.M., Measuring cell adhesion, John Wiley, New York 1990.

Chapter 3
Modelling Nanoparticle, Virus and Cell Adhesion

no free outgrowth of nerves in a fluid medium has ever been observed ... solids serve readily to support them

Harrison (1914)

As we start to look at nanoparticles, viruses and cells, it becomes apparent that the one parameter model described in the previous chapters breaks down. The reason is that the van der Waals forces act over a certain distance which becomes comparable in size to the particles themselves. The approximation that work of adhesion alone is sufficient to describe adhesion then becomes unsatisfactory and a new model is needed. This chapter is the main theoretical part of the book and contains many equations. By the end, however, we are confident that readers of all backgrounds will realize that nanoscale adhesion results from several electromagnetic terms which are relevant to biological systems.

It was evident to Isaac Newton[1] three centuries ago that two terms were needed in the theory of adhesion when he wrote in his book Opticks that 'where Attraction ceases, there a repulsive virtue ought to succeed'. In fact it has become clear recently that there are more than two terms in the basic equations because, although the repulsive term is straightforward, the attractions are more complex: they can be ionic, covalent and/or van der Waals. By adding these several components, a realistic model of adhesion between atoms can be produced, then extended to larger bodies by the addition of adhesive forces from all atoms throughout the bodies. This must then be moderated by the incessant motions of the atoms and particles as a result of thermal (i.e. Brownian movement), to give a dynamic picture of adhesive contact. Also, we cannot ignore the double layer electrostatic forces resulting from the ion clouds around the particles in solution. Finally we must take into account adhesion molecules on the surfaces. These must be squeezed out of the way to give oscillating repulsions as cells approach.

The main purpose of this chapter is to consider the theory of these forces between atoms, then to add the forces together to give equations for particles approaching contact. After that, the influence of small contaminant molecules like water adsorbing on the surface is shown to reduce adhesion considerably, especially if there are ionic double layers to generate electrostatic repulsions. Similarly, the influence of

K. Kendall et al., *Adhesion of Cells, Viruses and Nanoparticles*,
DOI 10.1007/978-90-481-2585-2_3, © Springer Science+Business Media B.V. 2011

longer surfactant molecules and larger surface structures such as adhesion molecule projections from the surface is to reduce adhesion. Finally, we argue how adhesion molecules can interact with adsorbed surface water to lower the energy barrier for contact so as to aid the entry pathways into cells.

The second objective of this chapter is to move away from old and simplistic ideas of 'lock and key' adhesion described by many authors[2], which do not fit Newton's logic nor the well-established concept of van der Waals bonding. Lock and key models have led to complicated concepts like antibody, receptor, ligand etc. Here we talk only of surfactants or adhesion molecules, which are molecules involved with water in the adhesion process. Whereas surfactants are small simple molecules like stearic acid which separate surfaces by about 2.5 nm, adhesion molecules are larger complex protein-based molecules which can push surfaces 10 or 20 nm apart, or even more when charged, greatly reducing van der Waals attractions. Of course, there are a several mechanisms by which these adhesion molecules can interact to alter adhesion phenomena and these will be explored in later chapters. But first we must consider the basics of adhesion between atoms.

3.1 Two Parameter Model of Atomic Interaction Force and Energy

Newton's idea that there were two terms in the expression for the force between atoms and molecules really began to take shape around 1900 when a number of theorists, like Mie for example, started to formulate mathematical equations to describe both an attractive term and a repulsive term[3] in Eq. 3.1 for potential energy V between two spherical atoms

$$V = -A/r^n + B/r^m \tag{3.1}$$

where r was the separation distance between the atom centres as shown in Fig. 3.1.

In this diagram, the potential energy between two isolated atoms is plotted on the vertical axis as the atoms are separated a certain distance, shown horizontally. Conventionally, the attraction is viewed as a negative potential given by the first term of the theory, and the repulsion is seen as positive, i.e. the second term. The two terms added together gave the total picture of the energy as shown in Fig. 3.1.

From this energy, the force F between the atoms can readily be calculated by taking the negative gradient of the top curve at any point. Thus the force is zero at the minimum of the energy curve. The minimum of the force curve is where the bond breaks as the atoms are pulled apart. This is the maximum tension which can be supported by the attraction.

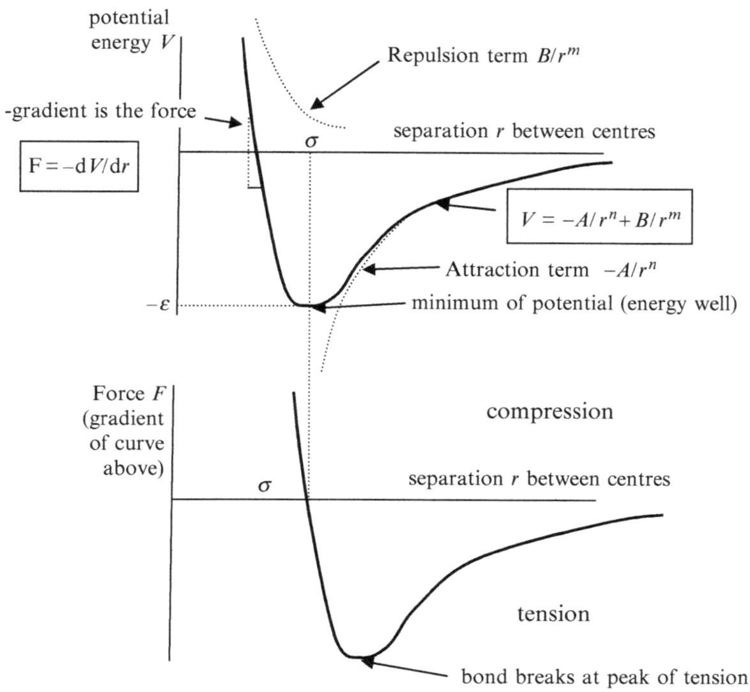

Fig. 3.1 Curves showing the two parameter model behaviour; the energy is given in the *top curve*, and the force in the *bottom curve*. The equilibrium point is at the minimum of the *energy curve* and zero of the *force curve*

The top curve shows that there is a minimum in the energy $V = -\varepsilon$ at a separation σ. This is the equilibrium point, that is the point at which the repulsion and attraction balance. It is an 'energy well' by analogy with a water well, with the atoms sitting at the bottom of the energy well. Lennard-Jones[4–6] applied this two parameter model in the 1920s to account for the properties of inert noble gases such as argon whose behaviour is dominated by van der Waals forces. With his colleagues Taylor and Dent[5], he realised that all known intermolecular forces are electromagnetic in nature. Argon does not experience the Coulombic forces between ions, nor the electron exchange force of covalent bonding, nor the dipole forces between polar molecules, so it is the weak instantaneously induced dipole van der Waals force which provides the attraction in this case. These adhesive forces, sometimes called London, London-van der Waals, dispersion forces, or simply van der Waals forces are always attractive because they result from instantaneous dipoles in one atom and their induced dipoles in a neighbouring atom. The Lennard-Jones equation can be written

$$V(r) = 4\varepsilon\left[\left(\frac{\sigma}{r}\right)^{12} - \left(\frac{\sigma}{r}\right)^{6}\right] \tag{3.2}$$

giving the energy V against distance r of the interaction as a function of the two parameters, ε the van der Waals bond energy and σ the bond length. By inserting values of these two parameters into the equation, Lennard-Jones could successfully calculate the properties of the atomic bond, especially confirming the measured viscosity of a rare gas like argon, its equation of state, the spacing of atoms in a crystal and the compressibility of the solid. The power of the repulsive term was between 9 and 20 and was typically 12. When this power was high, e.g. 20, then it approximated to a 'hard sphere' potential, such that the repulsion was very much like a hard wall. Later it was found that a whole range of two parameter models could be used to describe the experimental information. For example, Morse[7] fitted an exponential curve to both the attractive and the repulsive potentials. Buckingham had a power law attraction and an exponential repulsion. Born-Mayer, Eq. 3.3, fitted the alkali halides surprisingly well while ignoring the van der Waals force;

$$V = -ze^2/r + Ae(-br) \tag{3.3}$$

However, the power law expression of Lennard-Jones was most memorable because the attractive term corresponded very neatly with the London derivation of the van der Waals potential between two non-polar molecules, such as argon.

London[8] was working on a way to calculate the interaction between two atoms by solving Schrodinger's equation of quantum mechanics, starting from the known atomic structure and the fundamental constants. In practice this approach was too complex, so approximations and compromises were necessary. London took the bold step of replacing the molecule with a harmonic oscillator, allowing much simpler mathematical analysis to give his r^{-6} attractive energy which fitted experiments with inert gases and the Lennard-Jones equation.

3.2 Demands on the Model

The purpose of the model is to reconcile the known atomic forces above with the evidence obtained from adhesion experiments. This evidence is of two types: large scale peeling or fracture tests as shown in Fig. 3.2 below and atomic force microscope (AFM) probing of forces between nano-scale tips.

Figure 3.2 shows the results obtained by Tabor and his PhD students Winterton and Israelachvili on the attractive force measured between atomically smooth crossed cylinder mica surfaces brought carefully together into full contact from separations less than 10 nm between the surfaces. The results under dry conditions fitted the solid curve, showing on the right-hand axis the attraction building up to 40 μN at a gap of 2 nm. The surfaces then jumped into contact and it was difficult to measure the mN forces shown on the increased scale at the left of the diagram. But the main conclusion was that the results fitted the equation which Hamaker[12] had worked out from the van der Waals forces added together for all the atoms in the crossed cylinders.

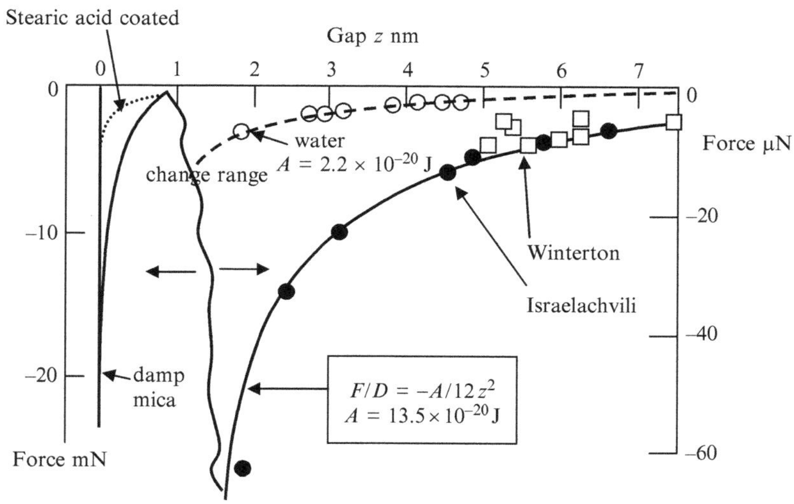

Fig. 3.2 Results of surface force apparatus tests on crossed mica cylinders[9–11] compared with Hamaker theory[12]

$$F/D = -A/12z^2 \tag{3.4}$$

where F was the attractive force, D the cylinder diameter, A the Hamaker constant and z the gap between the surfaces. The Hamaker constant is a material property which depends on the polarisability of the atoms, reflecting their ability to form instantaneous induced dipoles. The open circles are experimental points for the force under water, which reduced the Hamaker constant by a factor of about 6. It is clear that contamination of the surfaces reduced adhesion substantially, but that was beneficial experimentally because it allowed smaller gaps to be measured and less catastrophic jumping into contact.

The conclusions from these experiments are

(a) The atomic adhesion force falls off with r^{-7} (potential with r^{-6}) but when the atomic forces are added for spheres or crossed cylinders the fall is with gap squared i.e. z^{-2}.
(b) Absolutely smooth surfaces are needed because the gaps are less than 10 nm.
(c) Molecules on the solid surface lower the attractions by reducing the Hamaker constant typically from 13.5×10^{-20} to 2.2×10^{-20}J.

Further evidence from the surface force experiments on mica[13] showed that at very close approach under water, the attractions oscillated tremendously as shown in Fig. 3.3.

This curve looks very different from that of Fig. 3.2 because the van der Waals attraction shown as the broken curve is now dominated by two other effects: first the electrostatic (DLVO) repulsions caused by ionic charges on the surfaces and

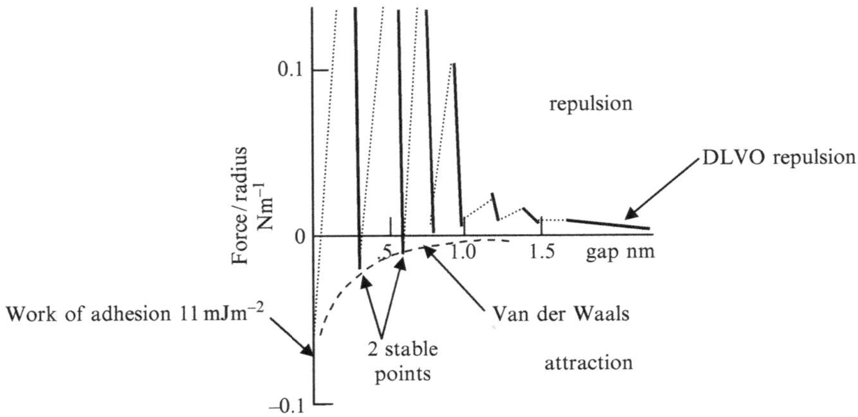

Fig. 3.3 Results for mica surfaces approaching in potassium chloride solution[13]

second by the large force required in steps to squeeze several layers of adsorbed water from the gap as contact is made. Clearly, the contaminating molecules and ions are having a huge moderating effect on the van der Waals adhesion. These influences are the key to understanding adhesion molecules as described later in 3.9–3.11. But before considering the effect of contaminant molecules, let us look at the adhesion of a clean nanoprobe approaching a smooth surface.

3.3 Modelling Simple Nanoparticle Adhesion

In realistic atomic models, which have been developed to describe adhesive behaviour of nanoparticles, four steps are necessary

(a) Defining an interatomic potential which is acceptable.
(b) Adding potentials together for all the atoms in the system.
(c) Calculating thermal motions of the atoms to take the dynamic nature of the system into account.
(d) Working out the equilibrium condition.

Here we start by describing models of simple ionic solids such as sodium chloride and magnesium oxide whose parameters are very well understood. The overall method is described as molecular dynamics and is carried out on large computers with around a million atoms under consideration. This is still not enough to model the smallest virus particle. However, the largest computers are able to calculate the movement of more than 10 billion atoms which could eventually be sufficient to simulate a whole virus.

A typical computational package is DL_POLY developed at Daresbury Laboratory by Smith, Forester and Toporov.[14] In recent papers,[15–17] a nanoparticle of sodium chloride was modelled as it approached a plane NaCl crystalline surface

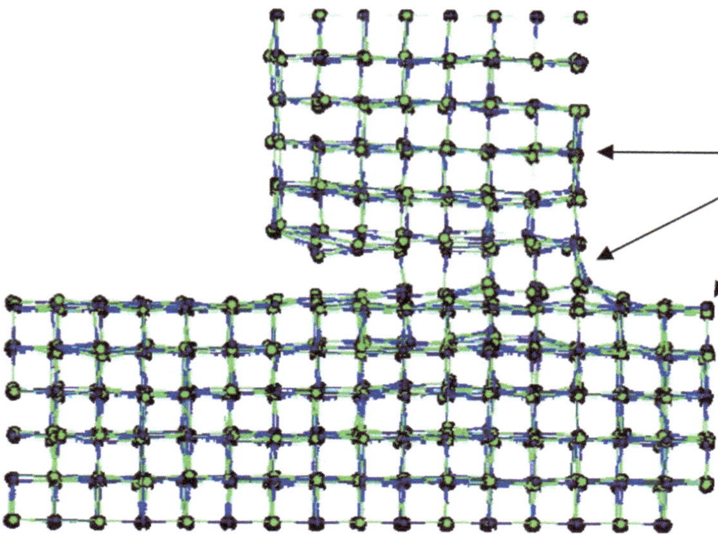

Fig. 3.4 Molecular dynamics model of a sodium chloride nanoparticle approaching a slab of NaCl to form new bonds with the surface atoms

and the equilibrium was calculated for each incremental step as the contact process occurred. A still picture from the movie is shown in Fig. 3.4, but this does not reflect the true situation because all the atoms are in continuous motion. A movie is available[18] showing all the atoms vibrating around their equilibrium positions as the nanoparticle approaches and then retracts from the NaCl plane.

It is evident from this model that there are no locks and keys necessary here for adhesion. In fact, as Newton postulated, atomically smooth surfaces adhere best.

The potential chosen for this simulation contained three terms, Eq. 3.3.

$$w = -e^2/r - C/r^6 + A\exp(-r/\rho) \tag{3.5}$$

where e is the electron charge, r is the distance between atoms and C, A, ρ are constants fitted empirically to the experimentally determined properties of NaCl such as elastic constants and crystal lattice parameters. The first term is the electrostatic attraction between Na and Cl ions, the second the van der Waals attraction and the third term is the repulsion between the spherical rigid ions, the last two comprising the well-known Buckingham potential. Later papers[19] have shown that the electrostatic and repulsion terms can show the main effects of 'jump to contact', elastic and plastic adhesion, giving Hertzian like behaviour for spheres above 10 nm diameter, but leaving out the van der Waals term is not acceptable. Continuum models with the insertion of Lennard-Jones potential have also been attempted[20] using finite element analysis, showing that a mix of continuum and molecular models could be beneficial.

For Fig. 3.4, the potentials were added pairwise, evaluating the long-range electrostatic potentials by 3D periodic Ewald summation. Then, atomic trajectories were solved by the Verlet leapfrog algorithm with a fixed time step of 0.5 fs. Initially, all moveable atoms were assigned initial velocities from a Gaussian distribution equivalent to a temperature of 300 K. The system was then allowed to equilibrate until a stable mean configurational energy was achieved, usually taking about 80 ps.

3.4 Sodium Chloride Nanoparticle Results

The point of the molecular model is that it overcomes the key problems of the continuum descriptions like JKR[11] which presume a one parameter model and perfect spherical shapes. In reality, at the subnanometre level, the crystal planes must be flat, the pressure distribution cannot be JKR and the potential must contain two or more parameters as in Eq. 3.5.

A 2-D model of soft Lennard-Jones discs had been modelled by Quesnel, Rimai and DeMejo to show that the brittle and ductile adhesional behaviour of nanoparticles could be varied by adjusting the range of the interaction.[21] To improve on this with a 3-D model and a real potential, a clean NaCl plane slab was set up in DL_POLY.[22] This slab consisted of 28×28 rows of NaCl lattice with an NaCl distance of 0.279 nm, six layers deep as shown in Fig. 3.5. Periodic boundary conditions were applied in the x–y direction to mimic an infinite crystal surface with the (100) surface oriented at the z direction. A small rectangular NaCl probe six layers thick was constructed and this was varied in cross-section from the smallest possible 2×2 to 18×18 atoms to simulate increasing size of the nanoparticle probe.

The layers of NaCl atoms were classified into three different groups as shown in Fig. 3.5. The first three layers of atoms, labelled free, near the contacting surfaces where commensurate contact was to take place were allowed to move freely to reach their equilibrium positions. No constraint was placed on these outer layers of atoms and this ensured that the natural outcome of the atomic configurations at the surface

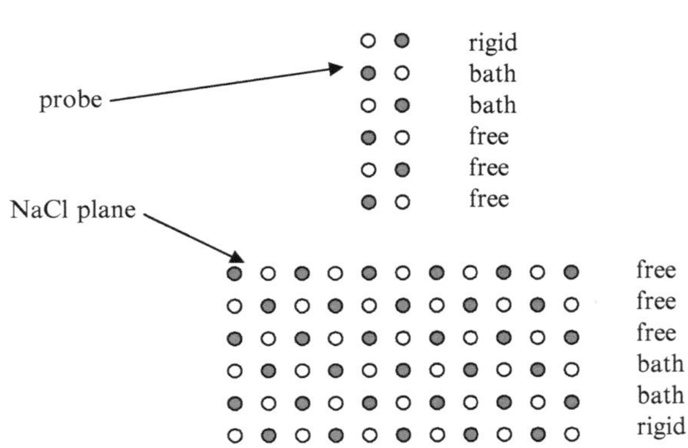

Fig. 3.5 Schematic of the smallest 2×2NaCl probe approaching an NaCl plane

was entirely due to the interatomic adhesion forces. However, the fourth and fifth layers of atoms were coupled to the Berendsen[23] heat bath to maintain the temperature of the whole system at 300 K. The final layer of atoms, labelled rigid, were held fixed and gradually displaced to push the nanoparticle towards the surface. The initial gap was 0.5 nm and this was gradually closed, giving increasing attractive force. A typical force versus distance curve for the 8 × 8 particle is given in Fig. 3.6.

Initially, the attractive force was low but, on further approach, a sharp change in F_z was observed, the 'jump to contact'. Further displacement caused the force to approach zero, the equilibrium contact position. Subsequently, the particle was compressed as the particle was pushed into the slab giving elastic indentation to the left of Fig. 3.6. The withdrawal was essentially reversible but the 'jump from contact' or fracture of the adhesive bond was slightly delayed, indicating adhesion hysteresis. By changing the size of the nanoparticle, the average work of adhesion W in joules per square metre could be calculated from the model and compared with the breaking stress (i.e. pull-off force/contact area) in GPa as shown in Table 3.1.

It can be seen that the work of adhesion remains almost constant, as assumed in the continuum theory of peeling of Chapters 1 and 2, but the stress decreases as the nanoparticle gets bigger as expected from fracture mechanics theory. In addition, it may be seen that the energy is dominated by the electrostatic forces for NaCl; a van der Waals particle would have W around 0.1 Jm^{-2}. Also the strengths are extremely high. Macroscopic crystals of NaCl break at stresses about 0.001 GPa because the adhesion force is proportional to size to the power 3/2 and not d^2, as shown in the plot of computed results from the model in Fig. 3.7. All these conclusions are consistent with the ideas on brittle adhesive fracture proposed previously.[11]

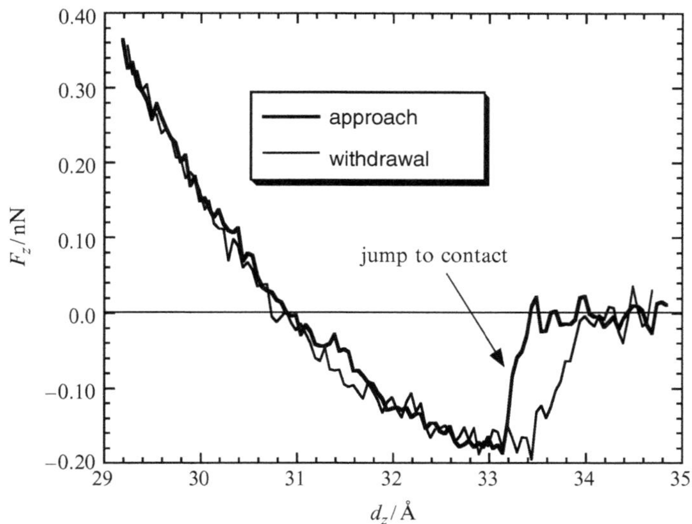

Fig. 3.6 Force per ion versus distance curves for approach from *right* to *left* and withdrawal of the 8 × 8 NaCl nanoparticle showing the jump to contact and the equilibrium at zero force

Table 3.1 Work of adhesion *W* and average adhesion stress for the different size nanoparticles

Contact size	Area/nm²	W/Jm⁻²	Stress/GPa
(2 × 2)	0.3114	0.47	4.03
(4 × 4)	1.2455	0.48	3.06
(6 × 6)	2.8023	0.44	2.55
(8 × 8)	4.9818	0.45	2.42
(10 × 10)	7.7841	0.45	2.38
(14 × 14)	15.257	0.44	2.08
(18 × 18)	25.220	0.43	2.03

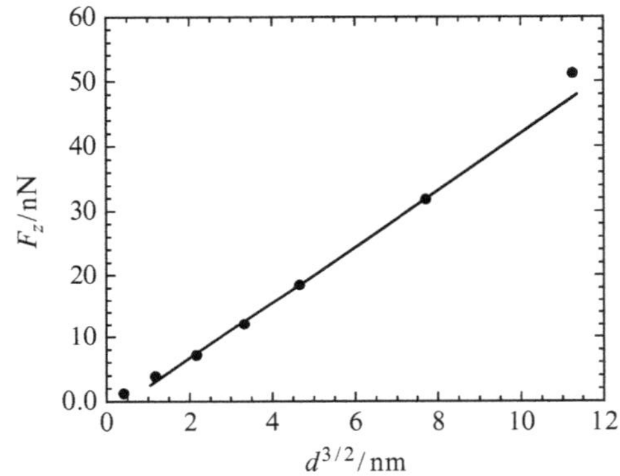

Fig. 3.7 Pull-off force versus nanoparticle size$^{3/2}$ from the model

3.5 Elastic and Plastic Adhesion

The results shown above are neat because the nanoparticle behaves elastically and separates cleanly from the slab despite the large attractive interatomic forces. More commonly, the forces rise so high as the separation occurs that the atoms in the nanoparticle are displaced from their normal positions and irreversible plastic deformation takes place.[22] Figure 3.8 shows the computed results for a stepped NaCl nanoparticle of contact size 4×4 for contrast with a stepped particle with contact 8×8. As stress is applied to the 8×8 contact (Fig. 3.8 top), separation does not occur, but the step acts as a defect which promotes plastic deformation, allowing the NaCl crystal to extend in line with the applied force. The smaller contact (Fig. 3.8 bottom) also begins to deform plastically, but separation also takes place at the interface such that the nanoparticle is released but extra energy is expended in creating a plastic defect hole in the particle. This model shows that there should be an elastic/plastic transition as the nanoparticles get larger,[17] making it more difficult to remove larger nanoparticles.

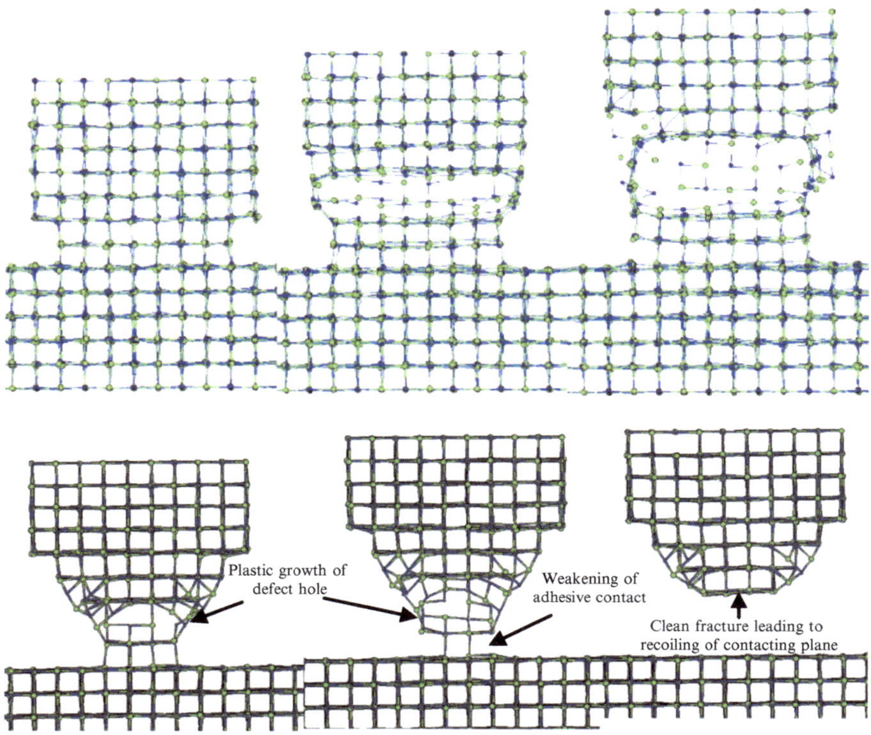

Fig. 3.8 (*Top*) Plastic deformation observed in the 8×8 contact model of the stepped nanoparticle. (*Bottom*) Computed results for 4×4 contact show clean detachment with some plastic deformation

There must be a limit to this effect because it is known that a ductile/brittle transition occurs for 1 μm particles where elastic effects again dominate for larger defect lengths.[24, 25] In conclusion, it is evident that a number of complex processes can influence nanoparticle adhesion including the nature of the atomic contact, its loading history, the shape of the particle and the propensity for plastic flow.[26]

3.6 Influence of Surface Contamination

The attractive forces above are seen to be very large, leading to plastic effects and irreversibility, with adhesive energies nearly 0.5 Jm^{-2}. Of course, this can only be true for atomically clean surfaces, as tested in high vacuum experiments with baked out materials.[27] In practice, ubiquitous contamination such as nitrogen or water adsorbs on the surface and this has a very large effect of reducing the adhesive force to much smaller values around 0.1 Jm^{-2} or even lower if thicker layers

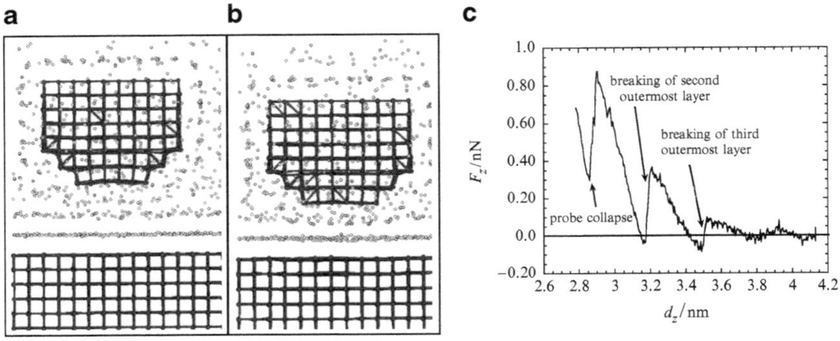

Fig. 3.9 (**a**, **b**) Stills from a movie of molecular dynamics model of MgO nanoparticle approaching an MgO slab coated with a monolayer of xenon molecules at 100K; (**c**) force distance curve approaching from *right* to *left*, showing steps as molecules are squeezed out[28]

of contamination build up. For example, Fig. 3.9 shows a pyramidal MgO model pushed towards an MgO slab in which the surface was covered by inert gas molecules i.e. xenon monolayer with van der Waals attributes.

The 6 × 6 MgO probe was surrounded by the Lennard Jones xenon-like spheres and it was immediately evident that these xenon molecules formed structured layers around the probe and upon the slab surface, with 3 distinct layers easily visible (Fig. 3.9a, b). Each sphere was weakly interacting with the surface atoms with an energy parameter of $\varepsilon = 0.02$ eV and $\sigma = 3.05$ Å at a temperature of 100 K. Figure 3.9c shows the response of the MgO probe in the presence of the contaminant, with the contact force increasing from right to left as the xenon in the gap was squeezed. The contamination prevented the jump-to-contact in this case but there were four significant steps; first at 0.02 nN where a slight bump could be seen, second at 0.1 nN as the third layer of xenon was penetrated at 3.5 nm; next at 0.35 nN as the second xenon layer was squeezed out at 3.2 nm; and finally at 0.9 nN as the probe collapsed plastically at high force because the first xenon layer could not be squeezed out by the probe.[28]

Because of this strong binding of the first contaminant layer, it is clear that this contaminated surface structure is much more likely to be found on earth than the ultra-clean model, because nitrogen, argon and other van der Waals molecules are readily available to adsorb. Also on earth, water molecules are abundant and these stick very well to the crystal surface and reduce the adhesion in the model. Indeed, the water layers can build up to much more than a monolayer, giving up to four monomolecular water films structured on the surface of the crystal. Figure 3.10 illustrates water molecules interfering with the approach of an MgO AFM probe towards an MgO slab.[28] The results indicate that the force oscillates as each layer of surface water is squeezed out of the gap. The final layer was too strongly attached to be shifted, and the model showed that the MgO probe was plastically deformed at stresses around 10 GPa, yet the last layer remained.

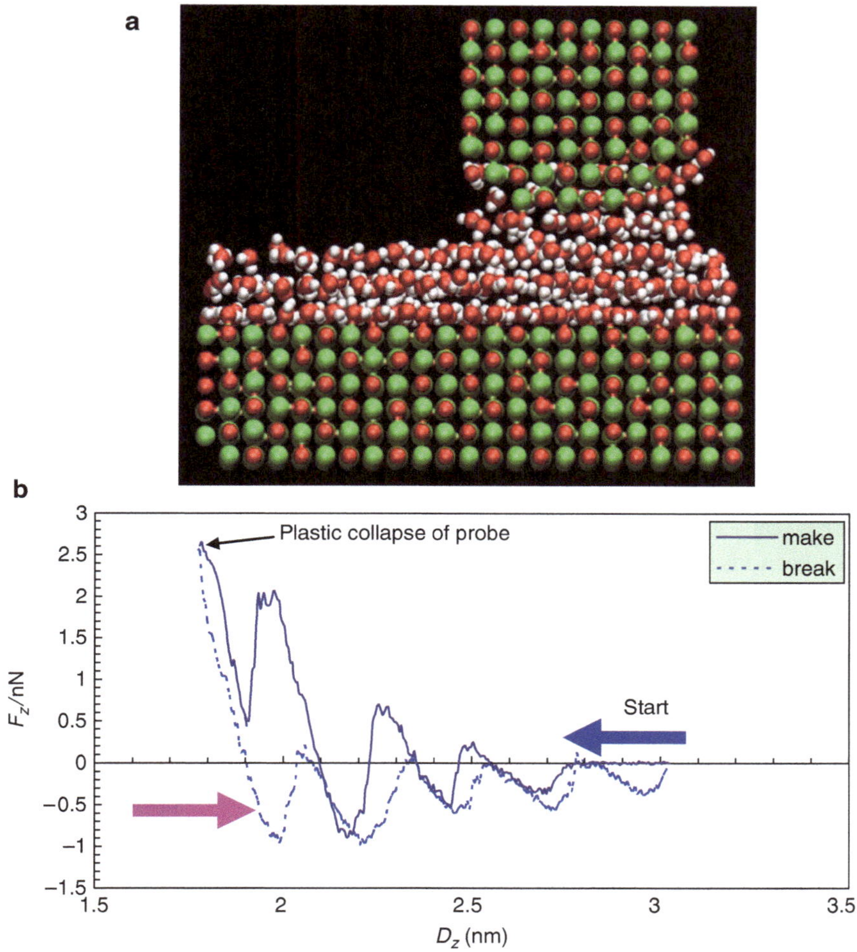

Fig. 3.10 (**a**) Molecular dynamic stills of movie from model of water on MgO nanoparticle contact. (**b**) Force vs gap showing the oscillatory prediction as water layers are squeezed out[28]

Of course, the interesting feature of this model is that there are now several adhesion energies at equilibrium, depending on the number of water molecule layers separating the nanoparticle against the kT of thermal interaction energy. This is shown by the presence of compression force required to squeeze the water layers out in Fig. 3.10b which illustrates that there are discrete equilibria of the adhesion at gaps of 0.25, 0.5, 0.75 and 1 nm, rather like the experimental results shown in Fig. 3.3. The question is 'How can the particles overcome these barriers to make contact?'

3.7 Effect of Surfactants

Water is not the only molecule which can reduce adhesion. It has been known since Langmuir's time[29] that molecules containing a hydrocarbon chain with an ionic group at one end behave as surfactants. In other words, such molecules tend to be attracted to solid and liquid interfaces and alter the adhesion and interaction between phases. Figure 3.11 shows two typical molecules, first the simplest stearate soap and second the lecithin (lipid) molecule which forms the backbone of cell membranes. Also shown is the schematic representation of a soap or lipid as a molecule with a polar head group which likes water (hydrophilic) and a long water-hating, hydrophobic tail. Soap made by boiling fats with alkali to form sodium stearate was the earliest detergent used for stopping dirt and grease sticking to clothes. Lecithin dissolved from egg yolk or extracted from soy bean oil by hydration is the classic lipid molecule which can phase separate from water to form interesting structures such as emulsions, cell membranes and liposomes.

The other major class of surfactants is dissolved polymer. Ever since Faraday[30] in his last paper described adding gelatin to his gold nanoparticle dispersions, polymers have been known to provide extra stability to dispersions by preventing adhesion. However, polymers can also cause flocculation, as shown in 1939.[31] In fact, certain polymers are used commercially both as stabilisers and as flocculants. For example, low molecular weight polyacrylic acid (Fig. 3.12e) is much used to reduce viscosity of clays and oxides in water by reducing particle adhesion, whereas high molecular weight polyacrylic acid is used as a flocculant for removing the same fine particles

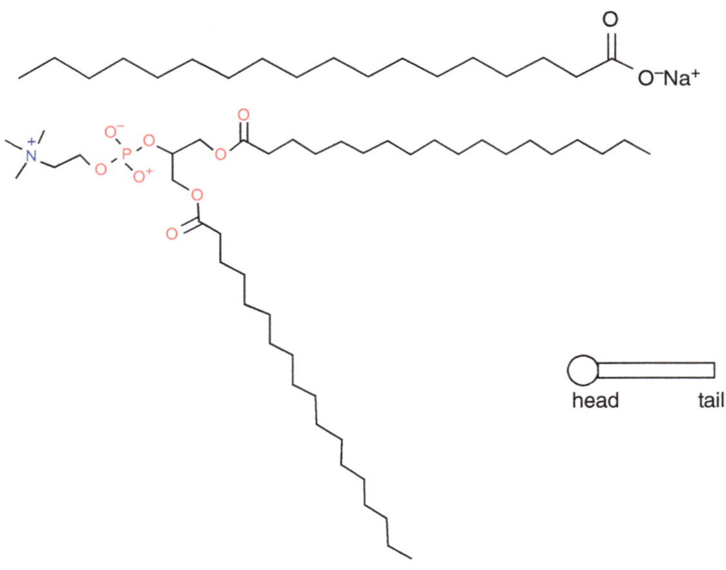

Fig. 3.11 Surfactant molecules; *top*, sodium stearate (soap); *bottom*, phosphatidylcholine (lecithin); *right*, diagrammatic model

during water treatment. We could view such high molecular weight polyacrylic acid as an adhesion molecule which binds silica particles together. In this model, the adhesion molecule is just an extended surfactant.

Figure 3.12a shows a typical vinyl monomer, acrylic acid, represented by a single hydrophobic chain with hydrophilic sphere attached (Fig. 3.12c), which forms long linear polymer chains of polyacrylic acid. Dissolved polymer molecules can enclose the polar groups in organic solvent (Fig. 3.12d) or behave as water-borne polar particles (Fig. 3.12e) which exhibit internal Brownian movement while also diffusing through the solvent like nanoparticles. The difference between a polymer particle and a silica particle is that the polymer can change its size and conformation because it is swollen by good solvents. Consequently, each segment can pulsate within the molecule, moving apart from other segments, and opening up to fill a larger volume. Also the polymer molecule can interact in a variety of ways with a solid surface. For example it can be strongly bound by one of its ends to the surface, or it can be weakly bound at several points along its length.

In the weakly bound state, shown in Fig. 3.13a, most of the polymer molecules coil up in solution to form particles which are considerably smaller than the extended chain length. A fully extended chain made up of N segments each of length l would be Nl long, whereas a polymer chain is known to coil up in a random configuration to give a distance $N^{1/2}l$ between the ends of the molecule.[32]

However, some of the molecules are adsorbed onto the particle surface, and these are in a dynamic equilibrium, constantly jumping on and off the particle.

The force acting between silica surfaces in polyacrylic acid solution has been measured by Atomic Force Microscopy,[33] giving the interaction curve shown in Fig. 3.13b. At large separations, the particles follow the van der Waals attraction curve as the silica

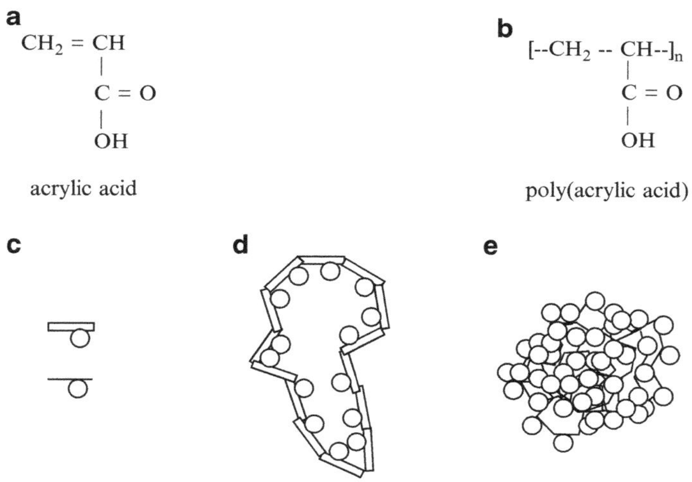

Fig. 3.12 (**a**) Acrylic acid monomer; (**b**) poly(acrylic acid) segment; (**c**) two representations of poly(acrylic acid) segment; (**d**) representation of polymer with 14 segments in organic solvent, molecular weight 1022; (**e**) polymer with 70 segments, M Wt 5110, randomly coiled in water

surfaces come closer together. But then the polymer molecules cause a repulsion as the polymer molecules resist desorption from the surfaces. This repulsion increases to a certain point but then the particles jump into a secondary minimum. As the particles are pushed further together, the polymer is squeezed out of the gap and eventually the silica particles make intimate contact in the primary minimum.

This curve is interesting because it shows a complex interaction with three minima, rather like the interaction of magnesium oxide particles with water molecules structured around the surface (Fig. 3.10). Typically, the secondary minimum occurred at a gap of 10 nm, indicating a swollen polymer particle diameter of 10 nm, and the tertiary minimum at a gap of 25 nm. This tertiary minimum was quite weak, with an attraction energy of about 10 *kT*, leading to soft adhesive behaviour, essentially a stabilized gel dispersion. However, the secondary minimum was very deep, around 30 *kT*, giving strong adhesion compared to Brownian movement. Once the silica particles get into this adhesive contact, they cannot break apart and are essentially strongly coagulated. Thus, polyacrylic acid can have both stabilising and flocculating effects in this system, depending on whether the particles are in the tertiary or secondary minimum (Fig. 3.13). The repulsions prevent strong adhesion of the particles, but the secondary minimum can produce aggregates with substantial strength. This aggregation can be viewed as the result of depletion of polymer molecules in the narrow gap between the particles, a gap so small that polymer tends to be excluded. The osmotic pressure of polymer molecules is diminished in the gap and so the particles are sucked together[34, 35] in steps depending on the size of the swollen polymer particle.

In this model of an adhesion molecule effect, the polymer is not showing any 'lock and key' behaviour. The molecule, swollen with water, binds weakly to the surface and reduces the van der Waals force between the silica particles. As the polymer is expelled from the gap, the natural attraction of the silica particles rises in steps, eventually forming tightly bound agglomerates.

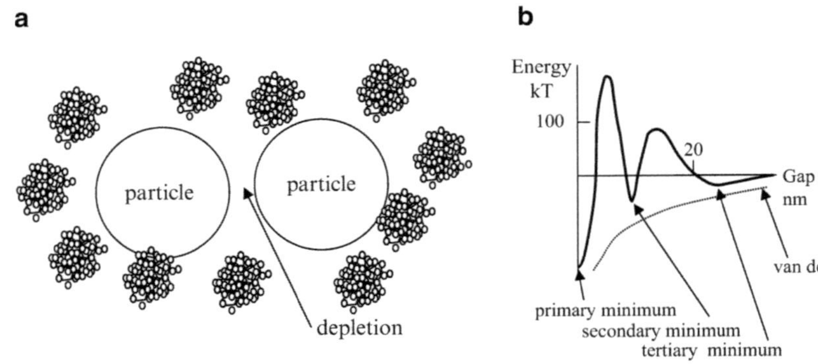

Fig. 3.13 (**a**) Representation of swollen polyacrylic acid molecules in water around two silica spheres; (**b**) interaction energy between particles in the presence of polymer compared with van der Waals interaction, showing the primary, secondary and tertiary minima

3.8 Proteins as Surfactants

Protein molecules can behave in a similar way because they are swollen polymer molecules which can interact with surfaces by weakly binding through their ionised groups.

Proteins are polymer molecules made from 20 available natural amino acid monomers (Fig. 3.14) linked by peptide bonds to form chains which can have complex and specific 3-dimensional structure because of the amine and hydrogen bonding between segments. An example is latherin,[36] a protein secreted on the skin of horses as they run to escape predators. This molecule at concentrations of 0.1% reduces surface tension of water from 73 to 55 mJm^{-2} and this allows spreading on the horse's skin to aid cooling. Perhaps the most important protein surfactant[37] for humans is lung surfactant which contains four surfactant protein components, Surfactant Protein A (SP-A), SP-B, SP-C and SP-D, mixed with 90% lipid. Without lung surfactant, the lungs would collapse and death would occur. SPA and SPD are members of the collectin subgroup of the C-type lectin family of adhesion molecules which feature in the immune response. SP-B and SP-C interact strongly with lipid molecules, especially DPPC (di palmitoyl phosphatidyl choline), to increase surface activity, giving exceptionally low surface tensions around 25 mJm^{-2}, but going to near zero under compression. By infusing these materials, or synthetic analogues e.g. Exosurf (a mixture of DPPC with hexadeconal and tyloxapol added as spreading agents) into the lungs of premature babies, many lives have been saved.[38]

The structure of such proteins has been widely studied.[39] The primary structure is organised into four regions: (i) a cysteine-containing N-terminus (required for disulfide-dependent oligomerisation) that is linked to (ii) a triple-helical collagen region composed of repeating Gly-X-Y triplets (associated with maintaining the molecules shape, dimension, stability and oligomerisation), followed by (iii) an α-helical coiled neck region (whose main function is protein trimerisation), and (iv) a globular structure at the C-terminus comprising a C-type lectin or CRD as in Fig. 3.15.

Protein based surfactants are also produced commercially because they are more biodegradable,[40] mild and biologically active than synthetic molecules made from petrochemicals. Although protein molecules naturally contain polar and hydrophobic groups along the amino acid chain, they can be made more hydrophobic by reacting

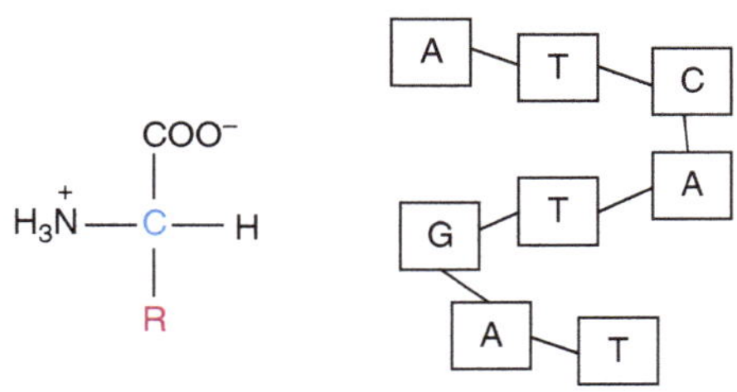

Fig. 3.14 *Left*; amino acid monomer, *right*; protein is linear chain[42]

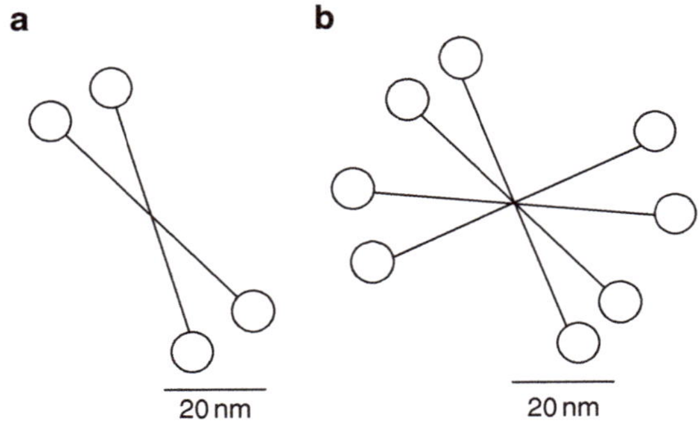

Fig. 3.15 (**a**) Schematic of single SP-D molecule (see ch. 11.8); (**b**) aggregate of SP-D molecules forming a globular surfactant

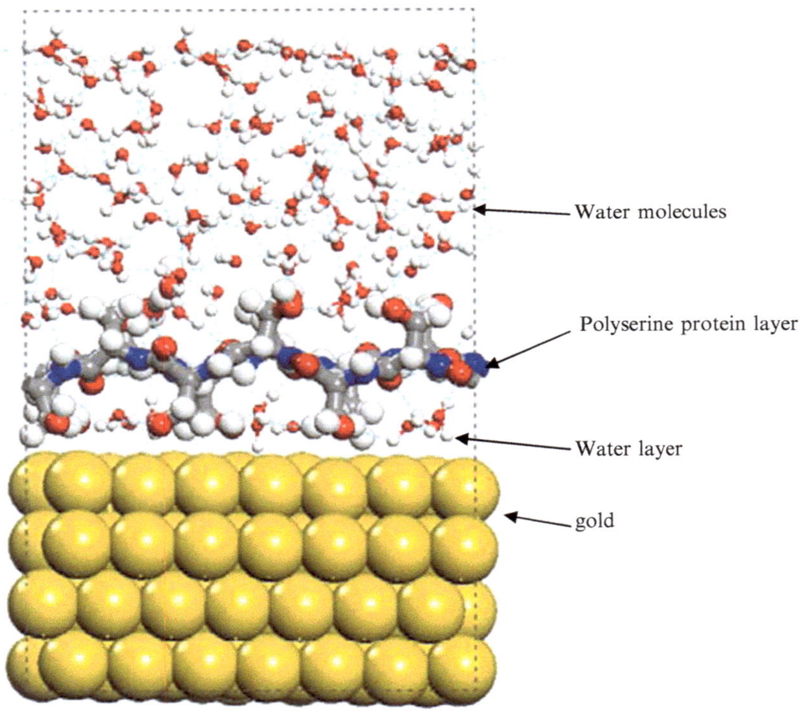

Fig. 3.16 Still from a computer simulation of a protein adsorbing to a gold surface in water[42] (Reprinted with permission of Prof Corni)

with various hydrocarbons including stearates to enhance their surface activity. Such products were first introduced in 1937 by the Maywood Chemical Company in the USA, based on condensation products of hydrolysed proteins and fatty acids to provide more hydrocarbon function. Typical chemical processes are acylation and enzymatic modification.

Proteins adsorbed at surfaces can be undesirable for membranes, paints, catheters and bloodstream implants because they are damaged by surface fouling. Desirable applications for protein covered surfaces are bone replacements and teeth implants where the protein coating enables cell adhesion to be enhanced.[41] Proteins are also known to coat gold surfaces and this effect means that gold nanoparticles can be used to understand protein interactions.

Although protein molecules are more complex than simple surfactants like stearic acid, molecular dynamics modelling can be used to visualise their adhesion. Figure 3.16 shows a computational model of a beta sheet of the protein polyserine formed from serine monomer adhering to a gold (111) surface under water.[42] There is evidently an interstitial water layer between the gold atoms and the polymer showing the interaction between polymer and water.

This brings us to the model of the cell because water, lipids and proteins interact to form complex structures at the surfaces of cells. Figure 3.17 shows the simplest representation of a sector from an animal cell. The lipid cell membrane is a bilayer of surfactant lipid molecules which can incorporate certain protein molecules and cholesterol within it. The lipid-protein cell membrane forms a physical boundary for the cell which can be dispersed in water and is negatively charged on its outer surface. Inside the semi-permeable membrane, the cell cytoplasm is contained. Our main interest is in the adhesion molecules which are proteins projecting from the cell wall, interacting with the extracellular matrix, that is the polymer layers which surround all cells and which also deposit on neighbouring solid substrates. Such a glycocalyx (literally sugar coating) can be microns thick and form protective and adhesive structures for biofilm and infective bacterial layers. These adhesion molecules are dealt with next.

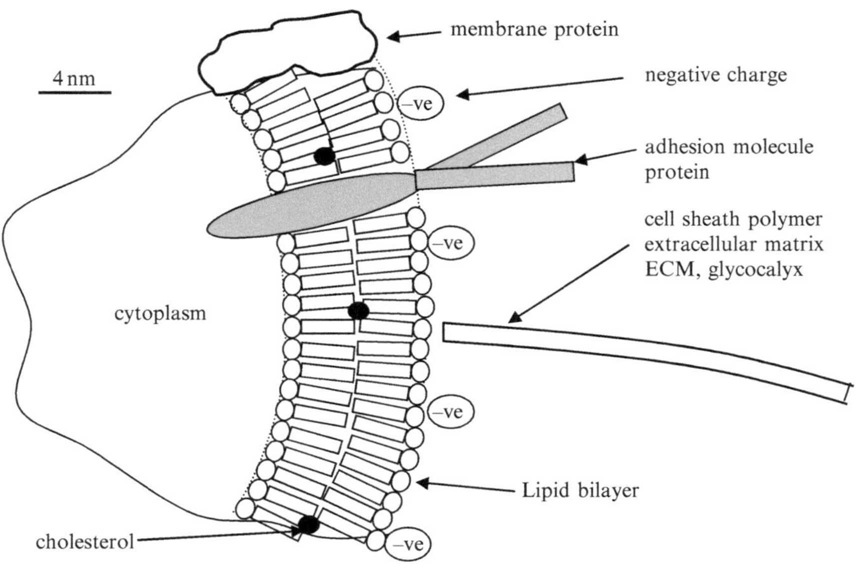

Fig. 3.17 Structure of cell membrane region

3.9 Models of Polymer 'Adhesion Molecule'

The question now is 'How does a large polymer protein adhesion molecule, typically 100 amino acids polymerised together and some 50 nm in length when fully stretched, influence the attraction between cells, causing aggregation; or between cells and solid surfaces, allowing cells to move around and reproduce?' From the arguments in adhesion molecule books,[43] it is suggested that the cell adhesion molecules (CAMs) form specific strong bonds on a 'lock and key' model to increase the adhesion. Of course this lock and key mechanism is untenable[11] because it is not consistent with the van der Waals theory described earlier in Chapter 1. The fact is that we would expect a large organic molecule interposed between cell surfaces to reduce the adhesion because the van der Waals attractive force drops rapidly with separation. In this section we suggest that a new mechanism of adhesion molecule function is required to understand cell adhesion.[44]

Adhesion molecules[45] come in several shapes and sizes as shown in the diagrams of Fig. 3.18. A cell adhesion molecule (CAM) is a protein sitting in the lipid membrane of animal cells that interacts with neighbouring cells or with the extracellular matrix (ECM). CAMs are also functional during growth and development in enabling cells to recognise each other and ensure correct cell–cell interactions. They also have other immune system functions and are excreted by leucocytes, the white blood cells, leading to the confusing name of leucocyte antigens.[46]

CAMs typically protrude from the membrane and interact with neighbouring molecules. There are several families of CAMs, the largest being the cadherins,

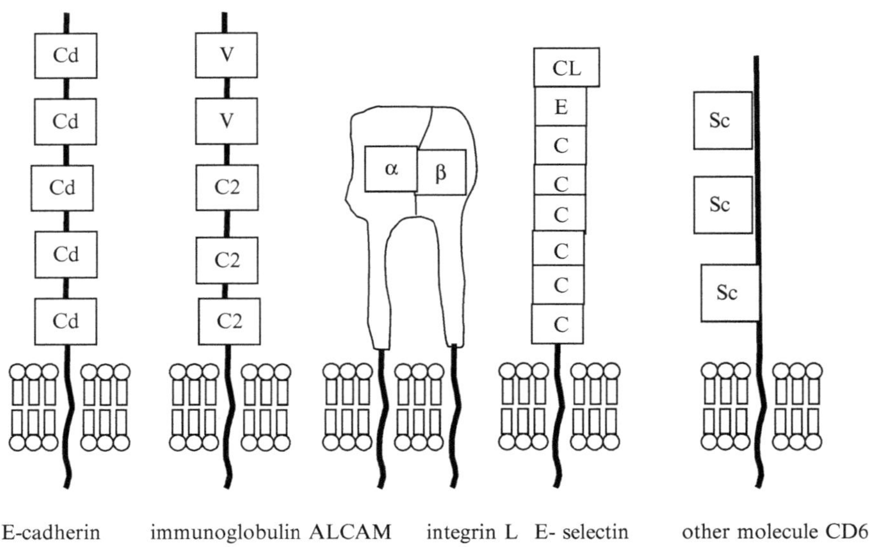

Fig. 3.18 Cell adhesion molecules in five families; *left* to *right*, cadherin; immunoglobulin; integrin; selectin; other molecules[45]

which are proteins found abundantly in junctions between epithelial cells. The second important grouping is the immunoglobulin superfamily (IgSF) which represents a large range of proteins with the common feature that they contain at least one immunoglobulin (Ig) domain outside the cell.

The integrins form the next largest family; they bind cells to components of the extracellular matrix (ECM), such as collagens and laminins, and can form cell–matrix junctions which act as 'contact spots' between cells and the ECM. Selectins and intercellular adhesion molecules (ICAMs) form the next family. These occur on the surface of endothelial cells lining blood vessels, where they help to tether passing leucocytes (white blood cells) at sites of inflammation. Finally there is a mixed bag of molecules which appear to be involved in cell adhesion, usually called the 'Other Molecule' family. The good news is that much chemical and structural information is now available on these complex molecules,[47] allowing them to be grouped by structure; the bad news is that there is information overload and no reasonable theory of adhesion molecule action to give a prediction of the adhesion effects. The paradox is that chemical structure and surface location can be measured, but the adhesion performance is not so easily analysed. Conventionally, the molecule structure is viewed as strung out while crossing the membrane into the cell. How these molecules get into the membrane is considered next.

3.10 Membrane Models

Recently, computer models of bilayer lipid membranes have been developed by Deserno and his colleagues[48, 49] based on Cooke's linked-bead model of a lipid molecule. Such models could explain curvature and elasticity effects in the membrane, which must influence the attachment of particles to the surface.[50] Typical membrane covered vesicles stick to glass surfaces with work of adhesion $W=10^{-8}$ to 10^{-5} $\mathrm{Jm^{-2}}$. Individual lipids were represented by one head bead and two tail beads. By means of simple pair potentials these lipid molecules self-assembled to a fluid bilayer state over a wide range of parameters, without the need for an explicit solvent. The model showed the expected elastic behaviour on large length scales, and its physical properties (e.g., fluidity or bending stiffness) could be widely tuned via a single parameter. In particular, bending energies in the experimentally relevant range were obtained, at least within 3–30 kT. The model was naturally suited to study many physical topics, including self-assembly, fusion, bilayer melting, lipid mixtures, rafts, and protein-bilayer interactions. A typical bilayer membrane being approached by a nanoparticle (which could be an adhesion molecule) is shown in Fig. 3.19. The cylindrical nanoparticle has on its side a white patch which is hydrophobic. This can interact with the hydrophobic tails on the membrane lipid molecules to give attachment to the membrane.

Molecular dynamics (MD) simulations were performed using the package ESPResSo.[50] The bead diameter was σ, its mass m and the adhesive energy well ε. The time step was set to $\delta=0.005\tau$ where τ was the time unit $\sigma(m/\varepsilon)^{1/2}$. Temperature

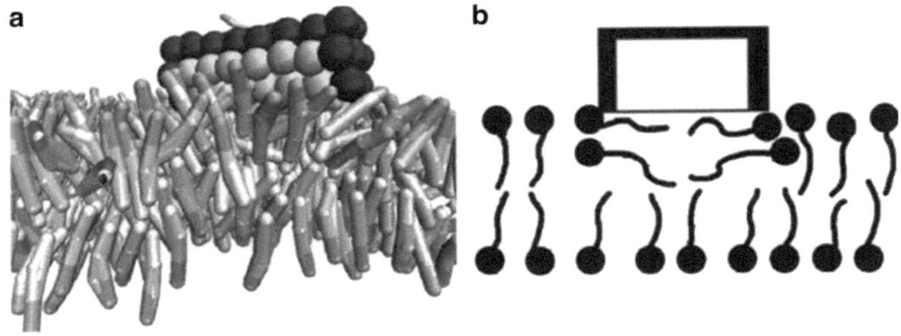

Fig. 3.19 Cross-section through a model bilayer membrane showing a nanoparticle attaching by hydrophobic interaction at the surface[50] (Reprinted with permission)

control was achieved by a Langevin thermostat with friction constant τ^{-1}. A cuboidal simulation box, with lengths $L_x = L_y$ and L_z subject to periodic boundary condition in all three directions, was used. Initially 4,000 lipids were preassembled into a flat bilayer spanning the square xy plane, giving a side length of $L_x \approx 50\sigma$. Then, 40 peptides were placed in a rectangular lattice at a distance of 4σ above the bilayer. The lateral tension was kept at zero via a modified Andersen barostat, allowing simultaneous box resizing in x- and y-dimensions, with a box friction $10^{-4}\tau^{-1}$ and a box mass $Q = 5 \times 10^{-4}$ m. Each calculated equilibrium could be presented as a movie snapshot as in the pictures presented here.

Measured results for observables (such as box sizes, pair correlation functions, etc.) were always averaged over three independent runs over the timescales indicated in the text. When discussing mechanisms of localised processes (e.g., peptide insertion), Deserno and his colleagues described a sequence of events that appeared typical, judged from the observation of repeated instances of such a process over three independent simulation runs.

A key simulation was to show how a bunch of nanocylinders with hydrophobic sides could penetrate the membrane to produce a channel. The sequence is shown in Fig. 3.20 which illustrates how nanoparticles approach the surface of the bilayer, adhere to it, then gradually rearrange to penetrate and form a pore structure. This model is revealing because the mechanism is cooperative, requiring several nanoparticles to produce the effect.

3.11 Three Components: Surface, Water, Adhesion Molecule

In this final theoretical section, we address the question of surfaces in water in the presence of protein chains. It would be exciting to consider a bilayer lipid membrane with a protein molecule at the surface. Some authors have presented pictures of protein molecules approaching a bilayer membrane. For example, Tsigelny and

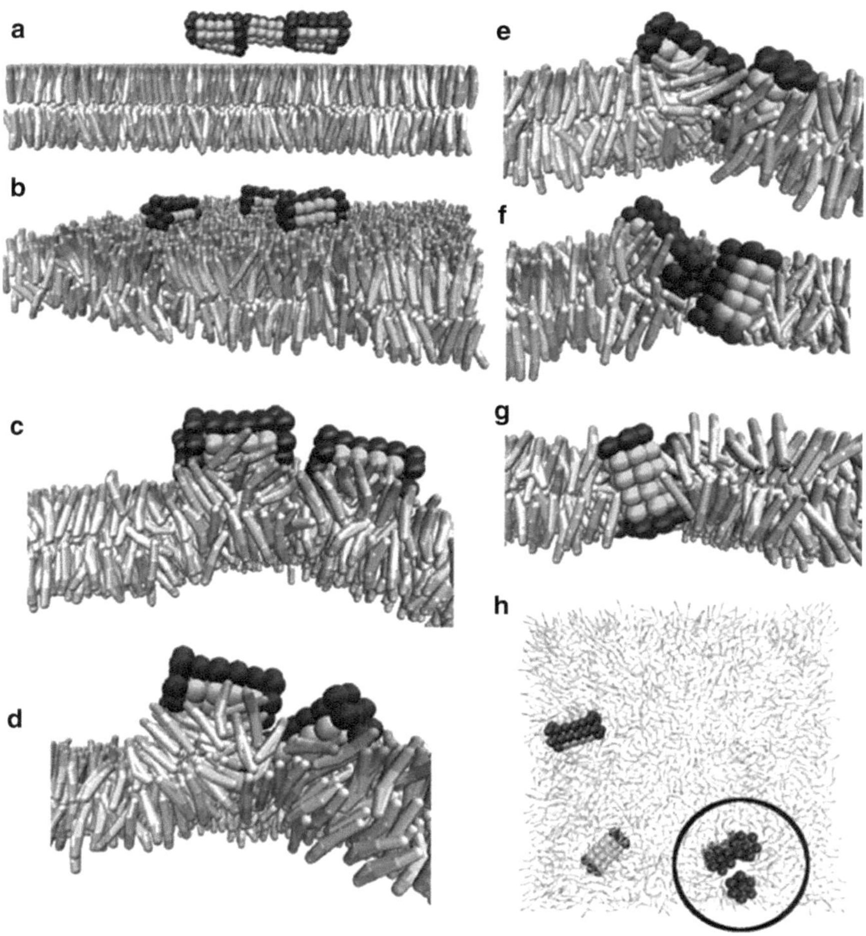

Fig. 3.20 (**a**) Three nanocylinders approaching the bilayer; (**b**) three particles adhering to the surface; (**c–f**) particles gradually penetrating the bilayer; (**g**) full penetration of the membrane; (**h**) plan view showing an aggregate of three nanoparticle forming a membrane channel, with two other particle still adhering to the surface (Reprinted with permission)

co-workers have modelled a-synuclein, a 140 amino acid protein associated with Parkinson's disease as it contacted a lipid bilayer membrane.[51] The problem with this approach is that proteins need water to give the correct swelling and conformational characteristics, so dry models like this do not consider all the factors involved.

A simpler approach is to compare a well characterised model surface like MgO with water already equilibriated at its surface as in Fig. 3.10, with a model in which swollen protein molecules are present in the gap between the MgO probes. Figure 3.21 shows an initial schematic of the theoretical result, with the broken line showing the

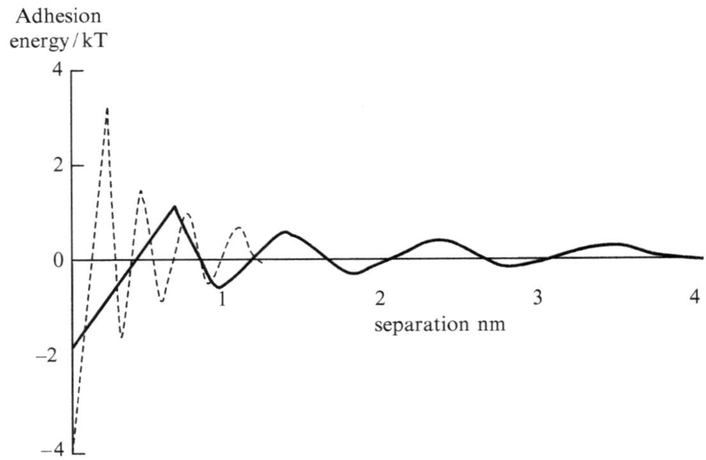

Fig. 3.21 Interaction of MgO probe with swollen protein molecule (*thick line*) for comparison with the pure water case (*broken line*)[52]

behaviour of water molecules alone. The last layer of water cannot be removed at the high force of 2.5 nN per atom because the MgO goes plastic at such large pressures. And the oscillation of the adhesion energy with the multilayering of the water is readily seen. However, when a three times larger molecule of protein swollen in water is adhering to the surface, the adhesion energy is less at full contact and the protein oscillations are much smaller in amplitude because the gaps are now larger and the van der Waals attractive energy falls rapidly as the gap rises. Thus several times less energy is required to remove the last layer of swollen protein, allowing the Brownian forces to spontaneously break through the energy barrier, thus leading to true contact. In other words, the protein acts as a catalyst to reduce the energy barrier to contact.

Molecular dynamics modelling is now proceeding to address the detailed nature of the protein-water structure at the interface, to calculate the equilibria as the gap between the MgO probes is compressed.[52]

3.12 Conclusions

From this discussion of adhesion modelling, it is clear that elucidation of the adhesion between nanoparticles, viruses and cells requires several steps. The starting point is the one parameter model for adhesion between simple spherical polymer nanoparticles, the so-called JKR model[11] which can explain how a spherical yeast cell attaches to glass. The JKR model is successful in explaining the circular contact spot for a spherical yeast cell, the deformation of the cell and the forces of adhesion in terms of one parameter W the work of adhesion.

However, the JKR model fails at the atomic level because nanoscale geometry is not describable by a spherical surface; also the interatomic potential has two terms, one for repulsion and one for attraction. But the attraction is usually more complicated and requires two or more constants because van der Waals is normally accompanied by electrostatic, covalent or other types of bonding.

Having got the basic shapes and potentials right, the energies must all be added pair-wise to obtain the overall interaction of the bodies to give an instantaneous equilibrium. There is then the Brownian motion which needs molecular dynamics for its statistical description. The picture is a movie rather than a still.

Finally there is contamination. Water is the major surface molecule which weakens the adhesion by a large factor, but the last layer of water cannot be squeezed off easily. Also, ions in the water can give electrostatic repulsions as in DLVO theory. Surfactants have an even more substantial effect in reducing adhesion. Adhesion molecules are larger, typically about 5 nm in diameter when the chain is random, and these are yet more significant in lowering the adhesion energy between the atomic surfaces which are held further apart, lowering the van der Waals force. However, because the water at the surfaces has to be squeezed out in discrete layers, there are several meta-equilibrium states of contact and adhesion. This has two effects; first the phase diagram is very sensitive to the several states of contact; second, the adhesion molecule can act as a catalyst allowing the Brownian impacts to push the nanoparticle through the energy barriers holding the water in place and make atomic contact.

The paradoxical conclusion is that the adhesion molecule lowers the equilibrium adhesion energy but drives the kinetics more quickly towards the fundamental atomic contact. In short, it can be viewed as an adhesion catalyst.

There is no need for 'lock and key' models of adhesion that cause confusion about the terminology of antibodies, receptors and ligands. Adhesion molecules can be defined as molecules, larger than simple surfactants, which act with water at cell and other interfaces to influence the adhesion mechanism.

References

1. Newton, I., Opticks, Smith and Walford, London 1704, reprinted Dover, New York, 1952, p. 395.
2. Zhang, C-Z., Wang, Z-G., Nucleation of membrane adhesions, Phys Rev E77 (2008) 021906; Washbourne, P. et al, Cell adhesion molecules in synapse formation, J Neuroscience 24 (2004) 9244–49.
3. Mie, G. Ann Phys (Leipzig) 11 (1903) 657
4. Lennard-Jones, J.E., The determination of molecular fields.1. From the variation of the viscosity of a gas with temperature, Proc R Soc Lond A106 (1924). 441–62 and 709–18
5. Lennard-Jones, J.E. and Taylor, P.A., Some theoretical calculations of the physical properties of certain crystals, Proc R Soc Lond, A109 (1925) 476–508; Lennard-Jones, J.E. and Dent, B.M., Proc R Soc Lond, A112 (1926) 230–234.
6. Borg, R.J., Diennes, G.J., The physical chemistry of solids, Academic Press US 1992, pp 101–110.
7. Morse, P.M., Diatomic molecules according to the wave mechanics. II. Vibrational levels, Phys Rev 34 (1929) 57–64.
8. London, F., The general theory of molecular forces, Trans Faraday Soc 33 (1937) 8–26.

9. Tabor, D. and Winterton, R. H. S., "The Direct Measurement of. Normal and Retarded van der Waals Forces", Proc. Roy. Soc. A312 (1969) 435–50.
10. Israelachvili, J.N. Intermolecular and Surface Forces, Academic Press, London 1985, pp. 198–201.
11. Kendall, K., Molecular adhesion and its applications, Kluwer, New York 2001, ch 6.
12. Hamaker, H C., The London van der Waals attraction between spherical particles, Physica 4 (1937) 1058–72.
13. Israelachvili, J.N., Pashley, R.M., The Hydrophobic Interaction is Long Range, Decaying, Exponentially with Distance., Nature 300 (1982) 341–342; Pashley, R.M.. and Israelachvili J.N., J Colloid Interface Sci 101 (1984) 511–23.
14. Smith, W. (ed) Molecular Simulation, 32 (2006) 933–1121.
15. Yong, C.W., Smith, W., Kendall, K., Surface contact studies of NaCl and TiO2: molecular dynamics simulation studies, J Mater Chem. 12 (2002) 2807–15.
16. Yong, C.W., Kendall, K., Smith, W., Atomistic studies of surface adhesions using molecular dynamics simulations, Phil Trans R Soc A362 (2004) 1915–29.
17. Yong, C.W., Smith, W., Dhir, A., Kendall, K., Transition from elastic to plastic deformation as asperity contact size is increased, Tribology Lett. 26 (2007) 235–8.
18. www.fuelcells.bham.ac.uk movie of NaCl molecular dynamics contact.
19. Miesbauer, O., Gotzinger, M., Peukert, W., Molecular dynamics simulations of the contact between two NaCl nanocrystals:adhesion, jump to contact and indentation, Nanotechnology 14 (2003) 371–76.
20. Du, Y., Adams, G.G., McGruer, N.E., Etsion, I., A parameter study of separation modes of adhering microcontacts, J Appl Phys 103 (2008) 064902.
21. Quesnel, D.J., Rimai, D.S., DeMejo, L.P., J Adhesion 67 (1998) 235–57.
22. Kendall, K., Yong, C.W., Smith, W., Particle adhesion at the nanoscale, J Adhesion 80 (2004) 21–36.
23. Berendsen, H.J.C., Postma, J.P.M.,Van Gunsteren, W.F., DiNola, A., Haak, J.R., Molecular dynamics with coupling to an external bath, J Chem Phys 81 (1984) 3684–90.
24. Kendall, K., The impossibility of comminuting small particles by compression, Nature 272 (1978) 710–11.
25. Kendall, K., Complexities of compression failure, Proc R Soc London A361 (1978) 245–63.
26. Kendall, K., Yong, C.W., Smith, W., Deformation of NaCl particle in contact at the nanoscale, Powder Technol 174 (2007) 2–5.
27. Maier, S., Gnecco, E., Baratoff, A., Bennewitz, R., Meyer, E., Atomic-scale friction modulated by a buried interface: Combined atomic and friction force microscopy experiments, Phys. Rev. B 78 (2008) 045432.
28. Kendall, K., Dhir, A., Yong, C.W., Strength by Atomic Force Microscopy: squeezing water from an MgO suface, Phil Mag 2010 in press.
29. Langmuir I, The adsorption of gases on plane surfaces of glass, mica and platinum, J Am Chem Soc 40 (1918) 1361–1403.
30. Faraday, M., The Bakerian Lecture: Experimental Relations of Gold (and Other Metals) to Light Phil. Trans. R. Soc. Lond. January 1, 147 (1857) 145–181.
31. Napper, D.H., Polymeric Stabilization of Colloidal Dispersions, Academic Press, New York, 1983.
32. Yamakawa, H., Modern Theory of Polymer Solutions, Harper and Row, London, 1971.
33. Milling, A.J. and Kendall, K., AFM of silica in polyacrylic acid solutions, Langmuir (1998); Milling, A.J. (ed) Surface Characterization Methods, Surfactant Science Series Vol 87, Marcel Dekker, New York, 1999, ch.3.
34. Asakura, S. and Oosawa, F., J Chem Phys 22 (1954). 1255–6.
35. Asakura, S. and Oosawa, F., J Poly Sci 33 (1958). 183–92.
36. McDonald RE, Fleming RI, Beeley JG, Bovell DL, Lu JR, et al., Latherin: A Surfactant Protein of Horse Sweat and Saliva. PLoS ONE 4 (5) (2009) e5726. doi:10.1371/journal.pone.0005726.

37. Thurlbeck, W.M., Churg, A.M., Myers, J.L., Tazelaar, H.D., Wright, J.L., Thurlbeck's pathology of the lung, 3rd ed.,Thieme 2005, p 52.
38. Whitsett, J.A., Weaver, T.E., Hydrophobic surfactant proteins in lung function and disease, N Engl J Med 347 (2002) 2141–8
39. Kishore U., Greenhough, T.J. et al, Surfactant Proteins SP-A and SP-D; Structure, function and receptors, Molecular Immunology 43 (2006) 1293–1315.
40. Nnanna, I.A., Xia, J. (eds) Protein-based surfactants, Surfactant Science Series 101, Marcel Dekker, New York, 2001.
41. Efimova, Y.M., Proteins at surfaces, Delft University Press 2006.
42. Corni, S., Calzolari, A., di Felice, R., et al Protein surface interactions mediated by water, Lecture at Edinburgh conference (2008) deisa.eu
43. Hortsch, M., Nott, P., New Cell Adhesion research, Nova Science 2009; Umemori, H., The sticky synapse: Cell adhesion molecules, Springer Berlin 2009; Garrod, D.R., Structure and function in cell adhesion, Portland Press 2008; La Flamme, S.E., Kowalczyk, A.P., (eds.), Wiley VCH Weinheim 2008; Cell junctions, Adhesion..., Cress, A.E., Nagle, R.B., (eds.), Cell adhesion and cytoskeletal molecules in metastasis, Springer, Dordrecht 2006; Beckerle, M.C., (ed.), Cell Adhesion, Oxford University Press 2002; Ley, K., (ed.), Adhesion molecules: function and inhibition, Birkhauser, Basel 2007; Reutter, W., Schuppan, D., Tauber, R., Zeitz, M., Falk symposium, Cell adhesion molecules in health and disease, Kluwer 2003; Coutts, A.S., Adhesion protein protocols, Humana Press London 2nd Ed 2007; Behrens, J., Nelson, W.J., Cell adhesion, Springer Berlin 2004; Collins, T., Leukocyte recruitment, endothelial cell adhesion molecules..., Kluwer Dordrecht 2001; Barker, J., McGrath, J., (eds.), Cell adhesion and migration..., Harwood Academic Amsterdam 2001; Guan, JL., (ed.), Signalling through cell adhesion molecules, CRC Press USA 1999; Hamann, A., Adhesion molecules and chemokynes in lymphocyte trafficking, Harwood Academic Amsterdam 2002; Pearson, J.D., (ed.) Vascular adhesion molecules and inflammation, Birkhauser, Basel 1999;
44. Kendall, K., Dhir, A., Du, S., Yong, C.W., Adhesion of viruses to particles, 2010, in preparation.
45. Isacke, C.M., Horton, M.A., The adhesion molecule facts book, Academic London 2000
46. Barclay, A.N. et al, The leucocyte antigen factsbook, 2nd ed. Academic Press, London (1997).
47. http://www.cell-adhesion.net; http://www.ncbi.nlm.nih.gov/prow/cd/index_molecule.htm
48. Cooke, I.R., and Deserno, M., Coupling between lipid shape and membrane curvature. Biophys. J. *91* (2006) 487–495
49. Cooke, I.R., Kremer, K., and Deserno, M., Tunable generic model for fluid bilayer membranes. Part 1. Phys. Rev. E Stat. Nonlin. Soft Matter Phys. *72* (2005) 011506.
50. Illya, G., Deserno, M., Coarse grained simulation studies of peptide induced pore formation, Biophys J., 95 (2008) 4163–4173.
51. Tsigelny, I.F., Sharikov, Y., Miller, M.A., Masliah, E., Simulation and modeling of synuclein-based protofibril structures as a means of understanding the molecular basis of Parkinson's disease, J Phys: conference series 125 (2008) 012056.
52. Kendall, K., Dhir, A., Yong, C.W., Adhesion J., Model of protein molecule equilibriating on an MgO surface, 2010, J. Adhesion to be published.

Part II
Mechanisms

Chapter 4
Macroscopic View of Adhesion for Nanoparticles, Viruses and Cells

it is doubtful if any... movements take place without the solid base

Harrison 1914

In Part I of this book we have considered the several elements required in a description of adhesion. Now let us use these ideas to describe adhesion observations on nanoparticles, viruses and cells in different circumstances.

At scales above 10 μm, where Brownian movement can be largely neglected, adhesion appears macroscopic, steady and static when viewed with the optical microscope. The range of van der Waals forces is negligible and we can use a single parameter W to describe the adhesion. In reflected light, an area of adhesion appears black, in contrast to the non-contacting and non-adhering areas which look brighter. Thus we can identify the *'black contact spot'* which represents a true molecular contact where the van der Waals adhesion between the surface molecules is found. Identifying the way in which this black spot changes in size is critical to understanding the adhesion process.

The purpose of this chapter is first to review the basic definitions of the black spot measurements to be made and the concepts which back them up, then to consider the different arrangements that have been used in experiments. Essentially we see that adhesion depends in general on three terms: the van der Waals force, the geometry and the elasticity. By using the equilibrium theory described in Chapters 1 and 2, we can obtain similar numbers for adhesion in all the wide-ranging methods which have been employed in the past. Finally, it is important to see how some interesting mechanisms such as alteration of the work of adhesion by surfactant addition, geometrical change or modulus change can alter the apparent adhesion by several orders of magnitude.

4.1 Fundamental Definitions

The most basic observation of adhesion at the macroscopic level is the *'jump to contact'*. Figure 4.1 shows how Roberts[1] studied the adhesion of optically smooth rubber to glass by reflection microscopy. When the two smooth bodies, the elastomeric

K. Kendall et al., *Adhesion of Cells, Viruses and Nanoparticles*,
DOI 10.1007/978-90-481-2585-2_4, © Springer Science+Business Media B.V. 2011

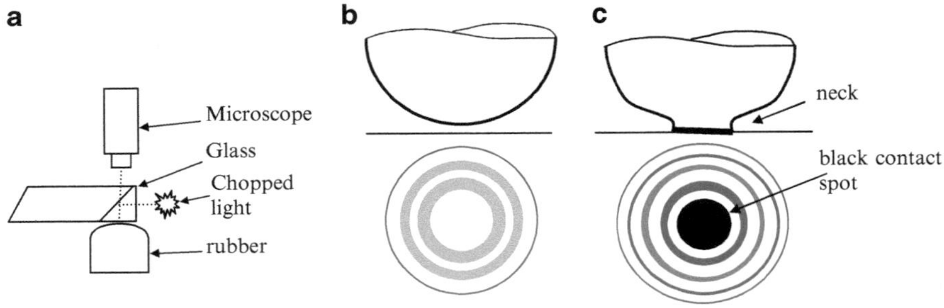

Fig. 4.1 (**a**) Roberts' apparatus for studying contact of rubber spheres onto glass; (**b**) Newton's rings before the jump to contact; (**c**) The black spot after the jump to contact, showing the rubber shape

sphere and a glass plate, were brought slowly together while observing them through the microscope in reflected light, interference fringes appeared as the gap between their surfaces approached 1 μm. These fringes were the familiar Newton's ring patterns, which gradually increased in diameter and spacing as contact was approached. Then, suddenly at a certain separation which depended on the stiffness of the system, the surfaces jumped together and the *'black contact spot'* appeared, spreading rapidly until it reached an equilibrium extent. This jump to contact is significant because it distinguishes van der Waals adhesion from familiar mechanical adhesion ideas such as 'lock and key' devices or Velcro, 'hook and eye' systems. Velcro does not jump into contact. It is not reasonable to consider atomic attractions in such simplistic lock and key terms.

As we saw in Chapter 2, once the 'black contact spot' reaches its final diameter, the adhesion is stabilised by the balance between the attractive van der Waals forces which pull the surfaces together with the elastic push of the squashed material flattened by the contact process. If we leave this system to its own devices, without applying any external forces, then we can equate the energies involved in the opposing processes of van der Waals attraction and elastic repulsion as described in Section 2.10 to produce the solution[2] first described in 1974 for the diameter of the black spot between two equal spheres.

$$d^3 = \frac{9\pi W D^2 (1-\nu^2)}{2E} \tag{4.1}$$

In this equation, d is the diameter of the black contact spot, W is the thermodynamic work of adhesion, that is the work in Joules required to separate reversibly 1 m^2 of contact area, D is the diameter of the equal spheres, ν is Poisson's ratio and E is Young's modulus of elasticity for the material.[3] This theory assumes that the spheres are much larger than the gaps, allowing the range of the van der Waals forces to be neglected.

The conclusion from this macroscopic theoretical argument is that the self-adhesion of particles can be explained in terms of the van der Waals attraction, embodied in the single parameter W the work of adhesion, the geometrical parameters involving D the sphere diameter and the elasticity described by E and ν. So three components are required in general to develop the adhesion picture: surface forces due to chemistry of the atoms, geometry which affects the form of the contact, and elasticity which allows deformation. Let us now consider these three aspects in more detail.

4.2 Theory Using Work of Adhesion *W*

The work of adhesion is a thermodynamic equilibrium quantity which adds up all the individual van der Waals bond energies across a surface, presuming that the bonds are extremely short compared to the macroscopic dimensions of the test apparatus. Since 99% of van der Waals bonding is below 1 nm gap, whereas most adhesion experiments are conducted above the 1,000 nm scale, this is a reasonable assumption.[4] But once the black spot shrinks to less than 1 nm in diameter, this assumption becomes invalid and we must move to the ideas described in Chapter 3 and later in Chapter 5.

Consider the situation shown schematically in Fig. 4.2a in which two semi-infinite plane rigid slabs of atoms approach each other to close the gap z between the surfaces. This problem of adding pairwise the individual r^{-6} atomic potentials was solved by Hamaker in the 1930s[5] to give the equation for attractive force as a function of z.

$$F = \frac{-A}{6\pi z} \tag{4.2}$$

Consequently, the adhesive force increases hyperbolically as the surfaces move closer together, as plotted in Fig. 4.2b. If this is stopped by a hard wall repulsive force as shown by the broken line where a is the atomic radius, then the work of adhesion is the shaded area under the curve and is the work done in moving the slabs apart from radius a to infinite separation at equilibrium. It is clear that the work of adhesion depends both on the radius a and on the Hamaker constant A.

Of course, geometry affects these forces as well. Hamaker showed that the force between spheres was different to that for plates made up of the same atoms. But he did not take into account the elasticity of the attracting bodies. The experiment sketched out in Fig. 4.2 cannot be carried out in practice because the plates stretch elastically as the van der Waals force increases, and this can be difficult to control. Such elastic stretching can be seen in the molecular modelling movie of Fig. 3.4 in Chapter 3 on the adhesion of a sodium chloride nanoparticle to a slab. The different effects of geometry and elasticity are considered next.

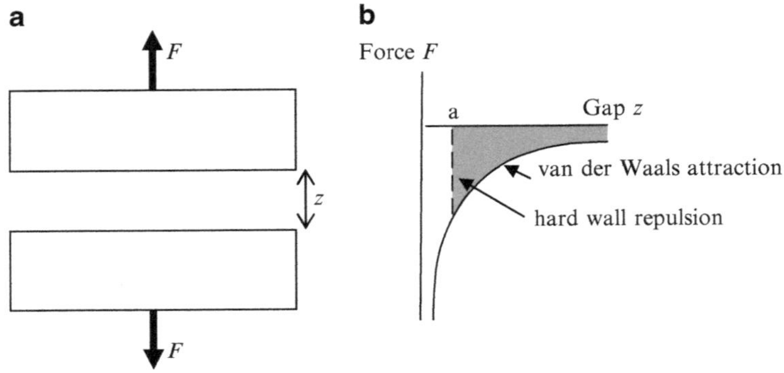

Fig. 4.2 (**a**) Idealised experiment in which two rigid van der Waals planes 1 × 1 m approach each other. (**b**) The schematic picture of the force curve; *W* is the area under this curve

4.3 Different Geometries; Wedging, Peeling and Spheres

In different adhesive geometries, the atoms at the surfaces remain constant, so the assumption is made that the Work of adhesion due to van der Waals force remains the same as the geometry is varied.

Take the wedging geometry first because this is familiar to us[4] when we chop wood with an axe as (Fig. 4.3). If the axe is well lubricated to give zero friction, then it can be pushed along the grain of the wood with a force F to separate the adhering cellulose fibres.

The crack ahead of the axe is length c and can be extended by an increment dc by pushing the wedge a distance dc further. By considering the energies involved in the process, a condition for the measured force on the axe, i.e. the wedging adhesion force, can be calculated.

The work done by the axe in pushing a distance dc is force times distance Fdc. All this work is presumed to go into separating the fibres against their adhesive van der Waals forces because friction is zero. The van der Waals bond energy created in the new surface area bdc is Wbdc. Because of energy conservation, the two work terms must be equal. Therefore

$$F \cdot \mathrm{d}c = W \cdot b \cdot \mathrm{d}c$$

Consequently,

$$F = Wb \tag{4.3}$$

This is the same equation obtained for peeling an elastic film from a plate, Eq. 2.1.

Several assumptions have been made in this derivation. The first is that the van der Waals forces do not stretch far away from the surfaces, such that once the crack has opened even slightly, no attraction remains. Secondly, the elastic deformation has been neglected, just as in the peeling argument of Chapter 2. The reason is that

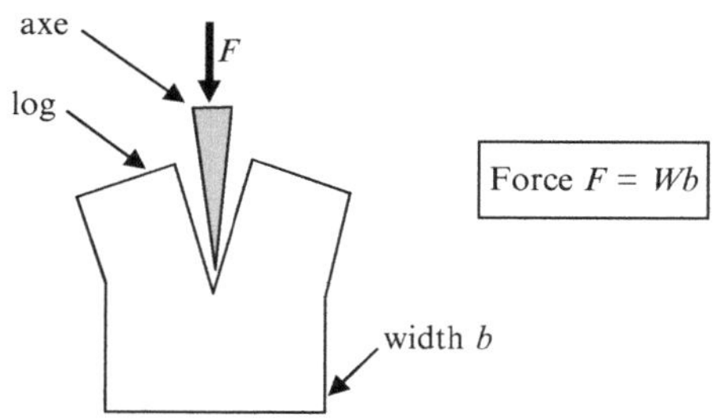

Fig. 4.3 A frictionless wedge separating adherent fibres in a block of wood

the elastic energy stored in the deformed wood remains constant as the wedge drives through. If there is no change in the stored elastic energy, then it cannot affect the crack. In this derivation, the basic argument of energy balance at equilibrium postulated by Griffith[6] is being used, i.e. fracture mechanics. Of course, if this system were truly at equilibrium as assumed in Griffith's theory, then the axe would be pushed back out of the wood by the van der Waals attractions once the force F was reduced. This effect was observed by Obreimoff as described in Section 1.12 but only happens when the split surfaces are atomically smooth so they can heal up, as in the case of mica and no contamination occurs.

The equation $F = Wb$ is surprising to many because it does not mention strength of the interface. However the concept of strength, or stress at failure, does not fit into the van der Waals theory of adhesion fracture. As shown in Section 3.4 on the theoretical modelling of van der Waals forces, the work of adhesion must remain constant as the geometry is varied, not the stress.

Similar equations for the adhesive force are obtained for peeling, $F = Wb$, and for sphere adhesion force $F = 3\pi WD/8$, neglecting the minus sign which is conventional for attractions. These forces are small, typically 1 mN to peel a centimetre wide strip of elastomer from dry glass. Because the failure is by cracking, the force is expressed per length of the crack line in N/m. It often seems surprising that such modest van der Waals attractions can account for the larger forces observed in many situations. One reason for this is that elastic deformation has a large influence and can increase the force substantially, as considered next.

4.4 Effect of Elasticity; Scraping and Stretching

The effect of elasticity is seen most readily when you are trying to start the axe into the wooden log. Once the crack gets running, the wedging is easy, but getting the crack started is more difficult. This is the problem Obreimoff studied with mica splitting. He showed (Section 1.12) that opening the crack against the Young's modulus E required a large force which then diminished steadily as the crack length c increased, as indicated in Eq. 4.4 derived by the energy balance theory.

$$F = b(WEh^3/6c^2)^{1/2} \tag{4.4}$$

where F was the force on the wedge, b its width, h the thickness of the split material and c the crack length ahead of the tip of the wedge. This equation shows that as the crack length c increases, the adhesion force decreases proportionally, eventually winding up at Eq. 4.3. Also, because E is such a large number for many materials, 2×10^9 N m^{-2} for polystyrene, the wedging force can be large for short cracks.

The simplifying feature of these tests, where the materials have to be highly strained to get the adhesive crack going, is that the parameter $(WhE)^{1/2}$ enters the equations, where W is the work of adhesion, h the thickness and E the elastic modulus of the material. This is revealing because it shows the dependence of adhesion force on the three factors: van der Waals bonds, geometry and elasticity. The conclusion is that, although adhesion bonding remains constant at the atomic level, i.e. W is constant, the line force of adhesion F/b can vary widely from 100 to 1,000,000 mN m^{-1}.

Similar considerations apply to the situation shown in Fig. 4.4 where an elastic film, perhaps of cells adhering to a plastic dish is pulled parallel to the surface by a force F.[7]

This elastic stretching action, often misleadingly called 'shear', drives a crack along the interface as the strip is pulled, eventually causing complete separation of the sheet from its substrate. In terms of the energy balance theory, the understanding of this test is simple because the energy terms are straightforward if we assume that the stored elastic energy in the stretched film is helping the crack, together with the potential energy of the force:

1. Adhesion energy in the peeled strip where c is the crack length — $w_1 = Wbc$
2. Elastic energy in the stretched film — $w_2 = \varepsilon^2 Ebhc/2$
3. Potential energy in the force F is twice this and negative — $w_3 = -\varepsilon^2 Ebhc$

where $\varepsilon = F/bhE$ is the strain in the stretched film, b is the strip width and h its thickness, E being the Young's modulus. Adding these three terms to give the total energy w and finding the equilibrium $dw/dc = 0$ gives the condition for crack propagation

$$d/dc\,(Wbc - \varepsilon^2\ Ebhc/2) = 0$$

and since $\varepsilon = F/Ebh$

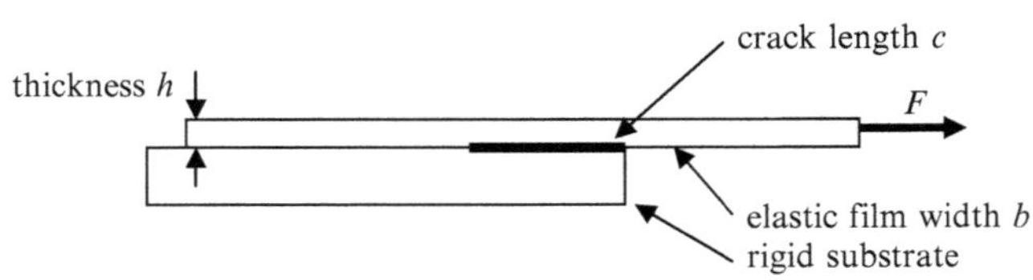

Fig. 4.4 Pulling an elastic strip parallel to the surface

$$\mathrm{d}/\mathrm{d}c\ (Wbc - F^2c / 2Ebh) = 0$$

Therefore

$$F = b(2WhE)^{1/2} \tag{4.5}$$

again showing the dependence of adhesion force on the three parameters, surface chemistry W, geometry h and E elasticity.

4.5 More Complex Geometries; Button Test, Tension and Probes

One of the simplest tests for measuring the adhesion of cellular material is to stick a cylindrical button onto a mass of cells and then to pull the button off with a vertical force as shown in Fig. 4.5.

This was first analysed by the energy balance method in 1971[8] when it was demonstrated that the same $(WEd)^{1/2}$ term entered the equation for the pull-off force F, but now d was the diameter of the button, not its thickness.

$$F = d\{\pi WEd / (1 - \upsilon^2)\}^{1/2} \tag{4.6}$$

This argument is essentially a modified form of the theory developed in the 1920s by Griffith[6], who first applied the theoretical idea that the crack line in a tensile fracture experiment separates the full contact area from the zero contact region. In other words, as the crack passes through an interface, the surfaces are separated instantly from molecular contact to full separation. The area swept out by the crack therefore demands a certain work of adhesion which is fully supplied by the energy fed in by the forces and elastic deformations. If the crack closes, the theory demands that the molecular adhesive forces can give all their energy back to the external system. Thus the Griffith theory is an equilibrium energy conservation argument as shown in Fig. 4.6. For such an edge crack, the equation for crack equilibrium was

$$F = 1.12\ b\ (WEh\ [h/\pi c])^{1/2} \tag{4.7}$$

Griffith thought that he had verified his theory experimentally for fracture of glass by measuring the cracking forces and showing they were consistent with the surface energy of molten glass, extrapolated down to the temperature of the cracking experiment. Unfortunately, Griffith made two errors: One numerical error was corrected in 1924[4] and the other mistake was to use glass as his ideal brittle material. It is now known that glass does not display reversible fracture, and that substantial energy losses occur in dissipative processes around the crack tip, stopping crack closure and preventing equilibrium: The cracks in glass were not truly in equilibrium.

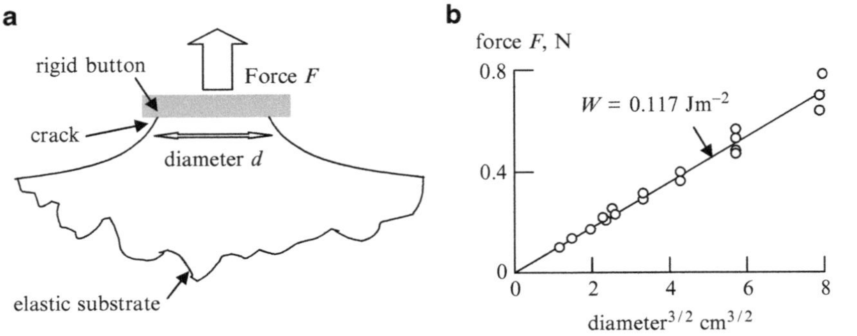

Fig. 4.5 (**a**) A rigid button being detached from an elastic substrate, showing the deformation of the substrate material; (**b**) Results for Perspex on gelatin solution[8]

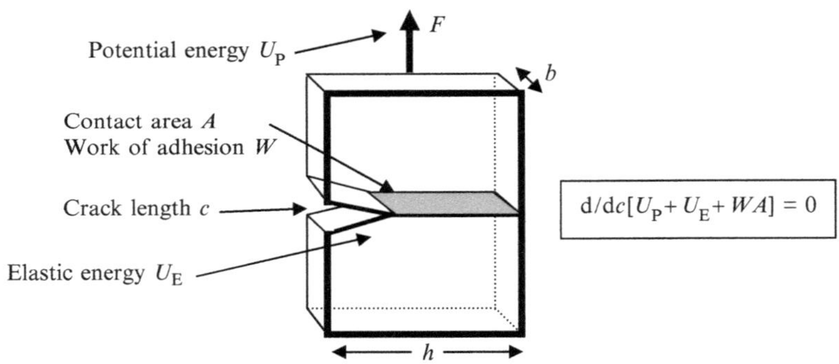

Fig. 4.6 Schematic view of the Griffith[6] tensile adhesion crack test

Similar arguments apply to testing adhesion of layers by probing them with the atomic force microscope as shown in Fig. 4.7. When a rigid probe is pushed into an elastic material, then the resulting deformations can cause the interface to break, with a crack propagating out around the probe tip. If the probe is then sheared across the surface, this detachment can propagate to detach the film, the basis of the scratch adhesion test for thin films.[4]

The conclusion is that the force applied to detach cells from a surface must depend on the work of adhesion, the geometry and the elastic properties, giving a wide range of adhesion forces. However, the measured adhesion forces may range wider still because the system is not at equilibrium as explained next.

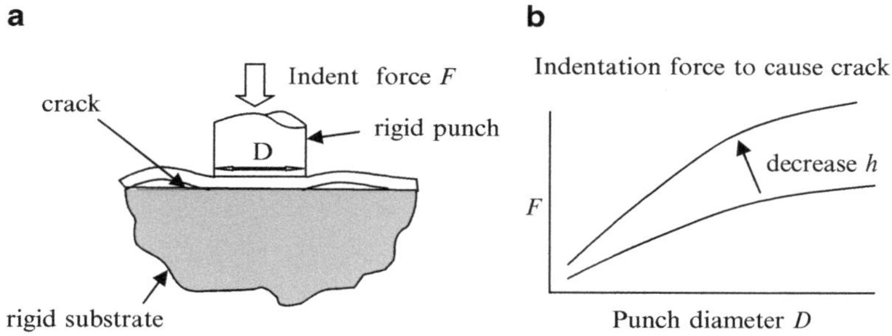

Fig. 4.7 (**a**) Crack at interface caused by indentation of an elastic film; (**b**) Results showing the cracking force changing with punch diameter and film thickness

4.6 Measured Values of Adhesion Energy

The fundamental problem of adhesion testing is that the energies W measured by the methods described above are too wide-ranging when compared to theory. Also, in many tests it is evident that equilibrium is not attained because the materials can be peeled apart but not easily healed up. Therefore it is necessary to distinguish the equilibrium value W from the measured adhesion energy R, which is defined as the experimental energy measured to break 1 m^2 of interface. Results then show that this adhesion energy can range from negative values, where joints fail spontaneously when immersed in ionic solutions, to very small values when colloidal particles remain separate and stable for long periods, to the very large values of 10^5 J m^{-2} needed in tough adhesive structures. The values of the theoretical molecular bond energies, i.e. equilibrium work of adhesion W, occupy only a small range from 0.1 to 10 J m^{-2}, and cannot possibly explain the full scale of measured adhesion. These values are plotted in Fig. 4.8, and we can see immediately that we need two logarithmic scales to describe the results, one for attractions and another for repulsion. How can this extraordinary range of experimental values be explained[9]?

It is evident from this diagram that adhesion of nanoparticles, viruses and cells cannot be explained by simple atomic or molecular bonding under clean conditions. van der Waals forces together with ionic and covalent bonding of clean surfaces can explain the range of adhesion energies from around 0.1 to 10 J m^{-2}. Contamination of the surfaces by foreign material such as water and adhesion molecules can then explain reductions in adhesion (i.e. repulsion as the van der Waals forces are shielded (see Chapter 3). Further contamination can also be the source of negative adhesion (i.e. repulsion) which pushes adhering bonds apart spontaneously, as in DLVO theory of colloids stabilised by ionic double layers or in steric repulsion due to polymer molecules.

To explain the adhesion variations shown in Fig. 4.8 we have to introduce even more mechanisms. Some of these are very obvious and have been mentioned earlier, for example, roughness which creates gaps between atoms on the surface, thereby reducing contact and the resulting adhesion immensely. However, there is

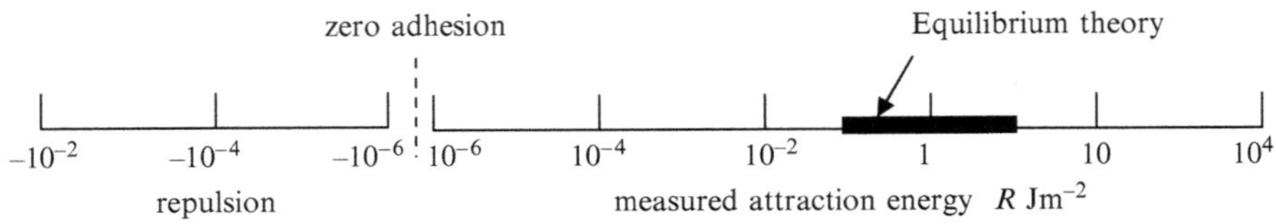

Fig. 4.8 Range of measured adhesion energies compared to theoretical clean molecular values

a significant problem explaining the high values of adhesion which are necessary to hold barnacles onto ships or to explain the adhesion of tendons to bone. We have to introduce some amplification mechanisms if we are to explain such values rationally. These amplification mechanisms can operate at the molecular scale or at the macroscopic level. It is the variety of these subtle mechanisms which makes the subject of adhesion so fascinating, both to chemists who wish to consider molecules, to engineers who think in terms of mechanics and to biologists who observe complex systems. Let us now consider changes in van der Waals forces, changes in geometry or changes in elasticity which can alter adhesion forces.

4.7 Reducing *W* by Adding Surface Active Agents

One of the earliest chemical effects observed in work of adhesion was obtained by contaminating adhering spherical elastomer surfaces with water[3] as explained in Section 1.10. The black spot was seen to diminish in diameter and the pull-off force dropped by a factor ten. By measuring the effect of applied load in the JKR test, it was clear by curve fitting that W had dropped from 71 to 6.8 mJ m^{-2} after contaminating the surfaces with water.

The interesting thing was that this drop in surface energy of the rubber surface due to the presence of contaminating water molecules was consistent with Young's equation for contact angle which was measured to be 66°C for water droplets sitting on the rubber, the first time that Young's equation had been verified experimentally in its 200 year history. When 0.01 M sodium dodecyl sulphate (SDS) soap solution was used to contaminate the rubber sphere contact region, the black spot contact size fitted the Hertz equation down to the lowest loads obtainable, showing that the work of adhesion was less than 1 mJ m^{-2}. This was consistent with the wetting behaviour of the soap solution on the rubber surface, and proved that the surfactant molecules were adsorbed to the elastomer molecules in the water solvent, producing much lower van der Waals attractions. The conclusion is that molecules attached to a surface, such as adhesion molecules, reduce the adhesion between the materials. They may increase the tackiness or stickiness of surfaces like a glue, but the adhesion force is reduced by these molecules.

Chaudhury and his colleagues at Dow Corning Corp and Lehigh University in Pennsylvania followed up these early results, especially looking at the surface treatment of the rubber to give a variety of surface species.[10,11] Poly dimethyl siloxane

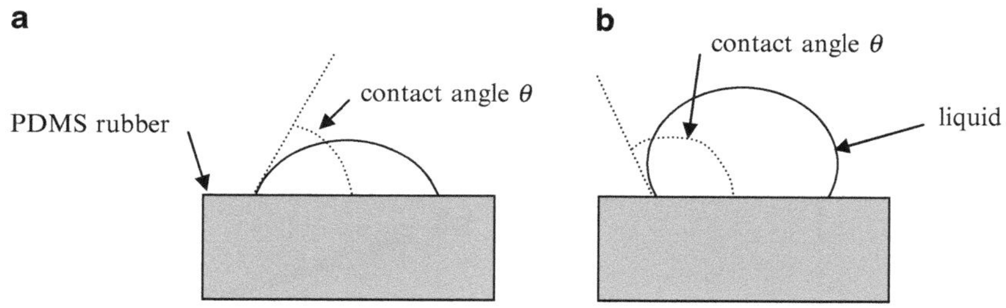

Fig. 4.9 (**a**) Wetting liquid in contact with rubber; (**b**) non-wetting liquid

elastomer (PDMS polymer or silicone rubber) was made by cross-linking a liquid polymer to give a very elastic rubbery network. In the first experiments, the adhesion of the PDMS rubber to itself was measured to give a work of adhesion of 44 mJ m^{-2}, significantly smaller than the ordinary hydrocarbon rubbers used previously.

Then a number of liquids were used to wet the rubber/rubber contact and the black spot size was measured under contaminated conditions. The results depended very much on how the contaminant liquid wetted the rubber surface, as shown in Fig. 4.9.

The wetting liquid caused the black spot to decrease whereas the non-wetting liquid caused the black spot to increase slightly in size. For example water, which does not wet PDMS rubber but has a contact angle of 102° as in Fig. 4.9b, gave a work of adhesion of 74 mJ m^{-2}. In this case the liquid makes the rubber adhere slightly stronger. But methanol, which wets the rubber, as in Fig. 4.9a, gave a greatly reduced work of adhesion of 6 mJ m^{-2}, illustrating the large contamination effect that reduces adhesion. The wettability of material may therefore be important when inhaled or ingested. For example during inhalation particles first become humidified introducing a water layer to the surface of particulate materials that will interfere with adhesion.

We can calculate the effect of wetting angle on W through the modified Young equation

$$W_{SL} = W_{SV} - 2\gamma_{LV} \cos\theta \tag{4.8}$$

Where W_{SL} is the work of adhesion of the solids measured immersed in liquid, W_{SV} is the work of adhesion measured in the vapour and γ_{LV} is the surface tension of the liquid/vapour (LV) interface.

When the contact angle is 90°, and the liquid droplet forms a little hemisphere on the surface, then no work is done by the wetting, and the work of adhesion in liquid is the same as in vapour. But if the liquid wets the surface then adhesion must be reduced. This is equivalent to Young's original theory of 1805, which he did not express in symbols but only in words.[4]

To check this law out more rigorously, Chaudhury and Whitesides[10,11] made 11 different mixtures of methanol and water, ranging from 5% methanol to 100%, and

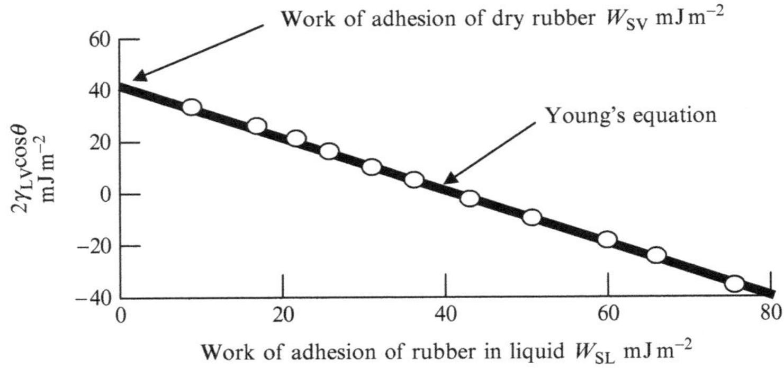

Fig. 4.10 Verification of Young's equation using PDMS rubber[11]

used them to verify Young's equation. First they measured the surface tension γ_{LV} of each mixture, then they measured the contact angle θ of each mixture on a flat surface of the PDMS rubber. This enabled them to calculate the term $2\gamma_{LV}\cos$ in Eq. 4.8. Plotting this as the vertical axis in Fig. 4.10 allowed a comparison with the work of adhesion of the rubber W_{SL} measured from the black spot in the methanol/water mixtures.

This graph shows the straight line plot expected from Young's equation, together with the experimental points which fall neatly on the theoretical line. This experiment was a convincing proof that the equilibrium work of adhesion could be measured over a wide range of chemical conditions. The solid rubber was found to have a surface energy very similar to the high molecular weight liquid, 42 mJ m^{-2}, suggesting that the methyl groups were sticking outwards at the surface.

Then Chaudhury and Whitesides found a way of modifying their silicone rubber surface to change its chemical character. The PDMS polymer was exposed to an oxygen plasma for a short period, as shown in Fig. 4.11, creating a thin layer of silica on the surface, about 3 nm thick. By treating this silica layer with molecules of siloxane, single molecular layers i.e. monolayers of particular structures could be formed at the rubber surface.

These monolayers are essentially adhesion molecules reducing the van der Waals force.

The method of applying the organic monolayer to the surface was simple. A solution of the silane molecule was made in paraffin oil, at a concentration of between 1% and 2%. This was evacuated in a desiccator containing the silica-surfaced-rubber samples. After a short time, the surfaces were found to be covered with a monolayer. Five long chain silane compounds were studied, as shown in Table 4.1 below. In each case, the chlorine atoms reacted with the

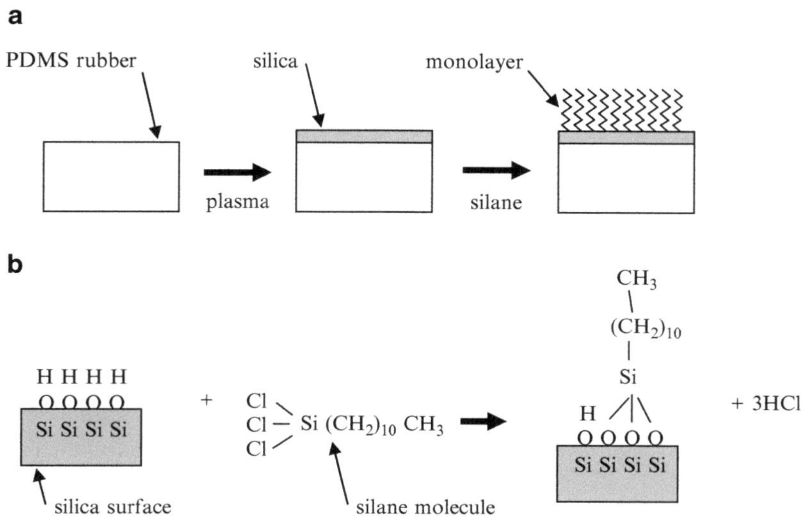

Fig. 4.11 (**a**) Process for modifying the surface of the silicone rubber to give specific molecular structure; (**b**) reaction of the silane with the surface of silica

Table 4.1 Values of work of adhesion W_{SV} for organic coated surfaces from dry adhesion tests[10]

Surface	Work of adhesion mJ m^{-2}
$-O_3\ Si(CH_2)_2(CF_2)_7CF_3$	32+/−1
$-O_3\ Si(CH_2)_{10}CH_3$	42
$-O_3\ Si(CH_2)_{11}OCH_3$	54
$-O_3\ Si(CH_2)_{11}OCOCH_3$	66
$-O_3\ Si(CH_2)_{11}Br$	74

hydroxyl groups on the silica to bond the long chains to the surface, as indicated in Fig. 4.11b.

These values show the remarkably sensitive changes in adhesion levels brought about by the changing chemical groups at the surface.

4.8 Measurements of Cell Adhesion by Probe Methods

Sirghi and colleagues recently used these ideas to measure the work of adhesion of living fibroblast mouse cells by indenting them with a clean silicon nitride AFM pyramid in growth medium where the cells could survive for several hours.[12] The arrangement is shown in Fig. 4.12.

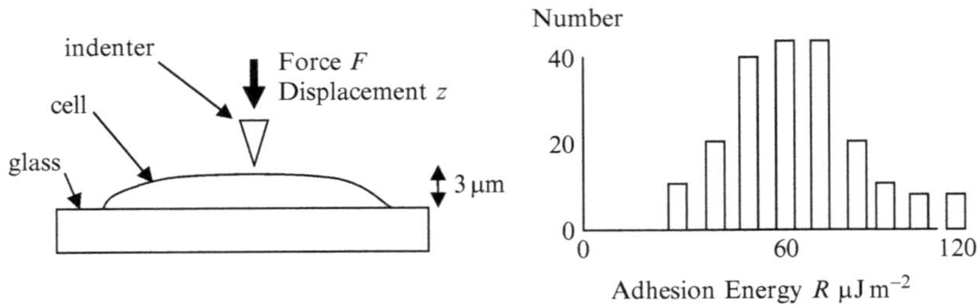

Fig. 4.12 *Left*, experimental scheme for indenting a living fibroblast cell immersed in growth medium[12]. *Right*, results for adhesion energy from 240 indentation tests on 12 different cells

The experiments were designed to measure both elastic modulus from the loading curve and the adhesion energy from the unloading part. By using a pyramidal indenter with an angle of 35°, the elastic loading curve with zero adhesion was planned to be linear, giving a benefit over the JKR sphere contact which leads to non-linear Hertzian curves. Two effects were overcome to obtain this linear elastic curve. First, the cells were found to be viscoelastic with the force decaying over time at a given indentation displacement. This problem was avoided by taking 5 s over each reading, by which time the force had levelled out. Second, the glass base was found to interfere with the readings at certain loading positions near the edges of the cells. Therefore, readings were taken near the middle of the cells where the cell thickness was 3 μm or more. The maximum probe penetration was held at 300 nm and the maximum force at 500 nN to reduce the substrate effect to a negligible level.

It was evident that the contact between probe and cell was not at equilibrium. The adhesion was developed mainly during pull-off and was not much apparent during the contacting process. There was no jump to contact and the pull-off adhesion energy was typically 60 ± 10 $\mu J\ m^{-2}$, about 1,000 times less than dry elastomer contacts. There was no experimental information on the peeling speed which must have had a strong effect on the adhesion energy, with significant adhesive drag.

The results showed higher adhesion than earlier fibroblast measurements. Zhu et al.[14] found by confocal–reflection interference contrast microscopy a value of $4 \times 10^{-8}\ J\ m^{-2}$ for fibroblast 3T3 cells at rest on gelatin-modified poly (lactide-co-glycolide acid) surface. This result indicates how much cell adhesion is reduced from a silicon nitride material to a polymer surface. The elastic modulus was also lower, measured at 0.85 ± 0.1 kPa. This was less than other reported values for fibroblasts ranging from 3–5 kPa[13]. In any case, the cells were 1,000 times more compliant than elastomers studied by JKR methods in macroscopic samples.

The most interesting result was found in a small number of experiments, 4% of total contacts, in which the smooth pull-off curve was interrupted by stiction events with much higher adhesion energy. A typical curve is shown in Fig. 4.13. The loading curve was uneventful and almost linear, indicating a low adhesion during the contact process. The unloading followed a smooth theoretically predicted path until three sudden jumps were observed. These were interpreted as polymer molecules being pulled off the cell surface, but could have other causes as described below. Previous

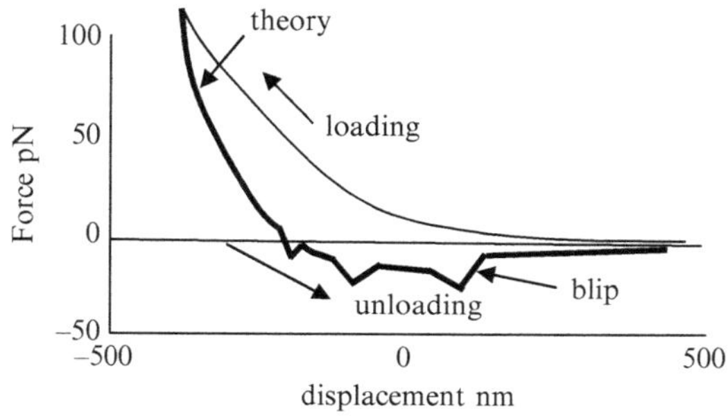

Fig. 4.13 Results showing three adhesive blips during steady detachment of the probe from the fibroblast cell. The loading curve was smooth but the unloading curve deviated from the theory line (E=0.96 kPa, R=67 μJ m^{-2}) in blips suggesting that smooth peeling was not possible

results had shown such blips[15] but the contact times had been longer in those earlier experiments, perhaps allowing more intimate contact to have been achieved.

In recent years, many AFM experiments have been carried out on force displacement curves using AFM with reviews showing hundreds of contributions.[16] A typical cell detachment experiment was by Bowen et al.[17] who showed that a yeast cell adhering to mica could be detached as in the experiments above. But they also described further adhesion which they interpreted as extracellular polymer being stretched out before complete separation occurred. This may be the correct mechanism but it is sensible to consider a range of other possibilities such as geometry change or elastic property change, because peeling of polymers from smooth surfaces can give stick-slip.[4]

4.9 Crack Stopping by Geometry Change

Adhesion does not just change with chemistry; it can depend on geometry in an interesting way. One particular example of an order of magnitude change in adhesion force, without any alteration in the atoms at the interface, nor in the elasticity of the system, was shown in 1975.[18] The adhesive crack geometry, shown in Fig. 4.14, was an adhesive peel crack viewed with a TV camera through a glass plate.

A rubber strip 10 mm wide peeling from glass was the experimental model. The crack speed was measured accurately as the crack approached an interface at which the thickness of the material was doubled by glueing another strip on top of the first strip. The adhesive crack stopped at this geometry change and the peel force had to be increased by a factor of eight to push past the step, after which the force dropped back to its original level.

In the first experiments, the crack behaviour for uniform rubber material was measured (Fig. 4.14a). As the force F applied to the crack increased, so did the crack speed. Identical behaviour patterns were observed for different rubber thicknesses, demonstrating the simplicity of the peel crack system in which the peel force F depends only on the adhesive fracture energy R at a given crack speed, as shown by the peel equation described in Section 2.5.

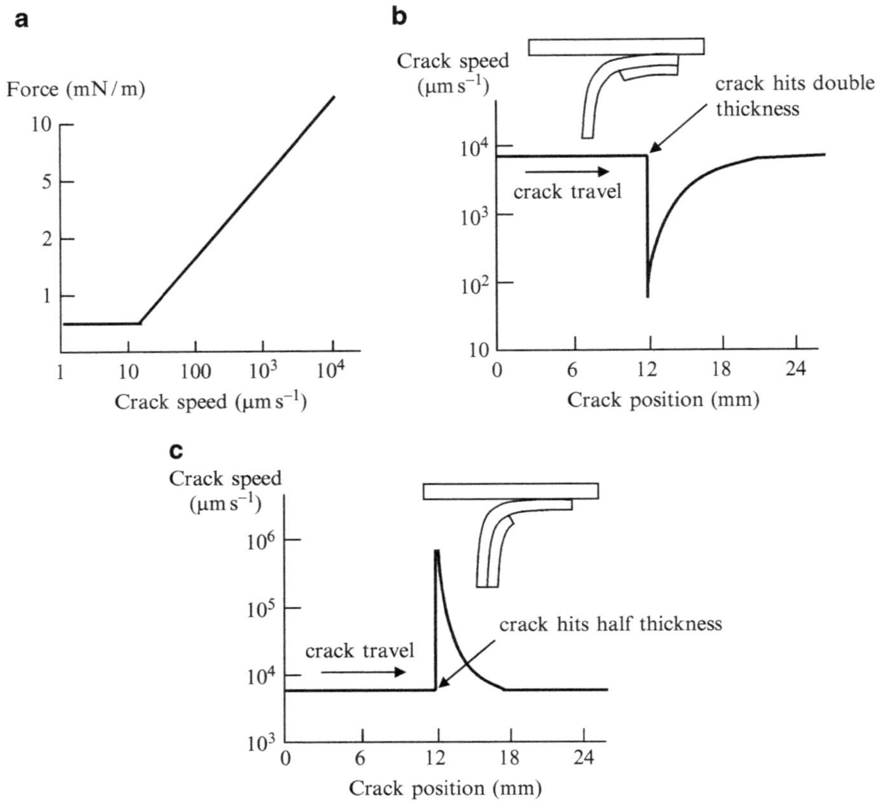

Fig. 4.14 (**a**) Peeling force versus crack speed for uniform rubber films of different thickness. (**b**) Crack slowing at an interface with double rubber film. (**c**) Crack acceleration at interface with half thickness rubber

$$F = Wb \tag{4.9}$$

Clearly, the film thickness itself does not affect elastic peel force according to this equation.

However, when the crack approached an interface where the strip thickness was doubled, the crack speed was seen to change substantially, as shown in Fig. 4.14b. At first the crack travelled at constant speed under the steady load, as expected from Eq. 4.9. Then, where the crack met with the thicker material, it slowed down by a factor of 100. Subsequently, as the crack passed the thickness change, it speeded up to regain its original constant speed after 15 mm of further travel.

When the crack moved from the thick film material towards the single thickness material, the opposite effect was observed (Fig. 4.14c). The crack travelled at the same constant speed as before at constant load. But at the point where the material

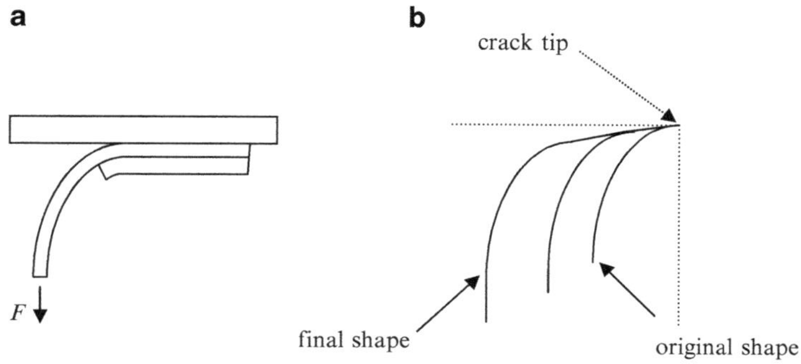

Fig. 4.15 (**a**) Shape of peeling film as the peeling just passes the thickening. (**b**) Change in shape as the peeling crack progresses further

was thinned down, the crack speeded up significantly, and then slowed back to reach its original constant speed after 5 mm of travel.

The theoretical explanation of these effects can be derived from the energy theory of fracture.[18] As a crack (Fig. 4.15a) tries to penetrate into the stiffer material, there is more elastic resistance to bending deformation and so the strip shape changes (Fig. 4.15b). These shapes were measured microscopically on peeling samples. The theoretical shapes were calculated from elastic beam theory. Since the thicker material is more resistant to bending, the peeling load cannot deflect so much and consequently does less work. But the interface fracture work remains constant. Therefore the peeling force must be raised to maintain the same fracture work and peeling speed.

Putting this theory mathematically, the beam deflection is

$$\delta = \frac{FL^3}{4Ebh^3} \tag{4.10}$$

where F is the force, L the beam length, E the elastic modulus, b the width and d the film thickness. To maintain the crack propagation at constant speed, this deflection must remain constant, i.e. F/h^3 is constant. Therefore, at the interface, the condition for cracking changes to

$$\frac{F_2}{h_2^{\,3}} = \frac{F_1}{h_1^{\,3}}$$

And since $F_1 = bW$,

$$F_2 = \left(\frac{h_2^{\,3}}{h_1^{\,3}}\right) bW \tag{4.11}$$

Thus, while it is clear that the thickness of a material may not affect its adhesion force as indicated in Eq. 4.8, a change in thickness at an interface toughens the material by a factor (h_2^3/h_1^3). This theory was used to calculate the full lines for comparison with experiment in Fig. 4.14.

This theory means that a cell could be peeling from a surface under a certain force, but could thicken itself in front of the peel crack to prevent further peeling. In other words, the cell need not change the adhesion molecules at the surface to prevent peeling; it could change its geometry instead.

4.10 Elastic Changes Affecting Adhesion: Shrinkage and Pre-stressing

In Eq. 4.10, because the elastic modulus E is equivalent to h^3 the crack-stopping effect described above can be equally produced by stiffening the cell from E_1 to E_2 just ahead of the peel crack, rather than thickening it. In that case, the extra force needed to peel the stiffened cell is given by

$$F_2 = \left(\frac{E_2}{E_1}\right) bW \tag{4.12}$$

So a cell that increases its elastic modulus will also increase its adhesion force.

Akin to this elastic mechanism of stiffening is the 'chewing gum effect' in which a flow process at the crack tip causes the elastic energy driving the crack to diminish. Although chewing gum does not have a strong adhesion to surfaces, it is extremely difficult to remove because it stretches extensively and the peeling energy does not get to the adhesive bonds but is dissipated in the stretching. That is why the best way to remove chewing gum from carpets is to freeze it to make it glassy. Then the gum comes off easily in a brittle manner.

A model for this chewing gum effect[4] can be built by taking an elastic rubber film peeling from a surface, measuring the crack speed as the crack travels along the interface. Then compare this with the same elastic film which has been cut in the middle, then the ends joined together with chewing gum or similar plastic material, as shown in Fig. 4.16.

The peeling stops at the point where the chewing gum is holding the cut film together. This shows that the sudden viscoelastic relaxation of the gum is sufficient to extract the energy from the crack and enhance the adhesion of the rubber. Quite clearly, the molecular adhesion at the interface remains unchanged; only the relaxation of the chewing gum can be implicated in the apparent increase of adhesion. It may be shown that the relaxation produces two distinct effects on the adhesion. The first is an amplifying effect on the adhesion energy caused by the effective fall in elastic modulus of the gum from E to E_t. The second is an additional resistance to cracking caused by the rate of change of elastic modulus.

Figure 4.17 shows these two effects graphically. Figure 4.17a gives the amplifying hysteresis factor E/E_t which magnifies the peel force by an order of magnitude.

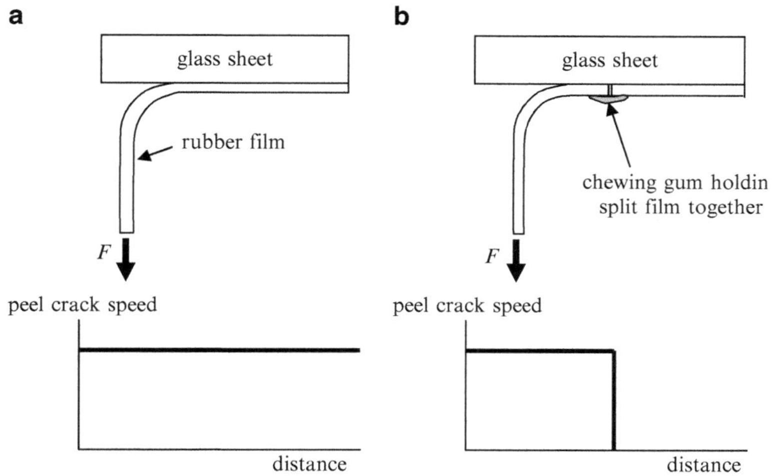

Fig. 4.16 (**a**) Steady peeling of rubber film from glass under elastic conditions; (**b**) sudden stopping of peeling at chewing gum holding cut film together

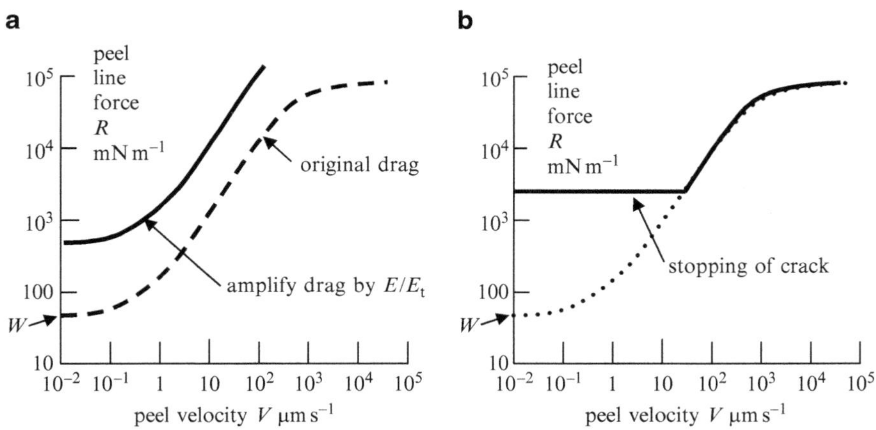

Fig. 4.17 (**a**) Amplification of adhesive drag curve by the modulus relaxation; (**b**) sudden crack-stopping effect caused by rate of relaxation

Figure 4.17b shows how the crack speed falls as a result of the second term, slowing hyperbolically with time. The crack can suddenly stop after travelling a short distance. Both these terms can give the sort of behaviour which is observed in real adhering systems where the adhesion force is much larger than expected from pure elastic concepts.

Another elastic effect on adhesion is shrinkage. If a cell shrinks and thereby stores elastic strain energy within itself, then this must reduce its adhesion force, even though the adhesion molecules at the surface remain constant. The equation first put forward for this effect in 1973 showed that the peel adhesion force F was reduced to[19]

$$F = Wb - \varepsilon^2 2Ebh \tag{4.13}$$

by the elastic strain ε which was released as peeling occurred. If the shrinkage strain was sufficiently large, then the material could detach without applying any force and in this case the condition for spontaneous detachment was

$$\varepsilon = (WEh)^{1/2} \tag{4.14}$$

The converse of this idea was that adhesion could also be enhanced by prestressing the material before contact was made and adhesion established. An apparent increase in force to failure by a factor 3 was possible by this mechanism.[20]

4.11 Stringing and Crazing

The mechanisms described in the previous sections are fundamentally molecular, but are enhanced by mechanical changes or losses in the bulk material. A particular mechanism which operates in the presence of polymers at the interface is stringing. It was mentioned by Rivlin,[21] who had observed the way in which sticky tape adhesive pulled out into fibres near the crack tip. In other words, the crack line did not remain straight but broke up into fibrous filaments which seemed to hold the crack faces together, as shown diagrammatically in Fig. 4.18.

In fact this type of separation process is bound to happen if a thin layer of elastic material, like sticky tape adhesive, is trapped between two more rigid layers which are pulled apart. The stress within the adhesive thin layer is a bulk stress because the material is constrained in all three dimensions. This bulk energy can be partly released by cavitation in the elastic body or by growth of adhesive cracks as shown in Fig. 4.19. The pressure then becomes two dimensional because one of the constraints is released, giving up energy which can drive the de-bonding. These adhesive cavities can often be seen in bullet proof glass laminates where polyvinyl butyral is used to glue thick plates of glass together. If the polymer is stressed, cracks penetrate along the interfaces to cause the characteristic crazed pattern. It has been suggested that this pattern is viscous in nature, the so-called Taylor instability. But it can happen with perfectly elastic rubbers, so viscous flow cannot be the cause. The thinner the polymer, the finer the craze cracks, as shown in detail by Chaudhury and his colleagues.[22,23]

Crazing mechanisms have been invoked to describe the attack of methanol on polystyrene,[24] to describe how plastic flow in a metal can retard a crack,[25] and to describe the effects of fibres in composites.[26] It can readily be proved that the brittle Griffith cracking mechanism is softened by the crazing phenomenon, making the adhesion appear tougher as shown in Fig. 4.20. The key factor is how much stress is supported by the strings to pull the crack faces together. If this crack-tying stress

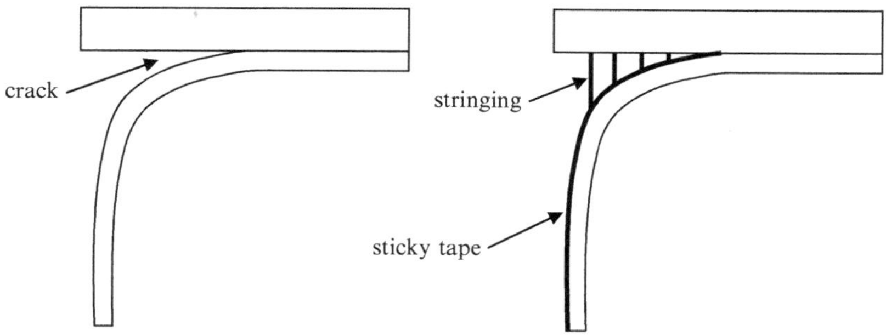

Fig. 4.18 Difference between a crack and stringing found with sticky tape

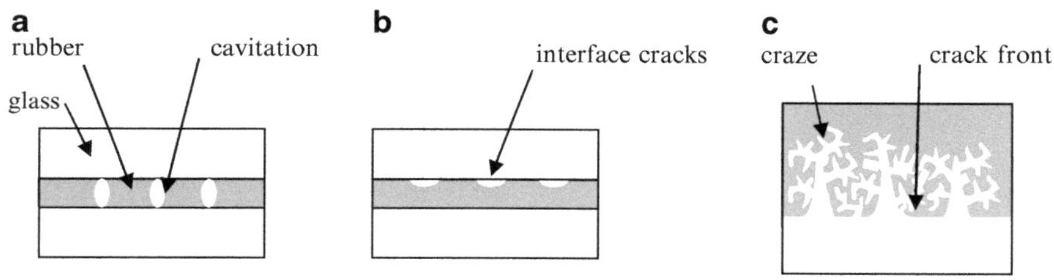

Fig. 4.19 (**a**) Cavitation in a polymer film stretched between rigid plates; (**b**) adhesion cracks giving same effect; (**c**) finger cracks develop into crazing

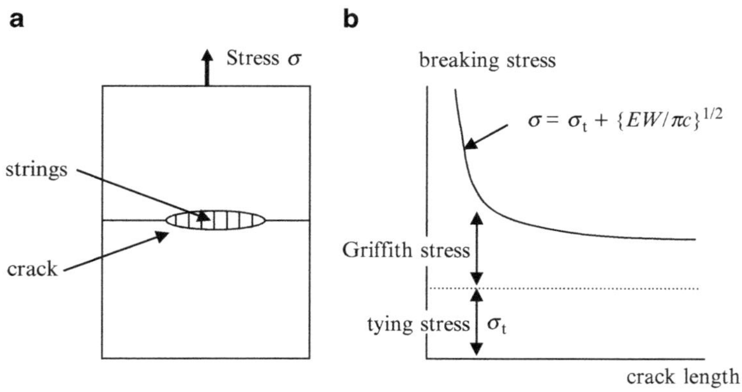

Fig. 4.20 (**a**) Interface crack with strings tying the faces together; (**b**) Stress required to break the tied joint is Griffith stress plus the tying stress

is high then the adhesive joint can appear almost ductile and fails more gracefully and predictably. The energy balance analysis applied to the interface crack shown in Fig. 4.20 proves that the stress applied to break the joint is the Griffith stress plus the tying stress. Thus the crazing mechanism means that the strength of the joint stays higher as the debonding propagates. Essentially, this is a Dugdale crack as described by Maugis in his book on fracture.[27]

4.12 Biofilm Adhesion

Perhaps the most ubiquitous illustration of macroscopic adhesion is that of the biofilm, the thin layer of material which is deposited on all surfaces surrounding living cells. A typical biofilm of modified E. coli is shown in Fig. 4.21, left, illustrating the cells within it and the surrounding matrix of extracellular polysaccharide fibres (glycocalyx) which sticks to the surface while providing an environment for cell proliferation. Glass slides coated with poly-L-lysine were placed in a bath of modified E.coli and agitated at 30°C in minimal M63 medium at 70 rpm in an orbital shaker incubator with a throw of 19 mm. After a time, the slides were removed and the thickness of the film was measured by interferometry, giving uniform growth over a period of 7 days to give an adhesive force to an AFM probe as shown in Fig. 4.21 right.

At first, the cells adhere weakly to the probe, but as the extracellular polymer (Fig. 4.21) begins to cover the cells after 5 days, the adhesion increases by two orders of magnitude because the polymer is stringing out as in Section 4.11. Costerton[28] has shown that most cells under normal circumstances like to grow on surfaces to produce a slimy film which can be found on pebbles in streams, on glass fishtanks, in sewage treatment facilities, on human tissues such as lung surfaces and generally on all solids. Indeed, Leeuwehoek had originally discovered cells by scraping the slime from his teeth and examining it in his primitive microscope. Of course, single cells can escape from these surface communities to form new colonies, but such free-swimming (planktonic) cells are in the minority, less than 0.1% of the population. The interesting observation is that antibiotics cannot easily kill

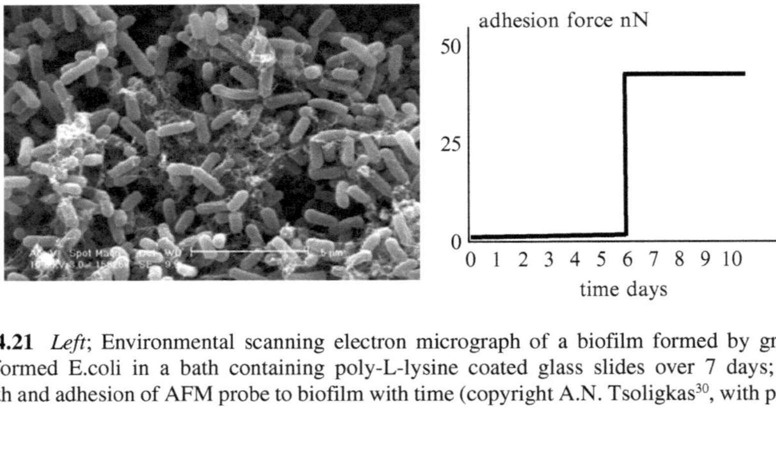

Fig. 4.21 *Left*; Environmental scanning electron micrograph of a biofilm formed by growing transformed E.coli in a bath containing poly-L-lysine coated glass slides over 7 days; *right*, growth and adhesion of AFM probe to biofilm with time (copyright A.N. Tsoligkas[30], with permission)

Fig. 4.22 Growth of biofilm; *left*, single cell attaches to surface, colony grows in extracellular polymer matrix; complex structure forms; *right*, cells break off and move away

the cells in the biofilm colonies because they are protected by their surrounding polymer coating. Thus biofilms are resistant to chemotherapy and require mechanical scraping for removal.[29]

The mechanism of growth is shown in Fig. 4.22. A free-swimming cell approaches the solid surface and adheres to it, subsequently excreting a layer of polysaccharide polymer. The cell reproduces and forms a colony in the extracellular matrix which hold the cells together. Several colony structures form a biofilm which can contain water channels supplying nutrients. Once the film becomes thicker, cells can break off to travel and reproduce elsewhere.

Such biofilms have often been considered to be a problem, for example in tooth decay, infection, prosthesis contamination and fouling of heat exchangers. However, in water treatment and biocatalysis, the films play a beneficial role through the formation of an extended reacting surface over which the nutrients flow.[30] The structure of this film may be enhanced by controlling its deposition through spin-coating to make an engineered biofilm.

Clearly, biofilm is a composite material with complex geometrical and viscoelastic properties which can affect adhesion by the mechanisms described in Sections 4.9 and 4.10.

4.13 Conclusions

In this chapter, it has become clear that there are three major factors influencing adhesion; Work of adhesion, geometry and elasticity. The fundamental theory of van der Waals forces shows that the energy per unit area of the adhesive bonds remains constant as the geometry and elasticity change. Therefore, a key parameter is *W*, the thermodynamic work of adhesion, the reversible work required to detach a unit area of adhesive junction. This term includes all the chemical effects such as the fall in adhesion under water and the influence of "surface contaminating" adhesion molecules which also reduce adhesion. Experiments on adhesion of living cells by wet-probing in the AFM show smooth elastic peeling adhesion with energy around 60 $\mu J\ m^{-2}$ but also show a number of local adhesion events which require interpretation. Changes in geometry from peeling to wedging, scraping or pull-off are effective in changing the adhesion force. When a crack meets a sudden increase in thickness, it can be arrested. Elasticity of the materials is similarly important. If a material stiffens up ahead of the crack, then the adhesion force is increased. Alternatively, plastic or viscoelastic dissipation around the crack can also raise the adhesion force, as also can stringing or crazing.

Whereas some adhesion tests depend mainly on *W*, like peeling or sphere adhesion, others are dependent on $(WEh)^{1/2}$ such as scraping, button pull-off or tensile debonding. Here *E* is the elastic modulus and h the geometry of the material, usually the thickness. Therefore it is vital to understand the adhesion test in detail if the adhesion force results are to be meaningful. Only then is it possible to analyse *W* the work of adhesion in order to assess the impact of adhesion molecules. AFM indentation and scraping can give an indication of the adhesion of both the cells and the biofilm colonies.

References

1. Roberts, A.D., Squeeze films between rubber and glass, J Phys D: Applied Phys 4 (1971) 423–432.
2. Kendall, K., Kinetics of contact between smooth solids, J Adhesion 7 (1974) 52–72.
3. Johnson, K.L., Kendall, K. and Roberts, A.D., Surface energy and the contact of elastic solids, Proc R Soc Lond A324 (1971) 301–313.
4. Kendall, K., Molecular adhesion and its applications, Kluwer New York, 2001.
5. Hamaker, H.C., The London-van der Waals attraction between spherical particles, Physica 10 (1937) 1058.
6. Griffith, A.A., The phenomenon of rupture and flow in solids, Phil Trans R Soc Lond A220 (1921) 163–98.
7. Kendall, K., Crack propagation in lap shear joints, J Phys D:Appl Phys 8 (1975) 512–22.
8. Kendall, K., The adhesion and surface energy of elastic solids, J Phys D:Appl Phys, 4 (1971) 1186–95.
9. Kendall, K., Adhesion: molecules and mechanics, Science 263 (1994) 1720–25.
10. Chaudhury, M.K. and Whitesides, G.M., Correlation between surface free energy and surface constitution, Science 255 (1993) 1230–32.
11. Chaudhury, M.K. and Whitesides, G.M., Direct measurement of interfacial interactions between semispherical lenses and flat sheets of poly (dimethylsiloxane) and their chemical derivatives, Langmuir 7 (1991) 1013–25.
12. Sirghi, L., Ponti, J., Broggi, F., Rossi, F., Probing elasticity and adhesion of live cells by atomic force microscopy indentation, Euro Biophys J 37 (2008) 935–45
13. Radmacher, M., Measuring the elastic properties of living cells by the atomic force microscopy, in 'Methods in cell biology' 68 (2002) 67–90, Academic Press NY eds. Jena, B.P. & Horber, J.K.
14. Zhu, A.P., Fang, N., Chan-Park, M.B., Chan, V., Adhesion contact dynamics of 3T3 fibroblasts on poly (lactide-co-glycoide acid), Surf Modified Photochem Immobil Biomacromol Biomater 27 (2006) 2566–76.
15. Sun, M., Graham, J.S., Hagedus, B., Marga, F., Zhang, Y., Forgacs, G., Grandbois, M., Multiple membrane tethers probed by atomic force microscopy, Biophys J, 89 (2005) 432–9.
16. Capella, B., Dietler, G., Force distance curves by Atomic Force Microscopy, Surface Science Reports 34 (1999) 1–104.
17. Bowen, W.R., Hilal, N., Lovitt, R.W., Wright, C.J., Colloids Surfaces A136 (1998) 231.

18. Kendall, K., Control of cracks by interfaces in composites, Proc R Soc A 341 (1975) 409–428.
19. Kendall, K., Shrinkage and peel strength of adhesive joints, J Phys D:Appl Phys 6 (1973) 1782–87.
20. Kendall, K., The effects of shrinkage on interfacial cracking in a bonded laminate, J Phys D: Appl Phys 8 (1975) 1722–1732
21. Rivlin, R.S., Paint Technol 9 (1944) 215–7.
22. Ghatak, A. Chaudhury, M.K., Adhesion induced instability in thin confined elastic film, Langmuir 19 (2003) 2621–31.
23. Ghatak, A., Chaudhury, M.K., Critical confinement and elastic instability in thin solid films, J Adhesion 83 (2007) 679–704
24. Kendall, K., Clegg, W.J. and Gregory, R.D., Growth of tied cracks: a model for polymer crazing, J Mater Sci Lett 10 (1991) 671–4.
25. Dugdale, D.S., Yielding of steel sheets containing slits, J Mech Phys Sol 8 (1960) 100–104.
26. Bowling, J. and Groves, G.W., Debonding and pull-out of ductile wires from a brittle matrix, J Mater Sci 14 (1979) 431–442.
27. Maugis, D., Contact, adhesion and rupture of elastic solids, Springer, Berlin 1999, p191–199.
28. Costerton, J.W., The Biofilm Primer, Springer Berlin 2007.
29. Costerton, J.W., Geesey, G.G., and Cheng, K-J., How bacteria stick, Scientific American 238 (1978) 86–95.
30. Tsoligkas, A.N., Winn, M., Bowen, J., Overton, T.W., Simmons, M.J.H., Goss, R.J.M., A new approach to generating catalytic biofilms, submitted 2010.

Chapter 5
Statistics of Adhesion at Nanoscale

Reaction to solids plays an important part in locomotion

Harrison 1914

As particles are smaller, towards the nanometer range, or when the gaps between surfaces approach the molecular level, it becomes necessary to consider not only the macroscopic i.e. average features of the adhesion, but also the Brownian movement and statistical nature of van der Waals bonding which is fluctuating and diffusing very significantly when viewed at the scale of atomic bonds.

As discussed in Chapter 3, we can think of a van der Waals attraction between two atoms, a single bond, but this can break in a Brownian collision, then later it can re-adhere. But each individual pair bond must be added together across the system to obtain the resultant overall force and energy. For large particles the fluctuations will average out but for nanoparticles there will be observable flickering of the black spot. The purpose of this chapter is to focus down to this fluctuating and diffusing molecular scale in order to understand what is meant by an adhesion molecule and how such an entity can influence the overall effects observed macroscopically.

In the first place, it is worthwhile to analyse the nature of the '*black contact spot*' more rigorously. This is clearly not constant across the black area because the adhesion increases with dwell-time and changes substantially with kinetic effects like peeling speed. Then we attempt to find true single bond adhesion between very small particles and come to the conclusion that this is revealed in aggregation tests.

5.1 The Flickering Black Spot; Large Contacts

When a contact between a macroscopic elastomer sphere and a surface is made, the black spot appears uniform in colour, except for dust particles and asperities which cause imperfections. Yet we know that the narrow gap between the surfaces can change in thickness substantially with time as fluid molecules are forced out of the contact region. Roberts[1] described these changes in 1971 when he studied interferometrically the approach of optically smooth rubber towards a flat glass surface through a liquid medium and showed that a number of different

K. Kendall et al., *Adhesion of Cells, Viruses and Nanoparticles*,
DOI 10.1007/978-90-481-2585-2_5, © Springer Science+Business Media B.V. 2011

effects can be seen depending on the molecules in the gap as shown schematically in Fig. 5.1.

The first effect is the squeeze film phenomenon noted by Stefan[2] and also studied by Reynolds[3] more than a century ago. Liquid is pushed out of the narrow gap against the viscous fluid resistance, a phenomenon that can be explained completely by fluid mechanics. With pure distilled water between rubber and glass, Roberts saw this happen quickly and then noted the jump to contact, with trapped islands of water in the black spot. These islands took a few hours to diffuse out, suggesting that the black spot contact was in fact separated by a few monolayers of water, about 1 nm thick, which allowed transport of the residual molecules to the edges (Fig. 5.1a). However, when 0.01 M sodium dodecyl sulphate (SDS) surfactant solution was squeezed out from between the rubber and glass, no jump to contact was observed (Fig. 5.1b). Instead the flattened rubber was held apart from the glass with a gap around 10 nm by repulsive double layer (DLVO) forces[1] (Fig. 5.2).

Two conclusions follow from these results: Firstly, the electrostatic forces depended strongly on the gap between the surfaces, increasing exponentially as the gap closed, like DLVO double layer forces in colloidal stability[4]; secondly, the effect of ions such as sodium chloride ions was to reduce the electrostatic repulsions, allowing the surfaces to approach more closely.

Roberts then went on to show that synovial fluid, that is the polymer solution which rests between the cartilage in knee joints, exhibited similar repulsive pressure on the rubber surfaces and suggested that this was important for lubrication of human joints.[5]

Later, Israelachvili and Adams[6,7] measured double layer repulsions in their surface force apparatus using crossed cylinders of mica. At pH around 6 in potassium nitrate KNO_3 solutions of various concentrations, the repulsion increased exponentially as the gap closed. For higher concentrations of salt, the slope increased as expected from DLVO theory. Results are shown in Fig. 5.3 for several concentrations of KNO_3, ranging from 10^{-4} to 10^{-1} M.

The results were found to fit the DLVO theory over a wide range of separations above 10 nm, when the Hamaker constant was taken as 2.2×10^{-20} J. At closer approach the repulsion increased faster. There was no clear cut evidence of the

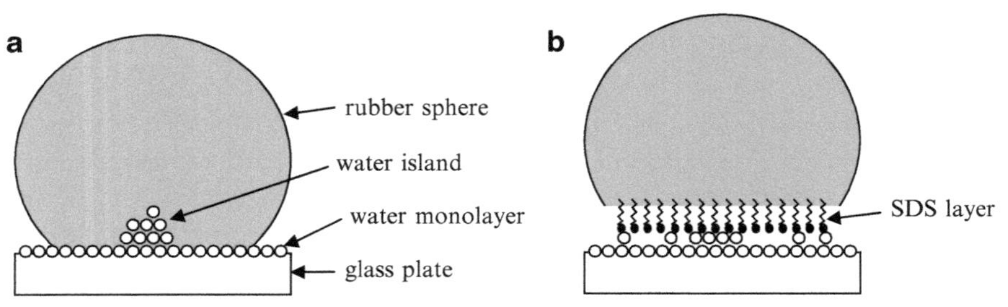

Fig. 5.1 (**a**) Water trapped between clean rubber surfaces at the contact spot; (**b**) Water containing SDS creates a repulsion due to the DLVO double layer

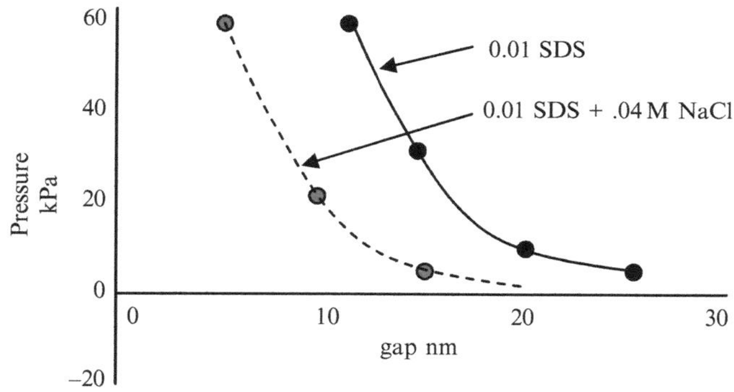

Fig. 5.2 Results of Roberts[1] for net repulsion of rubber from glass in 0.01 M SDS solution

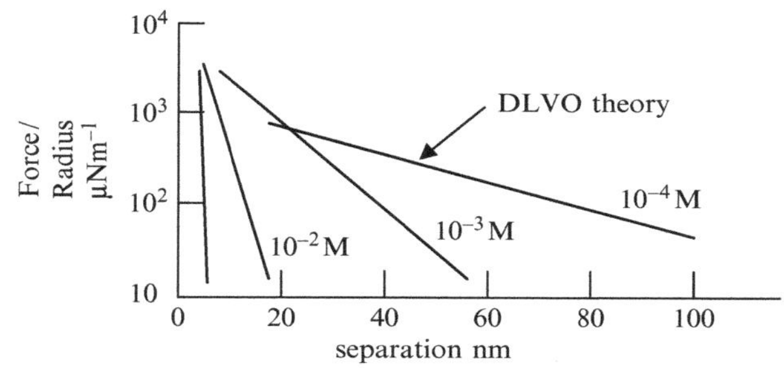

Fig. 5.3 Double layer repulsion forces measured in KNO_3 solutions[6]

primary minimum of attraction, which should cause jump to contact, to be expected below 5 nm when van der Waals forces dominate. However, it is known that with oxide surfaces, the repulsion can be continuous up to close contact, as a result of hydration layers, described in the theory of Section 3.6.

5.2 The Flickering Black Spot; Small Contacts

More recently, much work has been carried out on flickering contacts between small particles and surfaces, showing that colloidal particles can move statistically in and out of contact as they oscillate with Brownian movement.

The first test of this idea was carried out by Prieve and his colleagues around 1990.[8] They allowed a small sphere, 3–30 μm diameter suspended in water, to fall towards a glass surface under gravity, as shown in Fig. 5.4a. An argon laser beam prevented the particle from moving sideways. As the gap between the particle and glass reached about 200 nm, the particle began to scatter the evanescent light internally reflected from a helium neon laser shining laterally along the glass. Brownian movement of the particle up and down caused a flickering of this light which could

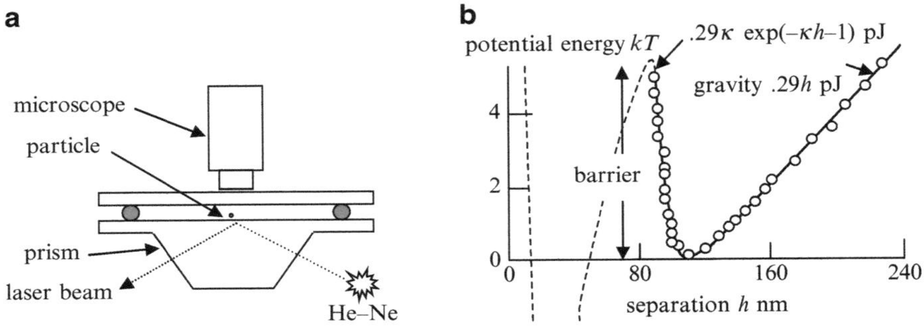

Fig. 5.4 (**a**) Prieve et al.[8] apparatus for measuring colloid repulsion; (**b**) results showing the gravitational potential and the double layer repulsion

be used to measure the particle position to within 1 nm. The probability $p(h)$ of finding the particle at a particular height h is given by Boltzmann's equation

$$p(h) = A \exp\{-\phi(h)/kT\}$$

where $\phi(h)$ is the potential energy of the sphere at height h. Thus the potential energy can be worked out as a function of height and shown to result from two forces; first the electrical repulsion pushing up due to the electrostatic double layer; second the gravitational force pressing down. Van der Waals forces were negligible at these large separations. The results fitted the theoretical curve, shown in Fig. 5.4b, demonstrating that this was a method with high force resolution, because it used kT as a gauge to measure 0.001 pN, almost a million times better than the surface force apparatus and a 1,000 times better than the AFM. The resolution of distance was 1 nm.

The flickering contact signal showed that the particle was statistically exploring the potential well at a distance of about 100 nm from the glass surface. The surprise was that a potential well could exist at such a large gap, because the energy landscape around kT has not properly been investigated until recently.[37] Of course, we wish to investigate much smaller gaps around 1 nm, as discussed in the theory of Chapter 3. These arise when jump to contact occurs between nominally clean surfaces such that adhesion develops. This regime has been more fully investigated by Zocchi and his colleagues[9] and Bevan's group.[36]

5.3 The Dwell-Time Effect

But the adhesion after jump to contact is not constant. It is observed to increase with the time of contact, the so-called dwell-time effect. Two surfaces were brought into contact and left for a time. The adhesion measured by peeling force was then found to have increased. Further time of contact led to further increase as shown in

Fig. 5.5. This effect could for example be a result of capillary condensation. When the surfaces are first in contact, the adhesion is low because roughness inhibits the short-range attractions. But as condensation occurs in the gaps, the adhesion rises with time, because the area of atomic contact is increasing rapidly with the extent of water filling the gaps.

Another possible cause of this effect is the creep of the interfacial contact caused by the gradual squashing of roughnesses. When two solids are placed in contact, the true atomic contact area tends to grow slowly with time because the material is not perfectly elastic. So, even when the atomic adhesion remains constant, more extended contact can occur from this junction growth. This has been measured particularly for polymers.[10]

When the contacting surfaces are very smooth, the above effects cannot arise so the cause of the dwell-time effect is the expulsion of contamination from the space between the surfaces, shown schematically in Fig. 5.5. In a particular experiment with elastomer,[11] shown in Fig. 5.6, natural rubber was mixed with 2.5% sulphur and cross-linked by heating at 145°C in contact with a smooth glass lens to give a spherical rubber surface. Immediately after de-moulding, this rubber was very smooth and adherent to glass surfaces Fig. 5.6a.

However, after the rubber sample had been in air for several days, the surfaces were no longer so smooth, and observation showed that tiny particles of sulphur had diffused out of the rubber to contaminate the surface, Fig. 5.6b. These particles were around 50 nm in size and reduced adhesion by an order of magnitude when tested with glass, Fig. 5.6c. After some hours in contact, the adhesion to the glass had increased back to its original high level, and no sulphur particles could be seen at the contact, Fig. 5.6d. The sulphur particles had been pushed back into solution in the rubber by the presence of the glass surface.

In conclusion, when contaminant molecules exist at a contact between two bodies, they will move towards their equilibrium positions by flow or diffusion. Adhesion will then increase as a function of time until a new equilibrium is attained. There

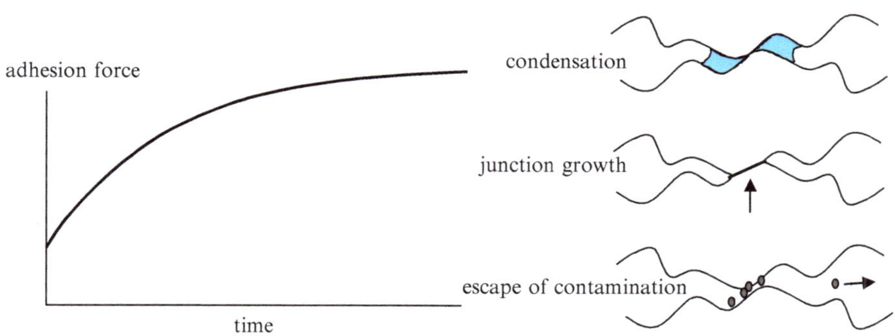

Fig. 5.5 The dwell-time effect, showing the increase in adhesion force with duration of contact, with three mechanisms; condensation, junction growth and escape of contamination

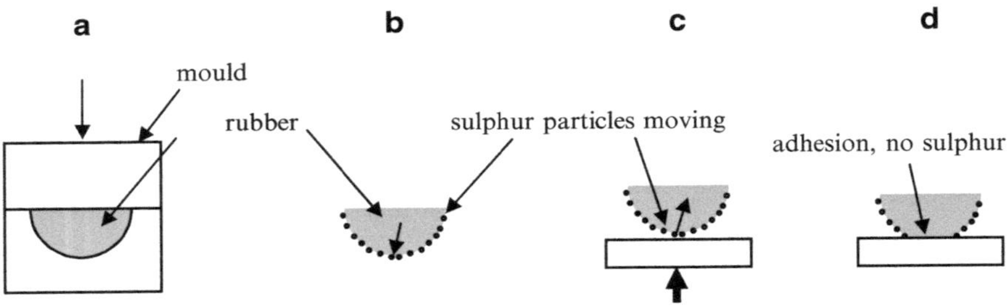

Fig. 5.6 (**a**) Moulding smooth rubber; (**b**) emergence of sulphur particles on surface; (**c**) contact with glass; (**d**) strong adhesion after sulphur diffuses into rubber

may be several different states of the contaminants between the bodies, leading to different values of adhesion.

5.4 Reaching Equilibrium

One of the best understood examples of these several adhesion states between contaminated surfaces is that of mica pressed into contact through water, as described by Israelachvili and his colleagues. Israelachvili joined the staff at the Australian National University in Canberra during 1973 and began to work with Adams in modifying the surface force apparatus previously built in Cambridge. The idea was to squeeze water and other molecules between ultra-smooth mica surfaces. A schematic of the equipment is shown in Fig. 5.7.

Two flakes of mica, about 1 μm thick, were cleaved and glued onto curved glass formers to give the crossed cylinder geometry. These mica surfaces could be brought together by a three stage mechanism; an upper coarse screw which positioned the lower mica surface; a bottom screw which adjusted the mica to 1 nm distance and the piezo tube which controlled the movement to 0.1 nm. Light was passed through the mica via a microscope objective to give the multiple beam interference fringes for measuring the gap between the mica surfaces. The lower mica flake was suspended on a cantilever spring of stiffness 100 Nm^{-1} so that the forces between the surfaces could be determined by the deflection. Special precautions had to be taken to purify the water to stop particles getting between the mica, inhibiting molecular contact.

Horn and Israelachvili[12] replaced the water with an inert liquid whose molecules were approximately spherical and of diameter 0.9 nm, octa methyl cyclo tetra siloxane or OMCTS. The idea was that this would not bond strongly to the mica surfaces (which anyway were contaminated with water and gas molecules), but would gradually be squeezed out of the way by the surfaces as they adhered together on close approach. The results shown in Fig. 5.8 indicated that the force of adhesion fluctuated significantly as each molecular layer was removed, revealing how the ordinary van der Waals forces were modulated by the molecular nature of the contamination

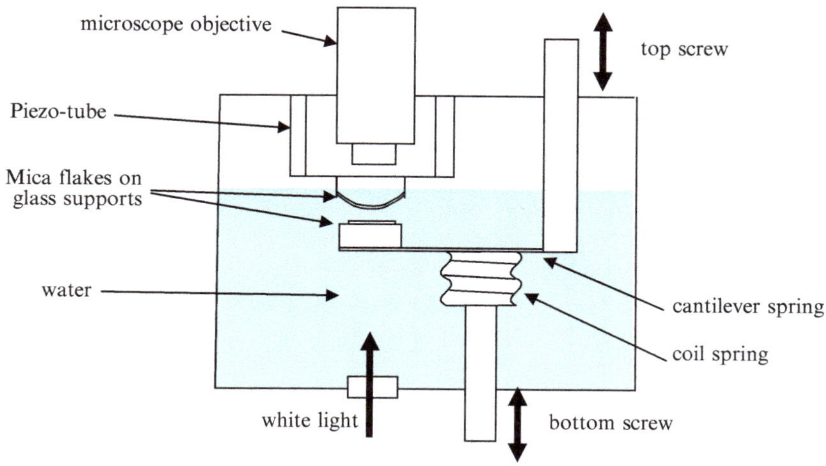

Fig. 5.7 Schematic of the surface force apparatus filled with water

(See theory of 3.6). In conclusion, the surfaces could not now jump together in one leap down the van der Waals curve; instead, the surfaces jumped together in a number of steps which depend on the size of the contaminant molecules.

In this case, the ultimate adhesion of the mica surfaces was weak, with an adhesion energy of 11 mJ m^{-2}. This was beneficial because no damage was then seen on the mica surfaces and the experiment could be repeated time and time again. However, this low adhesion was most likely due to contaminant water molecules strongly bonded to the mica in the first monolayer.

But the most significant result of this experiment, and of subsequent work by Horn and his colleagues[13,14] on other solvent molecules, was the demonstration that the system could now sit stably at several different stages of adhesion, shown by the minima in Fig. 5.8. In other words, there was not just one single adhesive state between two surfaces. Several different states of adhesion could exist depending on how many layers of molecules have been squeezed out from the gap. This was a suggestion first put forward by Kendall in 1973.[15] Obviously, the primary and strongest adhesion is when the surfaces are in intimate molecular contact. One layer of small foreign contaminant molecules will diminish adhesion by an order of magnitude. Larger molecules will have an even greater reduction effect. Further contaminant layers will diminish adhesion even more. It is clear from Fig. 5.8 that the size of the contaminant molecule is the most important factor in this argument, since the periodic jump distance is roughly equal to molecular diameter, larger molecules producing weaker adhesion.

These results were confirmed on a number of other solvent systems, including benzene, cyclohexane and carbon tetrachloride, which all behave as fairly rigid spheres.[16] Oscillations of attraction and repulsion were observed for up to ten molecular layers. With more flexible molecules, such as *n*-octane and 2,2, 4-trimethyl pentane, the oscillations died faster, after about four molecular layers.

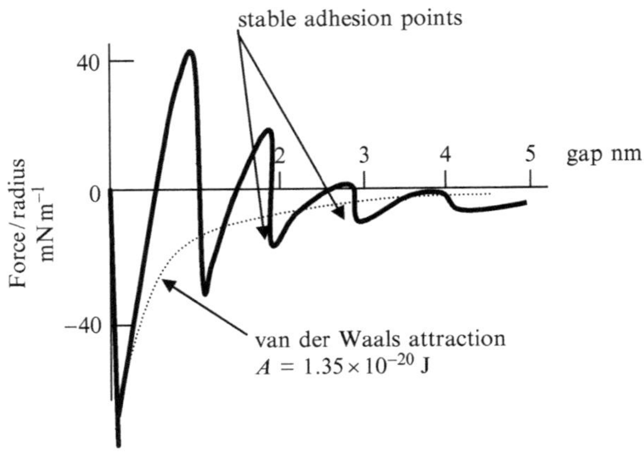

Fig. 5.8 Measured force between mica surfaces approaching each other through OCTMS[12]

Polar molecules, e.g. propylene carbonate, or methanol gave oscillations plus double layer repulsions. But water was the most interesting solvent, which must be considered in more detail.

5.5 Adhesion with Water Present at Surfaces

Pashley and Israelachvili[17,18] carried out a detailed study of mica surfaces approaching each other through water and dilute electrolyte solutions in an attempt to find 'hydrate crystal layers'. These layers had been inferred from the structure of damp clay, which is known to swell in water and to have a distinct lubricious surface, quite different from normal oxides. In the 1930s such clay had been investigated by the X-ray diffraction method, which showed that the clay plates moved apart in water to distances of 0.25 and 0.55 nm, about the diameter of one or two layers of water molecules, as shown schematically in Fig. 5.9.

Thus water appeared to be 'structured' near the surface of the oxide plates. Pashley at first could not see any structuring forces in his 1981 paper using 1 M KCl solutions. However, this was because of the behaviour of his spring system in bringing the mica surfaces together. Once he investigated the gaps below 1 nm more closely, especially with dilute KCl e.g. 10^{-3} M, then he found the stepwise jumping of the surfaces corresponding to removal of water molecular layers, as shown in Fig. 5.10.

The measurements of force could only be conducted in the regions shown by the black lines in Fig. 5.10. These measurements showed steep repulsions. If too great a force of compression was applied, then the mica jumped closer by removing one layer of water molecules, onto the next black line, which was again steeply repulsive. Eventually, the mica surfaces made contact to give a work of adhesion of 11 mJ m^{-2}. In addition to this molecular contact state,

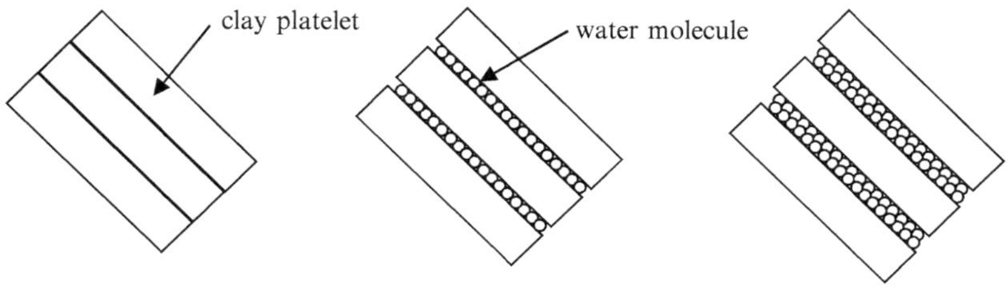

Fig. 5.9 Diagrammatic picture of clay platelets under dry, damp and wet conditions showing the water molecules between the layers

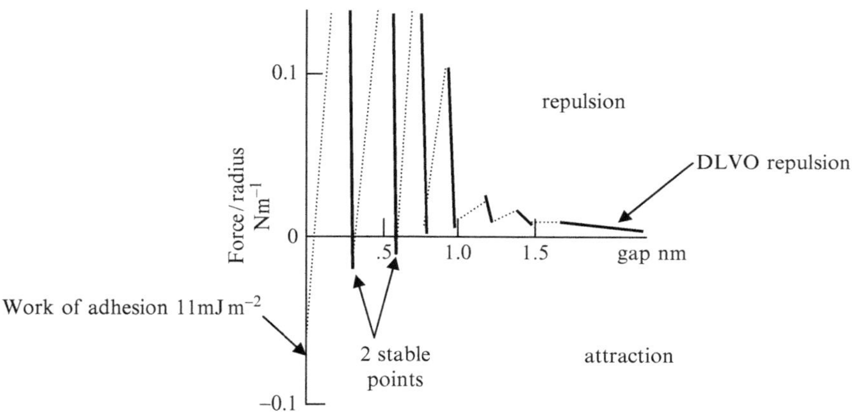

Fig. 5.10 Results for mica surfaces approaching in potassium chloride solution[17]

there were two other stable points of adhesion corresponding to one and two molecular layers of water between the mica respectively. In Fig. 5.10, these stable states are pointed out by arrows. These were located at gaps which agreed reasonably with the hydrated layers found in clays, around 0.25 and 0.55 nm.

When the mica plates were well separated, i.e. more than 2 nm apart, the results fitted the DLVO theory of double layer repulsions corresponding to a potential of −78 mV, which amounts to 40% coverage of the mica with positively charged potassium ions. These long-range forces were those previously measured by Adams and Israelachvili.[6,7]

Thus there appeared to be three types of force acting between the mica surfaces: the long range DLVO forces acting from 5 nm out; intermediate range repulsion acting from 2 to 5 nm; and finally the oscillatory jumping behaviour operating from contact out to 1.7 nm gaps. These close-in oscillatory forces in the presence of contaminant molecules have now been detected with atomic force microscopy but depend strongly on the tip geometry.[19]

5.6 Adhesive Drag

Once we have recognised that the contaminant molecules introduce an oscillating interaction energy between surfaces, as illustrated in Fig. 5.11, then we see that more complex adhesion effects must follow. For example, time effects must be observed because the contaminant molecules cannot get into position instantly. Molecules require time to diffuse into and out of the interface. Moreover, the contamination at the interface will depend on the force we apply to the joint, squeezing out the molecules. Also, we must consider the Brownian energy kT of the molecules which drives the diffusion process as the contamination escapes.

Consider as an example the situation shown in Fig. 5.11. The interface between the two surfaces can exist in the two metastable states with adhesion energies W_1 and W_2. Imagine first that the surfaces are in state W_1 and we wish to pull them apart into state W_2. To do this we apply a peeling force and this must be sufficient to overcome the energy barrier, with the help of the Brownian energy kT. This problem of molecules coming apart across energy barriers was first solved by Eyring in the early 1940s.[20] There are two forces required according to this theory; one to provide the reversible work of adhesion $W_2 - W_1$, and the second to overcome the energy barrier. Thus the total force can be expressed as the sum of two terms in the peeling equation given in Fig. 5.11. The second term is an energy loss term which appears as heat. Clearly this depends on the rate of peeling V, and also on the temperature variant constants A and B. The higher the rate of peeling, the greater the force required and hence the larger the energy dissipated. Also, the higher the temperature, the faster the peeling. Such behaviour is well-known for adhesive joints, such as those between silicone rubber and acrylic sheet as shown in Fig. 5.12.

In this example, at low speeds, the adhesion levelled off at a low value, corresponding to an apparent reversible work of adhesion of 0.3 Jm^{-2} at very low velocities of peeling.[15] We will study the precise nature of this equilibrium value in the next section. But at high speeds, the adhesion increased very strongly. This is the adhesive drag effect. Very similar results were obtained by Russian experimentalists

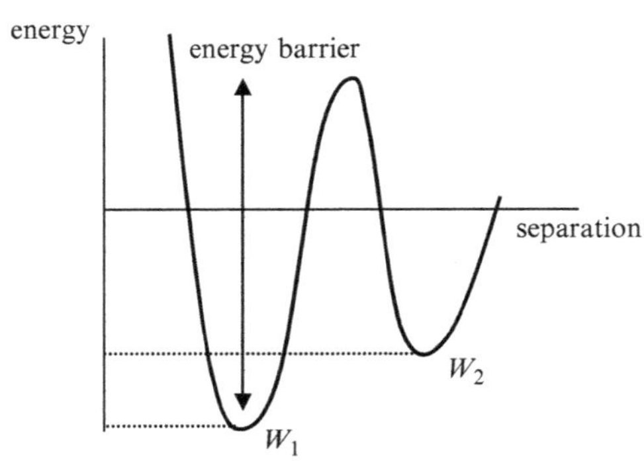

$$F/b = (W_2 - W_1) + A\sinh^{-1}BV)$$

Fig. 5.11 Schematic of the energy barrier causing adhesive drag, in separating a joint from state W_1 to state W_2

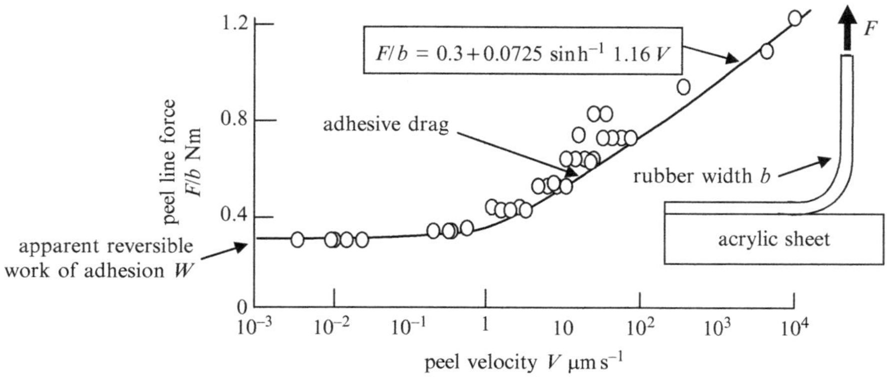

Fig. 5.12 Results for peel adhesion of silicone rubber from acrylic glassy polymer[15]

in the 1950s.[21] However, the Russian schools devised explanations of this adhesion behaviour based on charge separation or diffusion.[22,23] While there is no doubt that both charge separation and diffusion occur, overcoming the adhesion energy barrier at the interface should also be important.

The other significant aspect of adhesive drag is its relation to the surface contamination present on the surface. For example, an alkyd paint film was painted on a glass surface, cross-linked and then peeled off. For comparison, the same experiment was carried out on a glass surface coated with dimethyl dichloro silane, as shown in Fig. 5.13.

These results gave similar behaviour to that of silicone on acrylic, with an apparent equilibrium work of adhesion at low speeds, plus a velocity dependent peeling force at higher speed. However, there were two substantial differences; first the apparent work of adhesion was ten times too high at 4 Jm^{-2}; second, the presence of a silane coating had an enormous effect on the adhesive drag but not on the apparent equilibrium work of adhesion. This fall in adhesion due to one layer of molecules at the surface is akin to a catalytic effect: the monolayer is not changing the equilibrium, but is having a large effect on kinetics by reducing the energy barrier to peeling. Thus, the silane can be viewed as an adhesion molecule which catalyses the breaking of the van der Waals adhesion bonds (see theory of 3.11).

But adhesive drag is not the whole story because the measured energy of adhesion at low speed is too high. There must be other energy losses in addition to drag. Adhesive hysteresis is the term which describes these losses.

5.7 Adhesive Hysteresis

The problem of measuring adhesion, in general, is that the curves for peeling have a similar shape, with an apparent work of adhesion plus a large kinetic adhesion drag, but we are not sure exactly where the equilibrium is. So it is important to

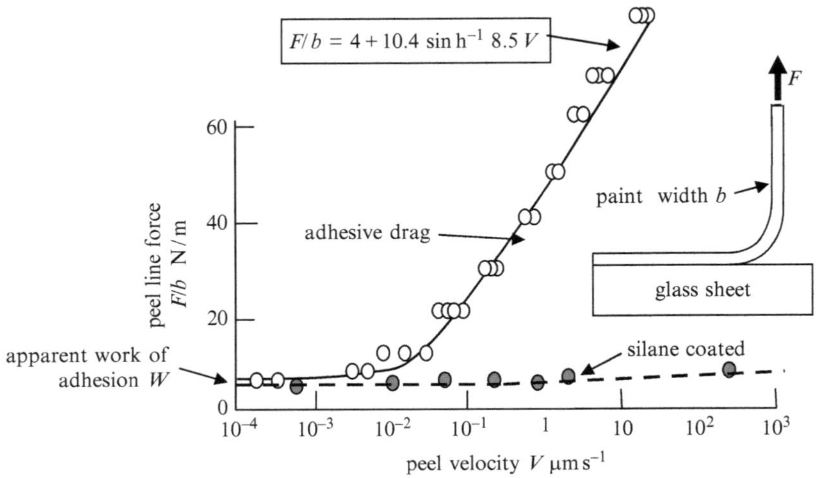

Fig. 5.13 Results for peel adhesion of paint from silicate glass[15], showing the effect of silane coating

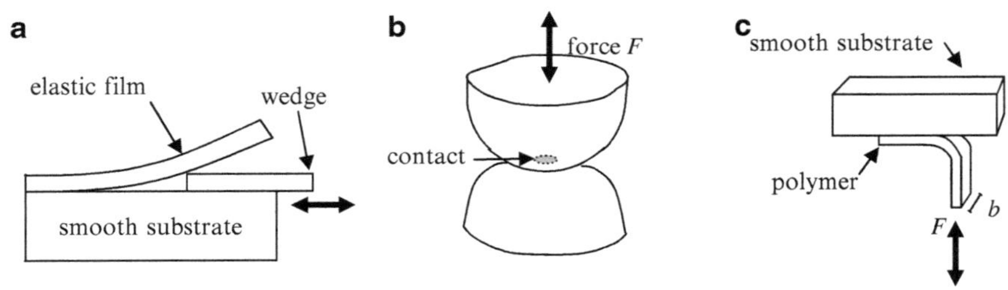

Fig. 5.14 Three experimental arrangements for studying equilibrium adhesion; (**a**) wedging; (**b**) sphere contact; and (**c**) peeling

devise experiments to study both making and breaking the joint in order to define the precise equilibrium point. Three typical experiments are shown above.

Figure 5.14a shows a wedging experiment, rather like that used by Obreimoff on mica.[24]

The film is detached by wedging, then the wedge is withdrawn slightly to allow healing. Figure 5.14b shows a sphere contact experiment, for example the JKR experiment,[25] in which a smooth sphere is allowed to make contact with a surface, then detached with a tensile force. Figure 5.14c illustrates a peeling film experiment in which the peel force is raised to peel the film, then lowered to heal the strip back onto the smooth substrate.[26] In each of these tests, the speed of movement of the crack front can be measured by observing the detachment line through the transparent materials, on both peeling and healing. The measured adhesion energy R, worked out from the force using the appropriate equation (e.g. $R = F/b$ for peeling), is then plotted against the crack velocity, on logarithmic scales as shown in

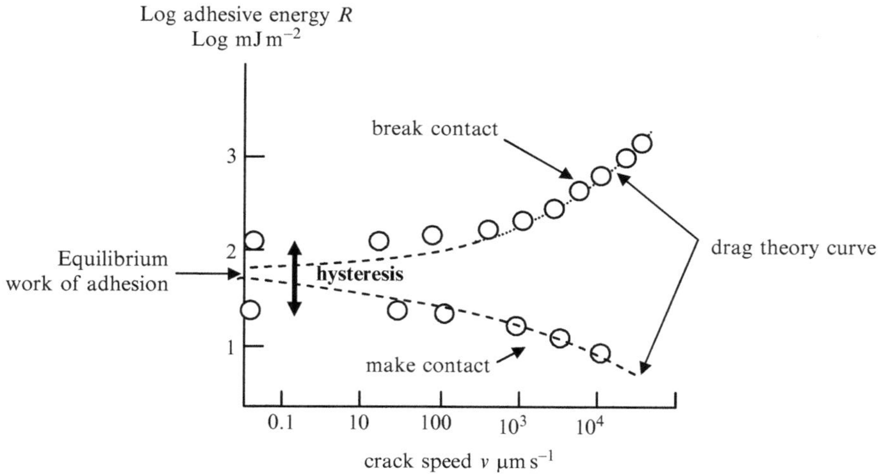

Fig. 5.15 Results showing the hysteresis for smooth cross-linked rubber on glass

Fig. 5.15. Both peeling and healing curves can be shown on the same logarithmic plot. This curve defines the adhesive drag on peeling and healing, and shows that equilibrium is not fully attained, but lies between the two asymptotes.

At very low speeds of crack propagation through the adhesive joint, to the left of Fig. 5.15, the peeling and healing curves should coincide. However, it was found experimentally that there was always a gap between the curves, which was small for silicone rubbers but larger for less elastic materials. This gap was defined as the adhesive hysteresis. The equilibrium work of adhesion was somewhere within this gap, around 70 mJ m^{-2}, but could not be found exactly in this experiment. Only by removing all the energy losses in the experiment would it be possible to attain true adhesive equilibrium. Such hysteresis energy losses could be caused by a number of mechanisms including roughness, impurities, inelastic deformation, etc.

However, one important energy loss which was explained was the effect of the visco-elastic behaviour of the polymer. This was studied by varying the cross-link density of the rubber, to alter the loss of elastic energy as the material relaxed. As the viscoelastic loss increased, so did the adhesive hysteresis, as shown in Fig. 5.16.

These results demonstrated that the viscoelastic relaxation in the rubber could stop the peeling to give an apparent adhesion much higher than the equilibrium value *W*. Essentially, this is the chewing gum effect discussed in Section 4.10. However, viscoelastic peel-stopping is somewhat different because it can be linear in time.

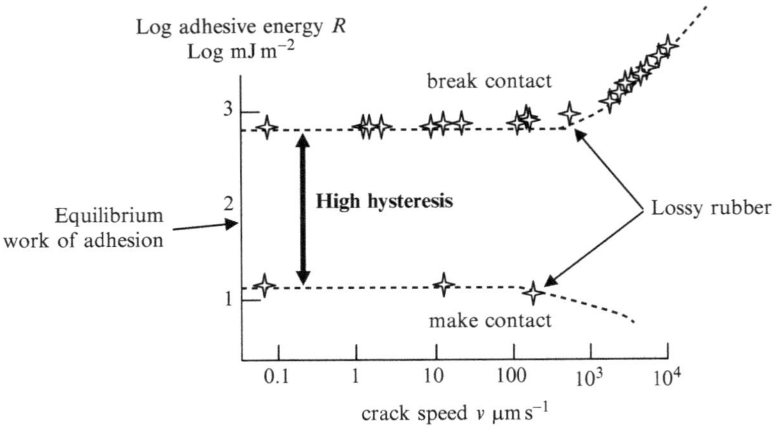

Fig. 5.16 Results showing the increase in hysteresis with viscoelastic loss

5.8 Peel Stopping by Viscoelastic Loss

Adhesive hysteresis was originally observed by Drutowski[27] in 1969. The effects were later studied systematically with smooth elastomer spheres using the apparatus shown in Fig. 5.17.[10]

Figure 5.17a shows a glass plate balanced on a pivot and gently moved down onto a smooth rubber sphere. The loading screw was adjusted until the Newton's ring fringes appeared. Then, suddenly, on further approach, the surfaces jumped to contact at zero applied load and the growing contact spot was recorded on the TV camera. The diameter increased rapidly for about 100 s and then remained constant for days. To measure the breaking contact curve, the glass plate was pressed momentarily into the rubber and then released. The large contact spot formed by the initial load then rapidly decreased in diameter for about a 100 s and then remained constant. The final contact diameter at zero load was almost twice that for making contact.

To obtain a more general explanation of these complex effects, Kendall[10,] separated out the drag and hysteresis terms. The drag was viewed as an interface reaction which was kinetically controlled by a surface energy barrier; whereas the hysteresis was viewed as a crack stopping effect brought about by the lossy relaxation of the inelastic material. The influence of these two separate phenomena was best demonstrated by measuring the contact spot size at various times and temperatures, as shown in Fig. 5.18.

The experiment was conducted as before, measuring the contact spot size d both making and breaking the contact, then calculating the adhesive energy R and crack speed v from the TV record using the equation derived from JKR

$$R = 2Ed^3 / 9\pi D^2(1-\nu^2) \tag{5.1}$$

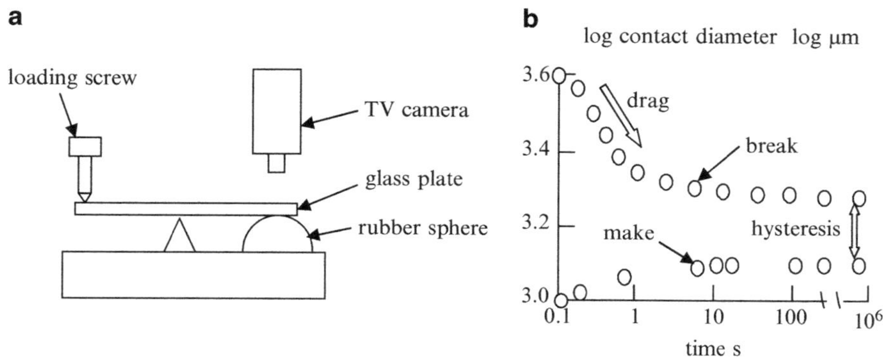

Fig. 5.17 (**a**) Apparatus for loading a rubber/glass contact for measurement of contact spot with time. (**b**) Results for sulphur cross-linked natural rubber, showing the difference between make and break, and indicating adhesive drag and hysteresis

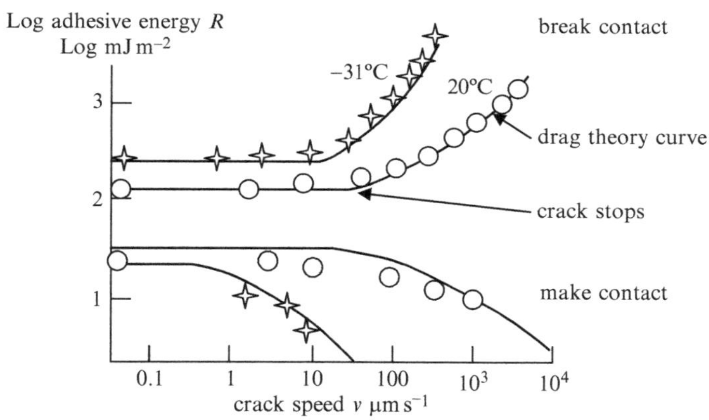

Fig. 5.18 Adhesive energy *R* calculated from contact spot size at various times and temperatures, compared with theoretical curves for drag and hysteresis

where *D* was the sphere diameter *E* its Young's modulus and *v* its Poisson's ratio. The measured adhesive drag at 20°C was fitted by a power law expression giving *R* as a function of crack speed

$$\log\{R / R_o\} = A \log \{v / v_o\} \tag{5.2}$$

and the temperature dependence was fitted to an Arrhenius type of curve. Then it was discovered that the crack stopping effect became more noticeable at lower temperatures as the viscoelastic loss in the rubber increased. The viscoelastic loss was quantified in terms of the relaxation constant *C* where

$$\mathrm{d}E / \mathrm{d}t = -C\, E / t \tag{5.3}$$

C was 0.01 for the rubber at room temperature, but increased at lower temperatures. The results were fitted to the following equation which was solved numerically:[28]

$$R(v) = \{2Ed^3 / 9\pi D^2(1-\nu^2)\}\{1 - Cx/vt\} \tag{5.4}$$

where x was the distance travelled by the crack and the term Cx/vt was a peel stopping term which increased with the relaxation of the rubber. The theory gave a reasonable description of the results as shown in Fig. 5.18, explaining why the adhesive drag and hysteresis of rubber both increase substantially at low temperature.

5.9 Rolling Resistance as a Measure of Adhesion Hysteresis, Drag and Dwell-Time Effect

One of the best ways to measure adhesive hysteresis is rolling. Consider a cell rolling on a smooth surface as shown in Fig. 5.19. Rolling in Fig. 5.19c is viewed as a combination of peeling and healing as in (b) and (a).

During rolling, both these peeling and healing processes are combined because one of the contacts is closing while the other is opening. Thus, we expect that rolling resistance for smooth cells should be dependent on the adhesive hysteresis and drag experienced in direct adhesion tests, less the dwell-time effect derived from the transient time of contact of the cell on the plate. This concept was tested[29] using silicone rubber in contact with glass surfaces as shown in Fig. 5.20. A cylindrical glass roller was placed on a smooth rubber strip and allowed to roll down under a controlled angle. The same optically smooth rubber was then peeled at various speeds from the glass to measure the break energy as in Fig. 5.20b. The dwell-time effect was evaluated by doing the peeling test after several periods of adhesive contact, from 10 to 100,000 s. Finally, the smooth rubber was allowed to heal back onto the glass by inclining the glass at a suitable angle. In all cases the speed of the crack moving through the adhesive interface was measured optically.

The results are shown in Fig. 5.21, indicating the significant dwell-time, drag and hysteresis effects. These were then compared with rolling friction. The energy required to break the bond was the most significant term, modified substantially by the dwell-time effect. At a rolling speed of 10 μm s^{-1}, the rolling contact (3 mm wide)

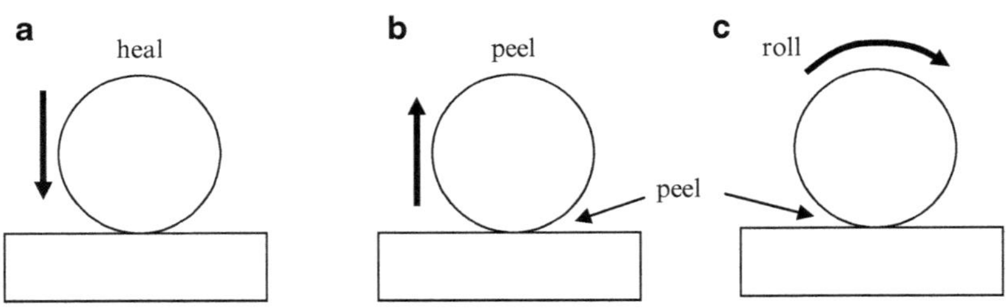

Fig. 5.19 (**a**) Cell making contact with surface; (**b**) breaking the contact; (**c**) rolling

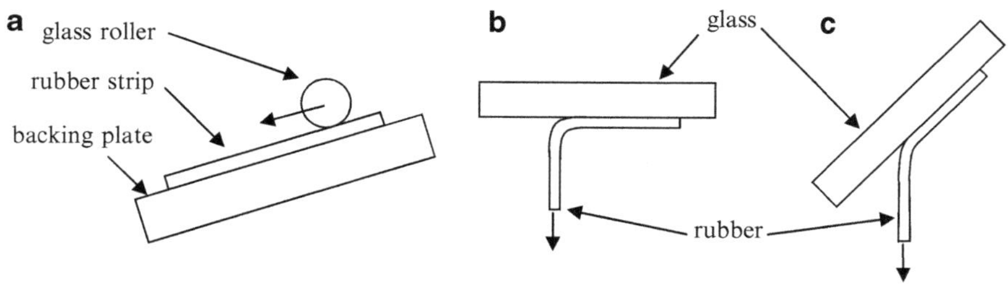

Fig. 5.20 (**a**) Rolling test; (**b**) peel test; (**c**) low angle heal test

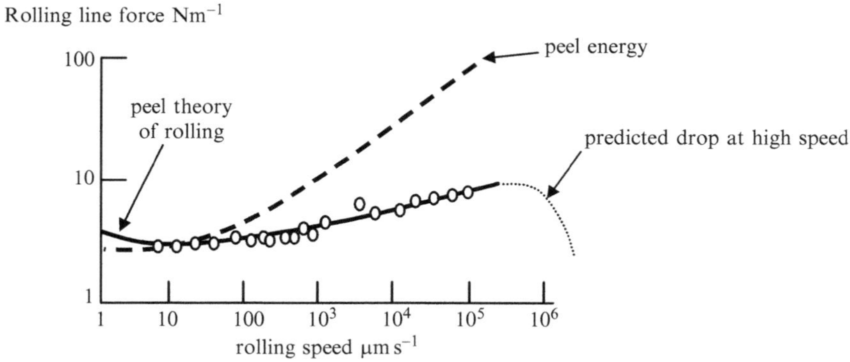

Fig. 5.21 Comparison of peel theory with rolling friction[29]

was bonded for 300 s which corresponded to the dashed line of Fig. 5.21. Faster rolling speeds gave lower dwell time so the rolling resistance was predictably lower than the peeling break curve. It was predicted that at high speed the rolling friction would decrease because dwell-time was too low. Also it was expected that dwell-time would dominate at low speeds, so that rolling resistance would increase. Rolling speeds below 10 μm s^{-1} gave a rapid slow down of the rolling as the dwell time effect took over, showing hyperbolic slowing of the roller with time.[29] It was noted that the rolling results could be predicted extremely well from the peel data.

5.10 Aggregation Statistics of Nano-Particles

Having considered the various mechanisms by which the contact spot can be affected by intervening molecules, which diffuse to give a complicated interaction between two adhering surfaces, it is now important to consider the situation where the contact spot is so small that only a small number of bonds are formed, or at the extreme, where there is only one bond between the particles, possibly by attachment of an adhesion molecule.[36,37]

One way to study the interactions between oxide surfaces in the presence of adhesion molecules is to use the Atomic Force Microscope (AFM). Milling with his colleagues performed a series of experiments using a 5 μm diameter silica sphere glued to the silicon nitride AFM cantilever, gradually brought towards a flat silica surface in solutions of polyelectrolyte, for example sodium polystyrene sulfonate or sodium polyacrylate. Well defined polymer samples were used to make sure all the molecules were of similar length, and the effects of polymer concentration and salt content were investigated.[30–32] There was little hysteresis in these measurements under dilute conditions.

A Nanoscope AFM from Digital Instruments was used with a wet cell arrangement. The spring constant of the cantilever, 0.06 Nm^{-1} was measured by the resonance method as different known masses were added. The silica was cleaned by brief boiling in ammoniacal hydrogen peroxide solution, followed by washing in ultra-pure water. Several molecular weights of sodium polyacrylate were used from 33,000 to 99,000. As expected, the higher molecular weight gave similar behaviour but at somewhat longer length scales (Fig. 5.22).

At a low polymer concentration of 39 ppm, there was an attraction with a minimum around 60 μm separation, together with repulsion at smaller gaps. However, at high concentrations, the minimum collapsed to a gap near 10 μm, with a much stronger attraction, followed by oscillatory behaviour at larger separations. This oscillating behaviour was similar to the structured hydration layers described earlier. It appears the polymer molecules were compressing as the concentration was increased, and the oscillations represented the force required to squeeze each layer of molecules from the gap. Putting electrolyte into the polymer solution had the same effect of collapsing the polymer molecules, again showing the oscillatory behaviour.[35] The AFM can only measure large forces. It seems possible that weak oscillations extending hundreds of nanometers from the surface could be detected

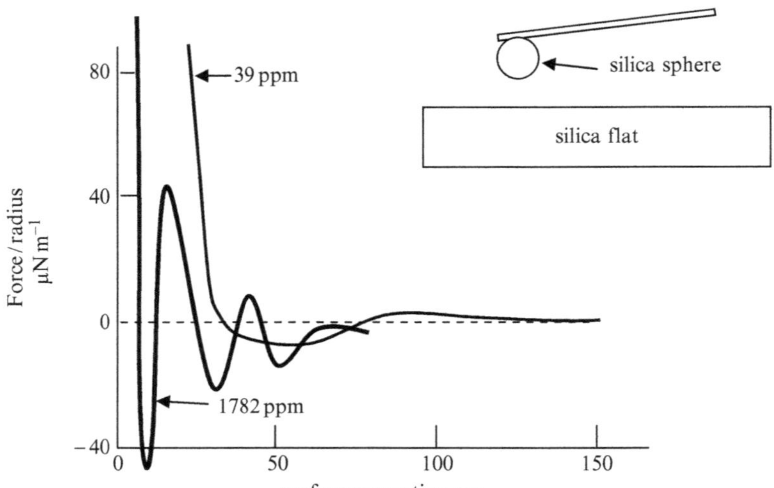

Fig. 5.22 AFM force-distance curves for silica sphere on flat using sodium polyacrylate solution at two concentrations

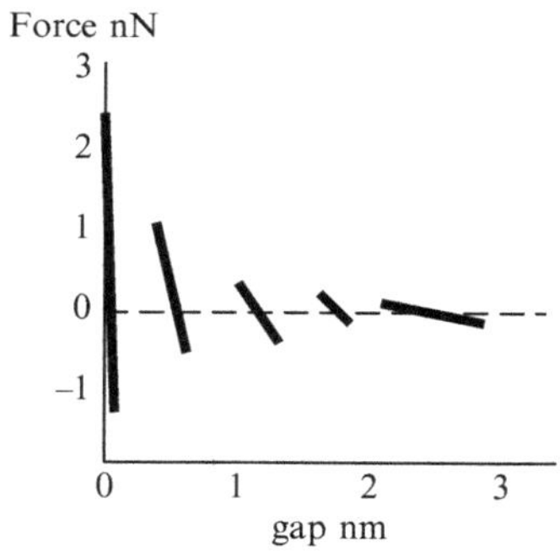

Fig. 5.23 Results for AFM Si tip on graphite with squalane molecules[19]

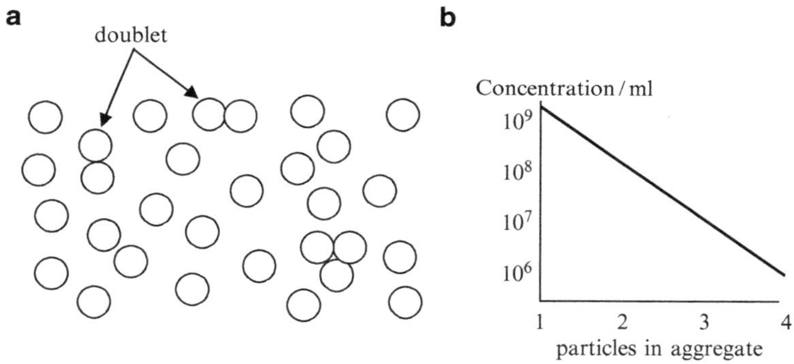

Fig. 5.24 (a) Particles adhering statistically; (b) Distribution of aggregates

by a probe sensitive to kT. For example, as shown later in Chapter 8, diffusing nanoparticles should detect such forces.[36–38]

The theory of these effects has been considered by a number of authors. A number of theorists have suggested structuring mechanisms in polymer solutions to explain the oscillatory forces.[33,34] It appears that polymer molecules can change the oscillatory behaviour of the attractions at the interface, perhaps improving the kinetics of the adhesion process.

Atomic force microscopy has now detected oscillatory forces in certain solvents but not yet water. For example, Lim and O'Shea studied squalane molecules on graphite with a silicon tip 110 nm in diameter and found that the adhesion at contact was –1.1 nN but with three more identifiable quasi equilibrium minima of –0.3, –0.2 and –0.1 nN with periodicity of 0.57 nm as shown in Fig. 5.23.

For a suspension containing many nanoparticles undergoing Brownian movement, the state of adhesion will be as shown in Fig. 5.24. There will be many single particles which are bouncing off each other because the adhesion energy is comparable with kT. There will be some doublets, triplets and higher aggregates where capture has occurred because of their slow collisions at the low end of the Boltzmann velocity distribution curve. Finally, there will be some doublets which have collided with high speed and sufficient energy to break through the molecular barrier to squeeze out the adhesion molecules to give the stronger adhesive contact. Measurements of these aggregates will be described in Chapters 7, 8 and 9.

The problem is that it is impossible to distinguish the weak doublets from the stronger ones merely by observing them through the microscope.

5.11 Conclusions

For small particles like cells, viruses and nanoparticles, the adhering contacts will not be steady and uniform as in the apparently static contacts described in Chapter 4 for macroscopic adhesive systems, but will be pulsating with Brownian collisions such that the dispersion consists of a range of states. At large gaps, say 100 nm, the DLVO forces and large polymer surface molecules dominate. But at smaller gaps, around 10 nm the polymer molecule squeezing forces come into play (i.e. adhesion molecule effects), and finally at 1 nm the hydration solvent layers begin to play a part. There are several possible adhesion states as the particles are pushed together by the Brownian movement, leading to adhesion drag, adhesion hysteresis and dwell-time effects. It is essential to understand these if the detailed adhesion process in the presence of adhesion molecules is to be predicted.

References

1. Roberts, A.D., Tabor, D., The extrusion of liquids between highly elastic solids, Proc R Soc Lond 325 (1971) 323–345.
2. Stefan, M.J., Versuch uber die scheinbare adhesion, Akad Wissen Wien Math Natur 69 (1874) 713–21.
3. Reynolds, O., On the theory of lubrication and its application to Mr Beauchamp Towers experiments, Phil Trans R Soc Lond 177 (1886) 157–234
4. Derjaguin, B., Landau, L., Theory of the stability of strongly charged lyophobic sols and of the adhesion of strongly charged particles in solutions of electrolytes, *Acta Physico chemica URSS* 14 (1941) 633.
5. Roberts A.D., Role of electrical repulsive forces in synovial fluid, Nature 231 (1971) 434–436.
6. Israelachvili, J.N. and Adams, G.E., Direct measurement of long range forces between two mica surfaces in aqueous KNO_3 solutions, Nature 262 (1976) 774–776.
7. Israelachvili, J.N. and Adams, G.E., Measurement of forces between two mica surfaces in aqueous electrolyte solutions in the range 0–100 nm, J Chem Soc Faraday Trans 74 (1978) 975–1001.
8. Prieve, D.C., Frej, N.A., Total Internal Reflection Microscopy: A Quantitative Tool for the Measurement of Colloidal Forces, Langmuir 6, 396 (1990) ; Prieve, D.C., Adv Coll Int Sci (1999).
9. Zocchi, G., Force measurements on single molecular contacts through evanescent wave microscopy, Biophys J 81 (2001) 2946–53.
10. Kendall, K., Kinetics of contact between smooth solids, J Adhesion 7 (1974) 55–72.
11. Kendall, K., Adhesion: Molecules and Mechanics, Science 263 (1994) 1720–25.

12. Horn, R.G. and Israelachvili, J.N., Direct measurement of structural forces between two surfaces in a nonpolar liquid, J Chem Phys 75 (1981) 1400–11.
13. Christenson, H.K. and Horn, R.G., Direct measurement of the force between solid surfaces in a polar liquid, Chem Phys Lett 98 (1983) 45–48.
14. Horn, R.G., Direct observation of the force between two lipid bilayers and observation of their fusion, Biochim Biophys Acta 778 (1984) 224–8.
15. Kendall, K., Peel adhesion of solid films-The surface and bulk effects, J Adhesion 5, (1973) 179–202.
16. Israelachvili, J.N., Intermolecular and Surface Forces, Academic Press, London 1985, pp.198–201.
17. Pashley, R.M. and Israelachvili, J.N., Molecular layering of water in thin films between mica surfaces and its relation to hydration forces, J Colloid Interface Sci 101 (1984) 511–23.
18. Israelachvili, J.N. and Pashley, R.M., Molecular layering of water at surfaces and origin of repulsive hydration forces, Nature 306 (1983) 249–50.
19. Lim, R., O'Shea, S.J., Solvation forces in branched molecular liquids, Phys Rev Lett 88 (2002) 246101–4
20. Glasstone, S., Laidler, K.J. and Eyring, H., Theory of Rate Processes, McGraw Hill, London, 1941, p. 339.
21. Krotova, N.A., Kirillova, Y.M. and Deryaguin, B.V., Zhur. Fiz Chim 30 (1956) 1921.
22. Derjaguin, B.V., Krotova, N.A. and Smilga, V.P., Adhesion of Solids, Consultants Bureau, London, 1978, ch.2.
23. Voyutskii, S.S., Autohesion and Adhesion of High Polymers, Wiley Interscience, New York, 1963, ch.1.
24. Obreimoff, J.W. The splitting strength of mica, Proc R Soc Lond A127 (1930) 290–97.
25. Johnson, K.R., Kendall, K. and Roberts, A.D., Surface energy and the contact of elastic solids, Proc R Soc Lond A A324 (1971) 301–13.
26. Kendall, K, The shapes of peeling solid films, J Adhesion 5 (1973) 105–117.
27. Drutowski, R.C., Hertzian contact and adhesion of elastomers, J Lub Techn Trans ASME 91 (1969) 732–7.
28. Kendall, K., Dynamics of slow peeling, Int J Fracture 11 (1975) 3–12.
29. Kendall, K., Rolling friction and adhesion between smooth solids, Wear 33 (1975) 351–8.
30. Milling, A.J., Depletion and structuring of poly (styrene sulfonate) at the silica-water interface J Phys Chem 100 (1996) 8986–93.
31. Milling, A.J. and Vincent, B., Depletion forces between silica surfaces in polyacrylic acid, J Chem Soc Faraday Trans 93 (1997) 3179–83.
32. Milling, A.J. and Kendall, K., Depletion, Adsorption and structuring of sodium polyacrylate at the water-silica interface 1. an atomic force microscope study, Langmuir 16 (2000) 5106–15.
33. Chattellier, X. and Joanny, J-F., Adsorption of polyelectrolyte solutions onto surfaces; a Debye Huckel theory, J Phys II 6 (1996) 1669–86.
34. Dahlgren, M.A.G., and Leermakers, F.A.M., Depletion zones in polyelectrolyte systems: polydispersity effects and colloid stability, Langmuir 11 (1995) 2996.
35. Ruckenstein, E., Manciu, M., Nanodispersions: Interactions, Stability and dynamics, Springer, NY 2010.
36. Fernandes, G.E., Beltran-Villegas, D.J., Bevan, M.A., Spatially controlled reversible colloidal self-assembly, J Chem Phys 131 (2009) 134705.
37. Bahukudumbi, P., Bevan, M.A., Imaging energy landscapes with concentrated diffusing colloidal probes, J Chem Phys 126 (2007) 244702.
38. Kendall, K., Dhir, A., Du, S., A new measure of molecular attractions between nanoparticles near kT adhesion energy, Nanotechnology 20 (2009) 0275701.

Chapter 6
Subdivision and Separation of Contact Spots

tissue came into contact with the cover-slip...and cells began to wander

Harrison 1914

In biological adhesion systems, a wide range of cell contact geometries has been found.[1–3] Where cells such as pollen or fungal spores need to be dispersed, the surfaces tend to be covered by spikes which prevent intimate extended contact between the particles, allowing the van der Waals force to be reduced to a low value. When adhesion needs to be maximised, for example where flies cling to the ceiling or lizards run up walls, the surfaces tend to split into close-packed hairy fibres or setai ending in flattened spatula tips. Figure 6.1 shows pictures illustrating these two cases.

These two opposing examples show that the subdivision of surface structures takes two basic forms: sharp spikes which hold surfaces apart to prevent van der Waals contact; and flat hairy closely-spaced contact spots which are multiplied and flexible to maximise the true van der Waals contact between rough surfaces. Whereas spikes must be well spaced out and have a very fine contact spot to diminish the true contact, flexible hairs must be as close packed as possible with a large flattened area at the end to maximise contact area.

6.1 Increase in Adhesion Force for Subdivided Contact Spots

The connexion between the scale of the contact hairs and the size of the creature was shown by Autumn and extended by Arzt and his colleagues.[2,3] A beetle with low gravitational pull had large contact spots at the 10 μm scale, whereas the heavy gecko had much smaller contacts, 100 nm in size as shown in Fig. 6.2.

At the same time, the number of contacts per unit area was seen to increase as the creature got heavier, as shown in Fig. 6.3 in a log-log plot. A small bug might have only one contact spot per 100 μm^2 to support its light weight whereas a gecko has about a thousand contact spot per 100 μm^2, with around half a million spots per foot. The straight line drawn through these results was described by the Equation.[2]

K. Kendall et al., *Adhesion of Cells, Viruses and Nanoparticles*,
DOI 10.1007/978-90-481-2585-2_6, © Springer Science+Business Media B.V. 2011

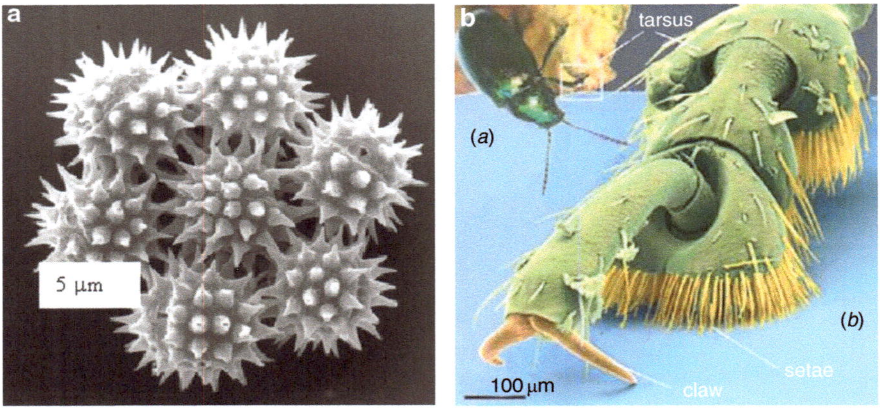

Fig. 6.1 (**a**) Spiky spores[1] (Copyright Rob Price, with permission); (**b**) Hairy insect foot showing subdivision into hairs (setai) with flattened tips (Copyright Stanislav Gorb, with permission)

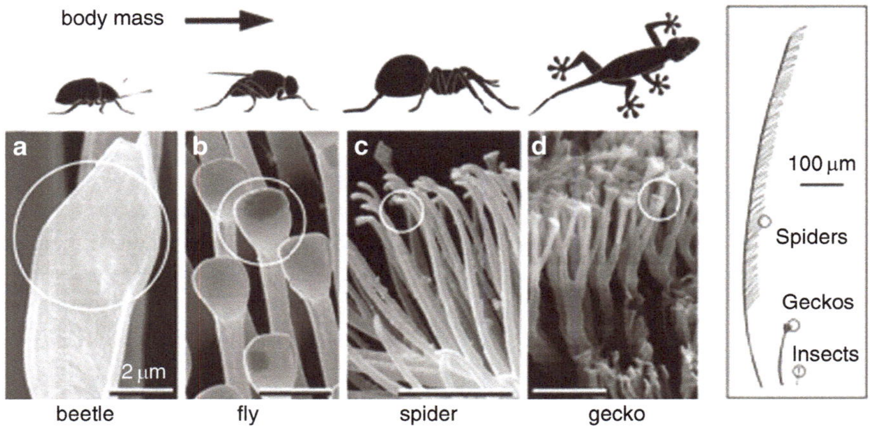

Fig. 6.2 Finer structure of adhesive setai as creature gets bigger[2] (Copyright National Academy of Sciences USA, reprinted with permission)

$$\log \cdot N_A \left(\mathrm{m}^{-2}\right) = 13.8 + 0.699 \cdot \log \cdot m\left(\mathrm{kg}\right), \quad R = 0.919 \tag{6.1}$$

where N_A was the number of contacts per m^2 and m was the creature's mass in kg.

The striking biological conclusion was that the varied attachment structures had evolved independently several times in the different species.[4,5] Various groups, such as spiders, lizards and insects have surfaces covered in patterns of asperities with different functions, such as hairs, hooks or pads. Not only the feet but also other areas of the bodies and legs were involved. The detailed geometries of the setai

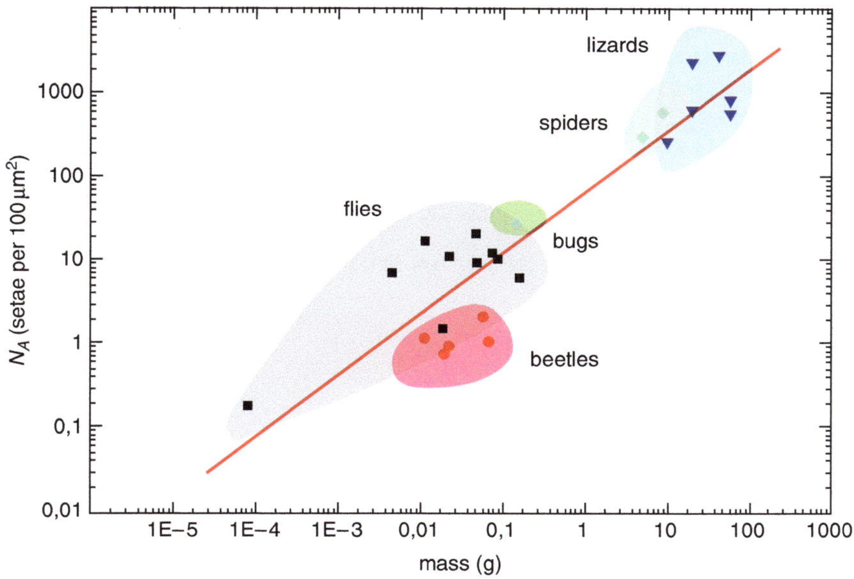

Fig. 6.3 Increase in the number of contact spots per unit area as creatures rose in mass[2] (Copyright National Academy of Sciences USA, reprinted with permission)

were different as also were the lengths of the hairs which ranged from mm to µm. It was clear from these observations that a general physical law was operating.

Some other groups have shown that this impressive correlation does not work within species. For example, 13 spiders studied by Keane et al.[2] did not show any change in setal concentration with mass over three orders of magnitude, confirming the results of Peattie and Full.

6.2 Force Versus Energy

Several attempts have been made to explain these results in terms of van der Waals forces, based on the knowledge that most spiders and lizards do not excrete sticky fluids so that dry contact of smooth polymer surfaces must be invoked to account for Eq. 6.1. Of course, the geometries are complex and an over-arching theory cannot be applied universally. However, contact mechanics offers a plausible solution to the problem.[6,7] Contact mechanics states that the energy of the contact is given by the contact area A multiplied by the work of adhesion W to give total adhesion energy AW, but the force is determined by the geometry and elastic properties as shown in Chapters 1 and 2. Thus, although a contact may be brittle because it is limited by the contact energy AW, the force for detachment at equilibrium may be increased by altering the geometry or the elastic properties. The weakest forces are for peeling, frictionless wedging and sphere pull-off, whereas the strongest forces are for rigid pull-off or lap shear because elastic modulus then enters the equations.[8, 9]

Imagine a creature as a perfect smooth elastic sphere, obeying the JKR theory described in Fig. 2.14, then it is evident that once the sphere exceeds about 6 mm diameter it cannot hang on a dry ceiling because the weight increases with L^3 where L is the creature's diameter while the adhesion only increases with L. The creature drops off when the weight just equals the adhesion, calculated from JKR i.e.

$$\pi L^3 \rho g / 6 = 3\pi W L / 4 \tag{6.2}$$

Therefore

$$L^2 = 9W / 2\rho g \tag{6.3}$$

So spherical creatures smaller than 6.7 mm will adhere to the ceiling (Fig. 6.4a), whereas creatures larger than L = 6.7 mm will fall, taking density to be 1,000 kg m^{-3} and W = 0.1 J m^{-2} for dry smooth polymer surfaces. One obvious strategy is for a creature to get larger by adding together a number of spheres less than 6 mm in size, extending along the surface as in Fig. 6.4b. The problem is that the biofilm is then limited in thickness to about 6 mm.

Another strategy is for the spherical creature to have smaller hemispherical asperities upon it. The JKR contact is then subdivided into a larger number of small contact spots. However, it can be shown that these are all dominated by the overall Hertzian pressure distribution, so that the adhesive force is actually smaller for this bumpy sphere-on-sphere configuration than for a single smooth spherical contact. Thus this more complex structure behaves as though it has been roughened and gives reduced adhesion, as in Fig. 1.10.

Arzt and his colleagues got around this difficulty by changing the geometry as shown in Fig. 6.5 from a single spherical contact of diameter L_1 into N close-packed spherical contacts of diameter L_2. If each smaller contact is independent of its neighbours, such that the contact forces can be added for each spot, then the overall adhesion force is now increased from the original JKR value

$$F_1 = 3\pi W L_1 / 4 \tag{6.4}$$

to

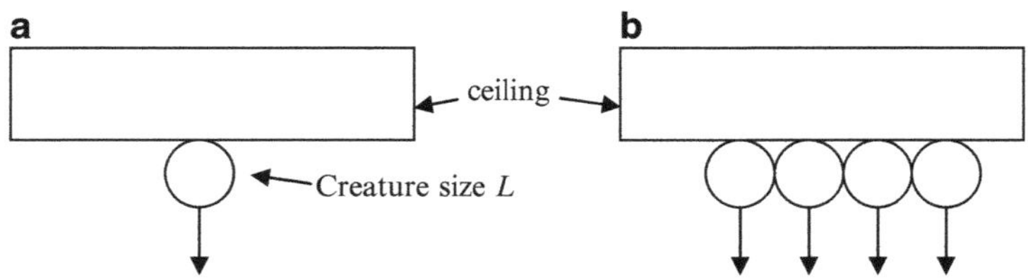

Fig. 6.4 (**a**) Smooth elastic spherical creature hanging from a ceiling needs to be less than 6 mm in size (**b**) Creature can grow by adding further spheres along the surface as in bacterial film

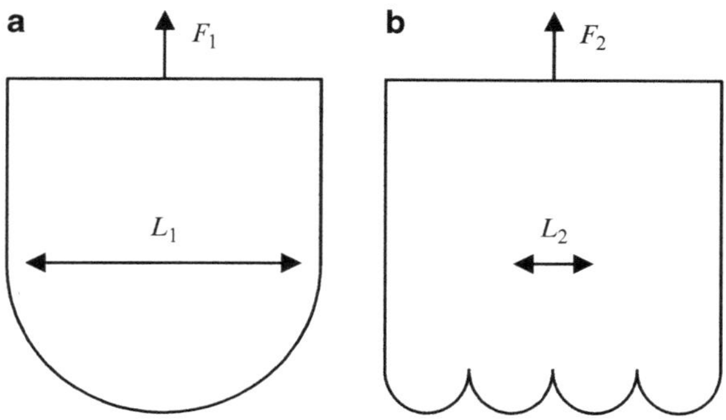

Fig. 6.5 (**a**) Insect foot with spherical end; (**b**) subdividing foot into N spherical setai

$$F_2 = N\ 3\pi WL_2 / 4 \tag{6.5}$$

And since $L_1/L_2 = N^{1/2}$then

$$F_2 = N^{1/2} 3\pi WL_1 / 4 = N^{1/2} F_1 \tag{6.6}$$

It is clear from this argument that increasing the density of setai in this way allows more load to be carried. So if an insect of diameter D with one contact spot, i.e. setal concentration $1/D^2$, doubled in size, then its mass would rise by D^3 i.e. by a factor 8, whereas the JKR force would only double and it would drop off. Four times more force is needed, which can be achieved by increasing the number of contacts from one to 4^2 i.e. 16 smaller contacts are needed to give the extra force because of the $N^{1/2}$ relation above. But its contact area goes as D^2 so there is four times more space for setai, giving a setal concentration of 16/4 = 4 times higher than the original insect. In other words, the setal concentration must rise with size squared or mass to the 2/3, fitting the curve shown in Eq. 6.1 and the line in Fig. 6.3. Setai have evolved according to the laws of adhesion, demonstrating one clear area where physical laws define biology.

6.3 Peeling Off the Gecko Foot

Consider how the adhesion force holding the gecko to the ceiling is large, but the force required by the gecko to lift its adhering feet is still low enough for the animal to walk around quickly and easily. This can be explained by the diagram in Fig. 6.6.

The diagram on the left shows the adhering gecko foot and the force transmitted to the gecko body by the leg which is pulling at a slight angle. If the gecko attempted to remove the adhesive foot by pulling in tension along its leg, then the force required would be very large, approximately equal to the full weight of the lizard which can be 50 g in mass, supporting 0.5 N force.

The quick release of the foot by the gecko as demonstrated by Autumn and his colleagues in movies, is facilitated by the lizard releasing its adhesion by peeling

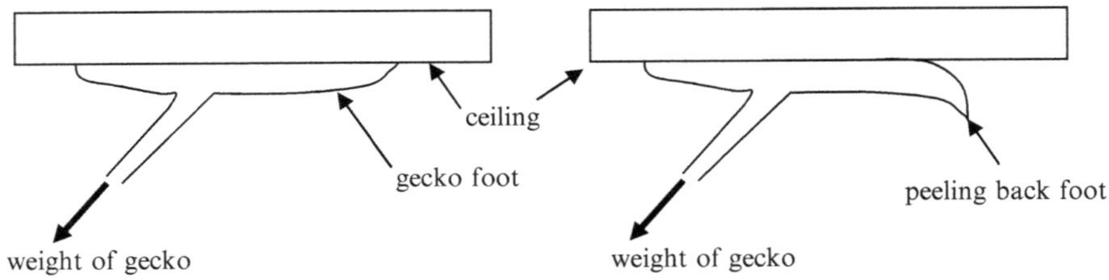

Fig. 6.6 *Left*; Force exerted on gecko leg is large; *right*; peeling force to bend toes back is low

the foot from its edge as depicted in the right hand diagram.[10] Release can occur by peeling in just a few milliseconds.

The peeling force at 90° is given by *Wb* where *W* is the work of adhesion, around 0.1 J m^{-2} and *b* is the width of the adhesive foot, about 0.005 m, leading to a peel force of 0.0005 N, a 1,000 times less than that required for instant detachment by pulling the leg to remove the foot vertically. Again this fits the concept that the normal adhesion force has been multiplied by $N^{1/2}$ where *N* is around one million for the large gecko foot.

This suggests that the gecko alters the attachment and detachment mechanisms to suit its requirements. The angle of the leg was also considered important based on the concept that the adhesion interaction of the setai with the substrate was a result of frictional traction of the foot against the surface.[11]

6.4 Measurement of Single Seta Adhesion Force

In order to prove that the multiple separated gecko foot contacts can together give a large force, Autumn et al in Nature (2000) reported the measurement of a single seta taken from a gecko as shown in Fig. 6.7a, which illustrates several lizard foot geometries. The picture shows the view through the glass plate which the lizard is climbing, followed by a detailed view of the toe pads which are clearly subdivided into a large number of hairs or setai shown magnified in Fig. 6.7b and c. A single seta is shown in Fig. 6.7d.

The direction of force *F* simulating a gecko climbing a vertical wall is indicated. The force was also measured perpendicular to the surface.

The initial efforts to attach a single seta failed to give any adhesion and only small forces as expected from friction were measured. But when a small preload was applied normal to the surface, the expected large adhesion forces around 40 μN were observed. It was shown that the seta had to be oriented properly to the glass surface, then a preload was necessary, then drag along the surface giving a 5 μm displacement was needed to generate substantial adhesion force of 200 μN, 32 times the force predicted by whole-animal measurements.[12] The discovery that maximal adhesion in isolated setae requires a small push perpendicular to the surface, followed by a small parallel drag, explained the load dependence and directionality of adhesion observed at the whole-animal scale by Dellit[13] and was consistent with the hypothesis that the structure of individual setae and spatulae is

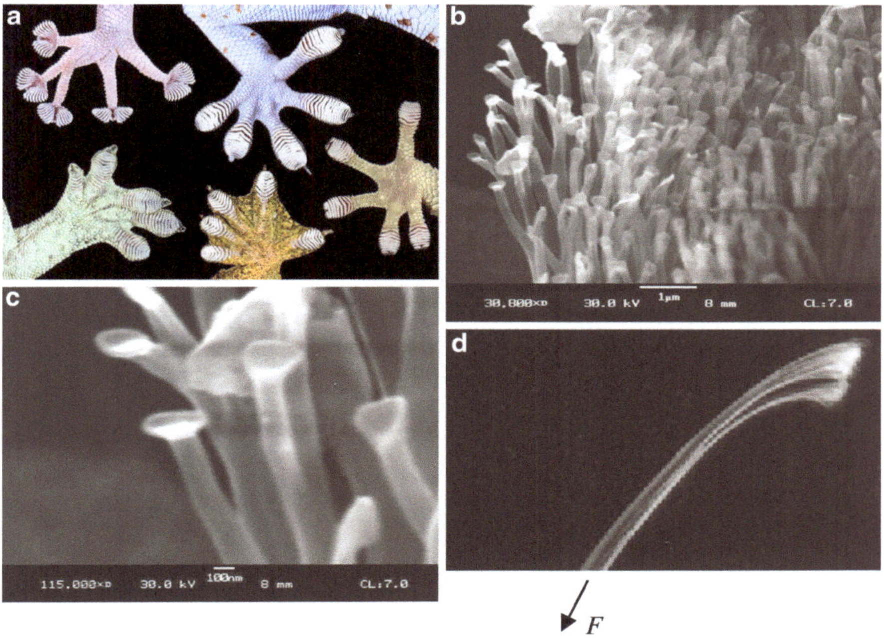

Fig. 6.7 (**a**) Pictures taken through glass plate of adhering lizard feet, showing the hairy pads in contact adhering to the smooth surface (Copyright Kellar Autumn, reprinted with permission). (**b**, **c** and **d**) Higher magnification pictures of the setai (Copyright Kellar Autumn, reprinted with permission)

such that a small preload and rearward displacement is necessary to engage adhesion.[14] The curvature of the setai was believed to be very important. The initial bending is flattened against the substrate by the preload such that improved molecular contact is made. The adhesion results are shown in Fig. 6.8 and demonstrate that the force achieved by touching is small at first, but the force rises by a factor of 4 as preload is increased. Once the force reached 200 μN, sliding occurred followed by fracture. The conclusion was that the contact of each seta was complex but the forces added together for millions of setai were large, around 100 N, many times more than needed to hold the weight of the lizard, around 1 N. Of course, the surface roughness and contamination in the real world could make this over-adhesion necessary because they would reduce the adhesion of each individual seta.

6.5 Possibility of Adhesive Dislocations

The above observations on geckos climbing walls illustrate the complex situation when a tension is applied to pull an adhering elastic strip from a smooth glass plate. The measurement on a single gecko seta is equivalent to experiments conducted in

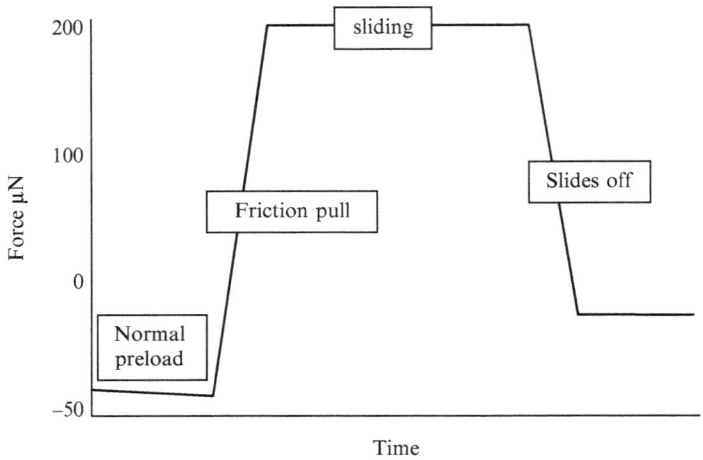

Fig. 6.8 Applying force to a single seta: first a normal preload pushes the seta onto the surface; then the frictional pull forms stronger contact; then sliding occurs; finally the seta drops off

the 1970s showing how an elastomer strip adhering through van der Waals forces could be pulled from a glass surface by a tension, as shown in Fig. 6.9.[15,16]

Figure 6.9a shows a single seta being pulled from the glass. This is essentially the same as reducing the peel angle of a peeling rubber film from 90° as described in Fig. 1.11. As Rivlin showed, the potential energy in the load is now changed from $-Fc$ to $-Fc\ (1-\cos\theta)$, so that the force must be raised to continue peeling. When the force is raised, the elastic film begins to stretch significantly, storing elastic energy F^2c/bEh in the uniformly extended elastic material, since elastic energy is a half of stress times strain times volume of the stretched elastic material.

The condition for equilibrium of the crack is then

$$(F/b)^2/2Eh + F/b(1-\cos\phi) - W = 0 \tag{6.7}$$

Showing that the result for peel force F is the solution of a quadratic equation. This theoretical calculation was compared with the experimental results for peeling of an elastomer from glass at various angles as shown in Fig. 6.10. As the angle was reduced, the peel force rose, but eventually levelled out at

$$F = b(2WEh)^{1/2} \tag{6.8}$$

which is the equation for lap failure of a flexible film in contact with a rigid surface, which applies also to shrinkage of films, to lap joints and to testing of composite materials.[9]

Equation 6.8, originally postulated and proved in 1973,[17] rather similar to the equation for fibre debonding of Gurney and Hunt[18] and Outwater and Murphy, has been ‘rediscovered’ regularly since that time.[19,20]

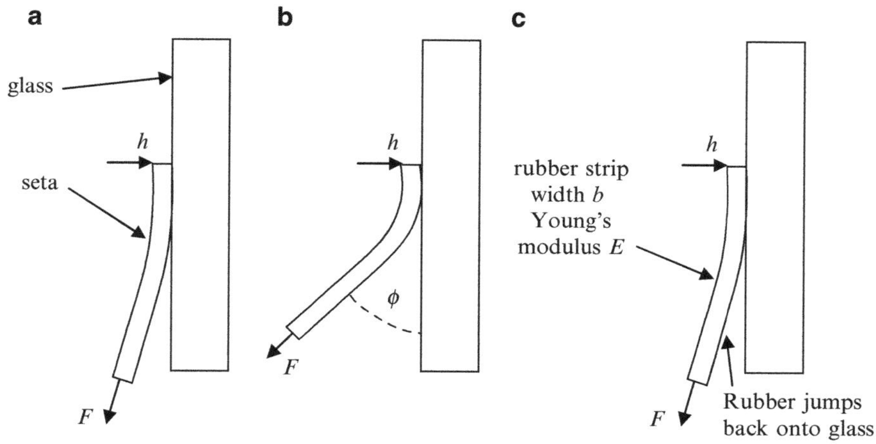

Fig. 6.9 (**a**) Gecko seta pulled from a glass plate; (**b**) rubber strip peeled at an angle φ; (**c**) rubber strip peeled at a low angle

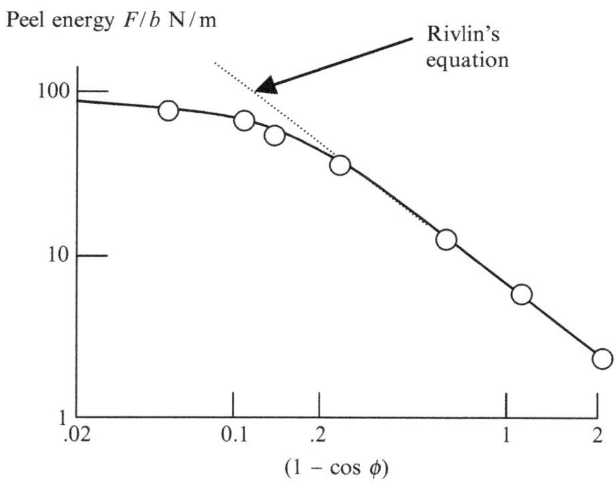

Fig. 6.10 Results showing how the peel force is reduced by the stretching mechanism for comparison with Eq. 6.7

When the peel angle was reduced to a small value, about 5°, an interesting phenomenon was observed.[16] The rubber peeled from the surface but the peeled material then jumped back into contact with the glass (Fig. 6.9c) and the peeling stopped. This was the formation of adhesive dislocations previously described in 1976.[21]

Schallamach first saw interface bubbles rippling across a frictional contact between smooth rubber and glass,[22] as shown in Fig. 6.11a. In that instance, there was no actual sliding of rubber on glass, merely the propagation of 'waves of detachment' i.e. interface dislocations through the contact region. In a related experiment, on the pull-out of a smooth rigid fibre from a rubbery matrix

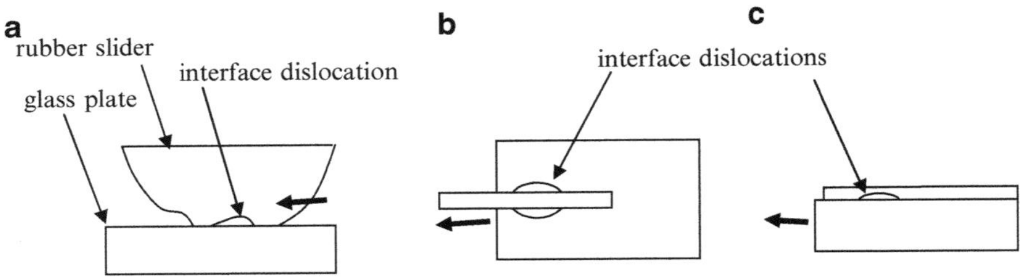

Fig. 6.11 (**a**) Schallamach waves at rubber/glass interface; (**b**) pull-out of smooth fibre from polymer; (**c**) failure of laminate

(Fig. 6.11b), interface bubbles were observed forming at the interface between fibre and matrix.[23] Further loading caused these to travel along the fibre, gradually getting smaller and disappearing. Additional dislocations then followed before gross failure occurred. Model dislocations were also seen in the failure of laminates (Fig. 6.11c).[24] Adhesive joints could be strengthened by a factor of three by the introduction of interface dislocations.

It seems likely that this same mechanism is operating in the gecko foot as indicated by Gao and colleagues recently.[34]

6.6 Subdividing Contact Spots Increases Adhesion

One of the most important reasons for having subdivided contact spots on an insect foot is the crack-stopping effect. This is familiar to us all in many composite materials, such as fibre-glass, and structures such as ropes. Whereas a sheet of glass can be broken by a single fracture propagating rapidly through it, fibre-glass breaks by the formation of a multitude of cracks, each requiring energy to initiate and drive through the material.[9]

Two mechanisms were shown to inhibit cracks by this subdivision process.[25,26] The first was the introduction of elastic modulus or thickness changes in a peeling material, causing the crack to slow down where the stiffness increased as described previously in Section 4.9. The second was crack deflexion along weaknesses which diverted the crack as in Fig. 4.16. Consequently, more force had to be expended to re-initiate the crack and get it moving again, even though the adhesion energy remained constant. This was the idea behind the Cook-Gordon mechanism of composite toughening, that subdividing a material by weak interfaces could give enhanced strength in a cracked specimen.[27]

The mechanism was proved by peeling a thin glass sheet from silicone rubber under various conditions shown in Fig. 6.12. The first experiment was carried out on a uniform rubber coating with silanised rubber to give reversible low adhesion.[28] Then the rubber surface was scored with a sharp blade to cut the rubber surface into three types of patterns, longitudinal along the crack direction, lateral parallel to the crack line, and cross-hatched i.e. both longitudinal and lateral. The results showed that the scoring increased the force of detachment but the cross-hatched pattern was most effective.

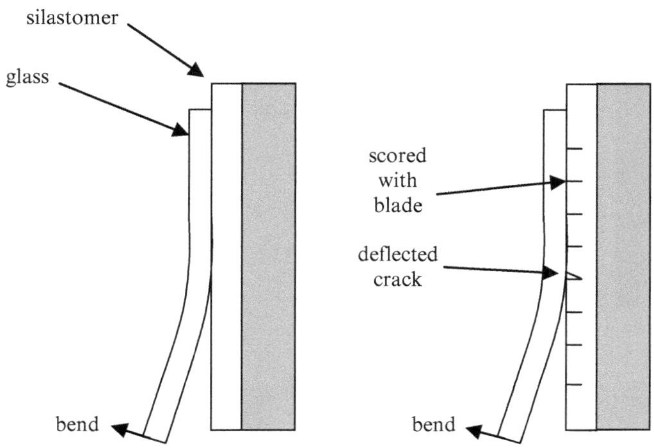

Fig. 6.12 *Left*; peeling a glass sheet from a silicone rubber surface; *right*; scored surface

The mechanism is clear from Fig. 6.12. The peeling crack proceeds smoothly along the interface until it meets the scoring, then the crack stops and has to restart, increasing the force and the losses, even though the adhesion energy is constant.

The pictures in Fig. 6.13 show that the scoring of the rubber was on a slightly larger scale to that of the bush cricket foot, i.e. 5 μm, which is also distinguished by its hexagonal geometry. Longitudinal and lateral scoring gave a slight increase in adhesion force. In contrast, the red curve for the cross-hatched geometry showed a factor 3 higher force and ten times more dissipation than the control. The pictures (c), (d), (e) and (f) show the differences in the detachment, seen through the glass, for the control, the lateral scoring, the longitudinal scoring and the cross-hatch respectively. The break-up of the adhesive bonds is clearly different for each case, indicating that adhesion force can be controlled wonderfully by altering the microstructure of the elastomer layer, while keeping the adhesion energy constant.

6.7 Gecko Tape

Such modifications of adhesive force have been discovered and patented in the search for bio-inspired adhesive tape; the so-called gecko-tape. The main aim of the research has been to produce hairy synthetic polymer tape which behaves like the gecko foot. A typical product was made from polypropylene (Fig. 6.14) and described in several papers.[29,30] Each polypropylene fibre was 15 μm long and 0.6 μm in diameter, comparable with some lizard setai

Kellar Autumn suggested seven properties which the tape must have to imitate the gecko structure fully:

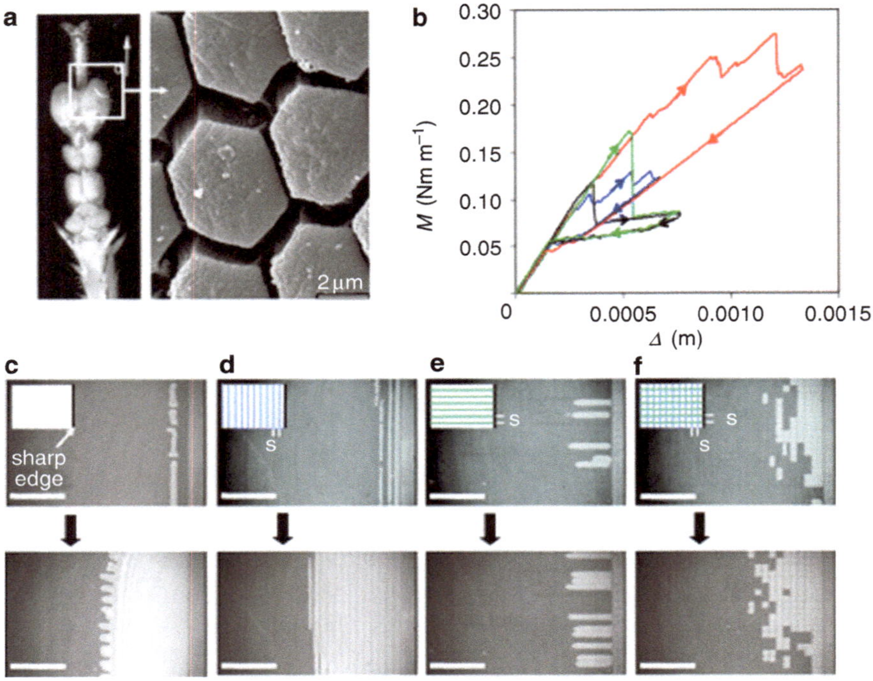

Fig. 6.13 (**a**) Scanning electron micrograph of the Tarsus of the leg of a bush cricket (Reproduced with kind permission from S. Gorb (Scherge and Gorb 2000)). (**b**) The peeling moment, M, as a function of the plate displacement, Δ, for differently incised films made of PDMS of shear modulus 0.9 MPa and thickness 305 μm, using cover plate of rigidity $D = 0.05$ Nm. All incisions were made to a maximum depth δ of 100 μm with the lateral and longitudinal spacing (s) of 250 μm each. *Black curve*: no incision, edge crack; *blue curve*: multiple lateral incisions; *green curve*: multiple longitudinal incisions; *red curve*: multiple crosswise incisions. (**c**)–(**f**) Sequences of the crack opening modes: (**c**) edge crack, (**d**) lateral incisions, (**e**) longitudinal incisions and (**f**) crosswise incisions. The scale bars in (**c**)–(**f**) are 2mm. In all cases, the crack propagates from *right* to *left* (Ref [28], copyright Royal Society, reprinted with permission)

1. Directional attachment; i.e. do not stick when applied vertically, but need transverse pulling to adhere strongly.
2. High pull-off force to preload; i.e. a slight initial loading triggers a large ultimate force.
3. Low detachment force; i.e. in normal pulling, the adhesion is very low.
4. Anti-self-matting; i.e. the fibres must not stick to each other but remain independent.
5. Non-sticky material; i.e. the polymer used is fairly hard like keratin and non-tacky.
6. The structure should also stick to rough surfaces.
7. Tape must be self-cleaning.

The tape shown in Fig. 6.14 satisfied the first five of these conditions and further work is proceeding on items 6 and 7.

Fig. 6.14 *Left*; Polypropylene tape structure; *right*; setai from anolis lizard (Copyright Kellar Autumn, with permission)

The application to a small robot and the mechanism of operation are illustrated in Fig. 6.15. The initial vertical contact of the tape to the glass under a slight preload gives a little adhesive area, but tractional movement has to be applied to pull more fibres into contact, increasing the downward adhesion substantially. Release is rapid once the downward traction is stopped because the fibres then lose their contact with the surface (communication from R. Fearing, Berkeley).

Many papers have been written on this interesting mechanism for adhesion of fibrous structures similar to gecko feet.[31–35] For example, very fine carbon nanotubes gave a similar effect but with ten times higher forces, around 100 N/cm^2, because the nanotubes were finer at 15 nm diameter and shorter at 150 μm, with ten billion fibres per square centimetre.[31] These nanotubes were grown in a vertical array by a low pressure chemical vapour deposition (CVD) process on a silicon/SiO_2 wafer.

The modelling of the scaling effects has also proved fascinating because there is a hierarchy of structures from the toe which is at the mm scale to the spatualae on the end of the hairy setai which are nanometers in thickness.[32]

At the bottom level of the hierarchy, the spatula pads have a tiny thickness of approximately 5nm, allowing them to be easily contacted onto a solid surface (Fig. 6.16). The peeling angle can then be adjusted by the lizard to give high adhesion at a peeling angle about $\pi/6$ and low adhesion at an angle $\pi/2$. In other words, the animal can release itself easily after lowering the adhesion force by a factor 10.

A slender hairy structure can not only provide a much higher adhesion force than expected from the van der Waals energy but also highly magnifies the difference in adhesion energy between attachment and detachment by altering the angle. Limiting the diameter of the seta below a critical value ensures uniform stress distribution in the structure.

By rolling in and pulling on the toe for attachment while rolling out and peeling the toe for detachment, the difference in adhesion forces between the two states has been further magnified at the scale of the toe. In this way, the gecko attains an adhesion force much higher than its body weight with displacement-controlled pulling and a detachment force much lower than its body weight with peeling at a large angle.

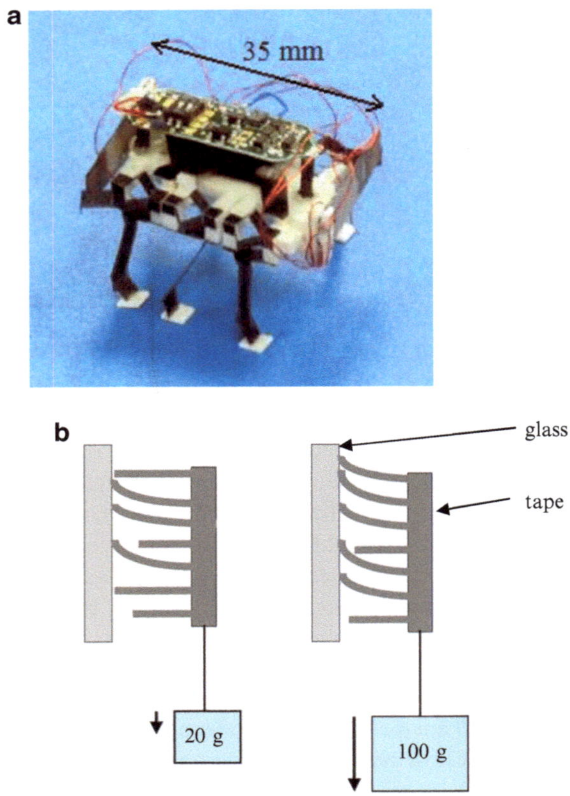

Fig. 6.15 (**a**) Biomimetic robot insect; (**b**) diagram showing the mechanism of contact as the gecko-tape is pulled along the surface (Copyright Ron Fearing, with permission)

6.8 Micro-Patterning

A different type of gecko-tape was made by creating a micropatterned structure using PDMS silicone rubber.[33] The structure, shown in an electron micrograph in Fig. 6.17 was made by moulding the liquid rubber onto a silicon chip which had been deeply ion etched with uniform square section holes, followed by cross-linking of the polymer, which was then peeled off and fixed to a glass slide support.

A thin film of PDMS was then coated on top to produce a 4 μm thick surface membrane which readily adhered to a metal or glass surface. When this structure was indented in a standard JKR test, the patterned sample was nine times harder to pull-off because the crack was trapped by the varying compliance of the contact, as shown in Fig. 6.18.

However, it is not so easy to see how this can be released easily on demand.

Alternative micropatterning methods can also be envisaged.[35] For example, the pattern could be made from adhesion molecules such as laminin or fibronectin on a glass or polystyrene surface by soft lithography which can create patterns of 1 μm

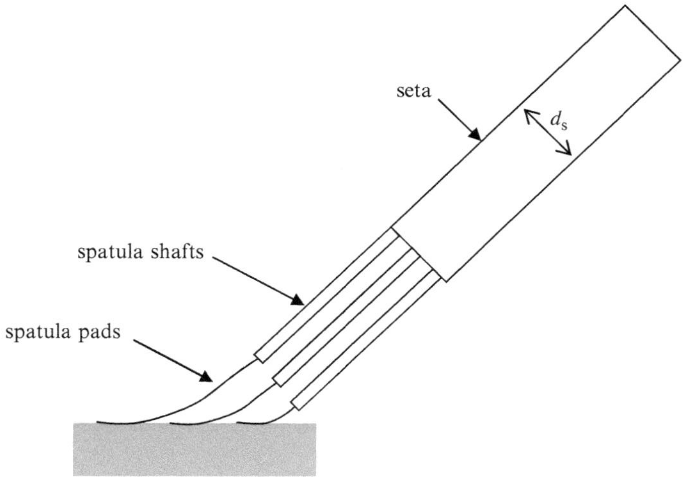

Fig. 6.16 Schematic of a seta with spatula shafts. Under an applied load, the distribution of forces among the spatula shafts is uniform only if the diameter d_s of the seta is below a critical value. Above this critical value, the adhesion with the substrate can fail via crack-like non-uniform stress distribution [34]

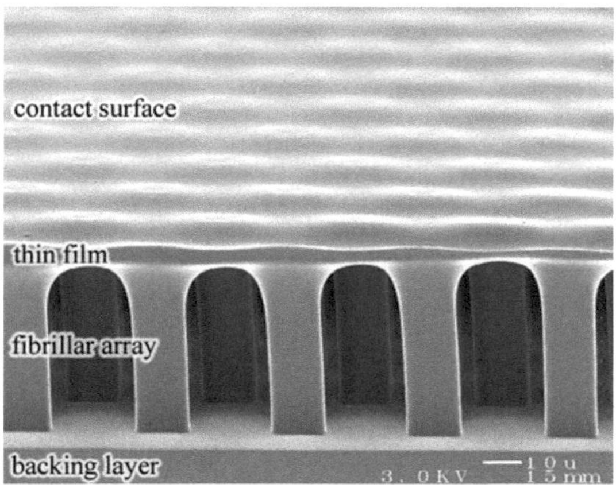

Fig. 6.17 Structure made from silicone rubber with 10 μm wide, 50 μm long pillars covered with a smooth film on a rigid backing [33] (Copyright American Academy of Sciences USA, with permission)

resolution. Another method is to print agarose walls which inhibit dispersion of the cells. Additional approaches have exploited the cell adhesive properties of polyamine (positively charged) silanes over the cell-resistive properties of alkylsilanes (neutral) through the use of photoresist and/or ablation with ultraviolet radiation.[36] The original methods go back to Whitesides and his colleagues who pioneered the soft lithography techniques.[37]

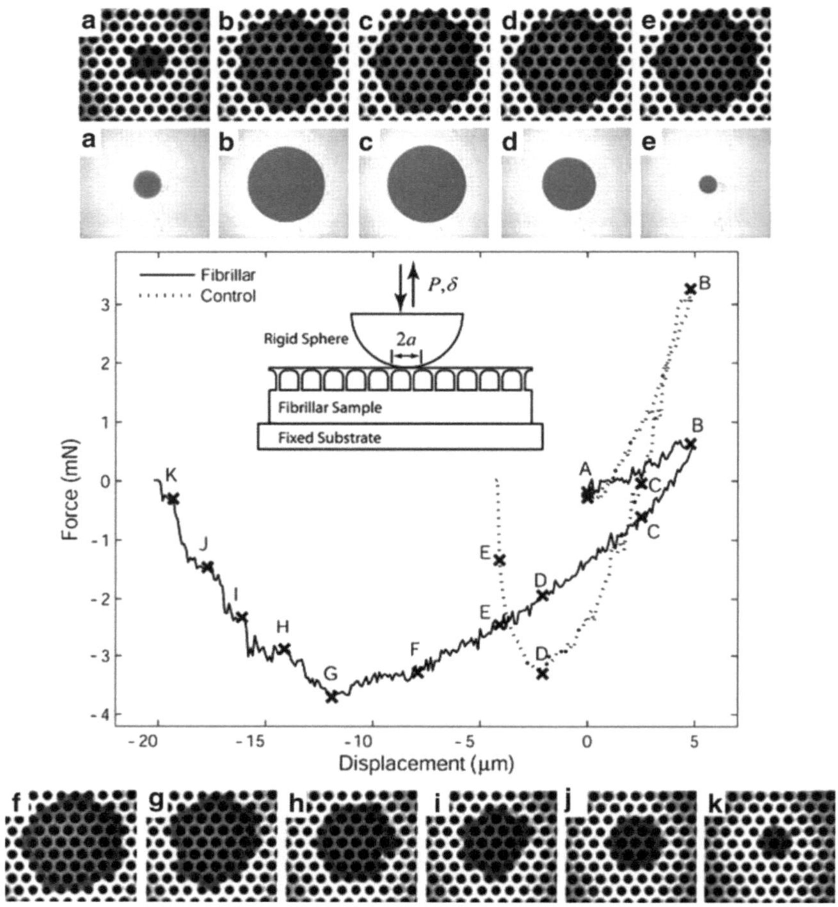

Fig. 6.18 *Top*; patterned JKR contact results compared to control contacts from A to E. The patterned contact remained larger during pull-off and required larger displacements[33] (Copyright American Academy of Sciences USA, with permission)

6.9 Stiffness and Adhesion of Multiple Contacts

One of the key issues relating to subdivided and separated contacts is how the force of adhesion can vary with the number and spacing of the contact spots. In the examples described above, the assumption is generally made that the contacts are independent. Then the force for each contact can be added to give the total force. Equation 6.6 depended on this assumption. But if the contacts lie close together, this assumption may be untenable. This question was tested in the model experiment shown in Fig. 6.19 by placing a metal disc on top of a block of elastic foam rubber and measuring the movement when the disc was loaded.[38] This is equivalent to observing the movement of an adhered disc being pulled vertically from an elastic

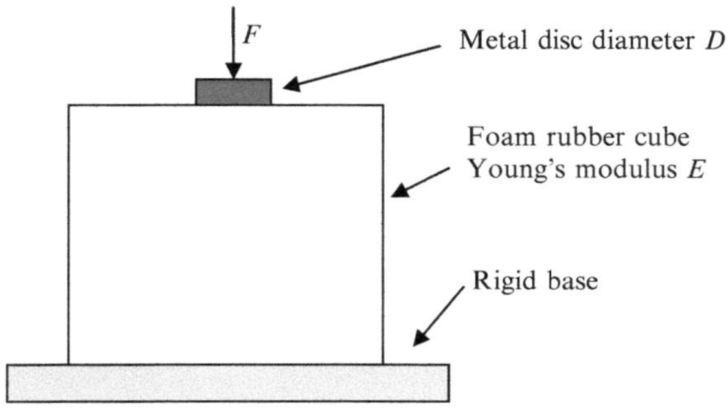

Fig. 6.19 Apparatus for measuring the model stiffness of contact spots

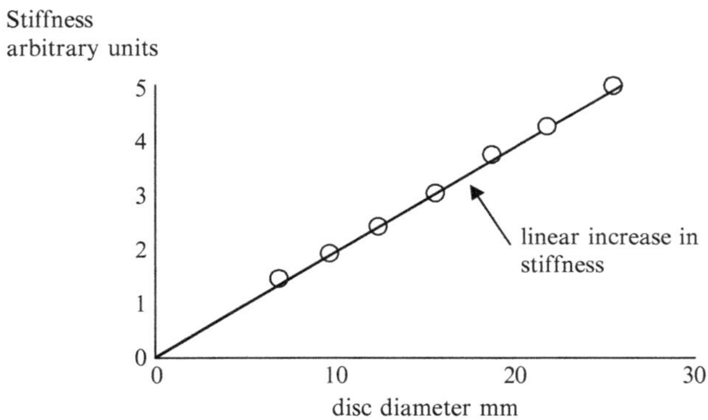

Fig. 6.20 Stiffness of disc contact on rubber as diameter increased[38]

surface as discussed in Fig. 4.5. According to Boussinesq elastic theory, the displacement z of the rigid disc into the elastic material under a load F is given by

$$z = (1 - \nu^2)F / ED \tag{6.9}$$

where E is the Young's modulus of the elastic half space and ν its Poisson's ratio. Thus the stiffness of this contact is

$$S = \mathrm{d}F / \mathrm{d}z = ED / (1 - \nu^2) \tag{6.10}$$

So as the disc diameter is increased, the stiffness should increase in proportion to D. Figure 6.20 shows the results confirming this equation for a 100 mm cube of foam rubber with Young's modulus of 0.36 MPa.

To investigate what happened as the disc was subdivided and separated, the single large disc was compared with seven small discs, each one third the diameter i.e. $D/3$, either close together or well separated as in Fig. 6.21.

If the seven discs are independent, then the forces should add so that the stiffness S_N for N disc contacts of diameter D would be

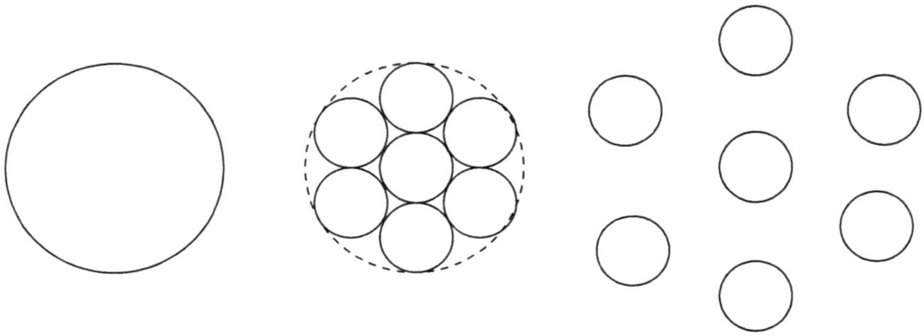

Fig. 6.21 *Left*; single large contact; *middle*; close-packed small contacts; *right*; spaced-out contacts

$$S_N = N\,ED/(1-\nu^2) \tag{6.11}$$

Whereas if the N discs are so close together as to behave like a single larger disc as in Fig. 6.21 (middle), then the stiffness will rise with $N^{1/2}$

$$S_N = N^{1/2} ED/(1-\nu^2) \tag{6.12}$$

as in Eq. 6.6. This idea was tested by placing arrays of discs with different numbers and spacings on the foam rubber to measure their stiffness. The results, given in Fig. 6.22, show that the close-packed discs fitted Eq. 6.12 but the discs separated by three diameters approach the independent disc curve for N less than 14.

Of course, the stiffness is important because, in the theory of brittle fracture, the crack propagation force must rise with stiffness at constant contact area. We can find how the adhesion force will depend on N by considering the energy balance for the two extreme cases shown in Fig. 6.22, i.e. the independent contacts where the force is given by

$$F = N\,KD_z$$

where K is a constant and the case where the contacts are interacting to give

$$F = N^{1/2} KD_z$$

For the independent contacts, taking the surface energy as $NW\pi D^2/4$ and the elastic energy as $F_z/2$ i.e. $F^2/2NKD$.

Then differentiating the total energy and making that zero to satisfy the condition of energy conservation.

$$\mathrm{d}U/\mathrm{d}D = 0 = NW\pi D/2 - F^2/2NKD^2$$

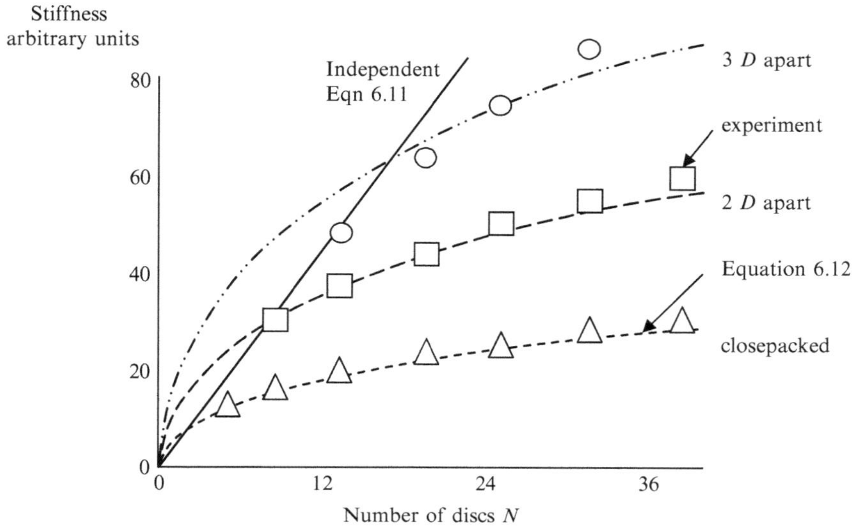

Fig. 6.22 Plot of stiffness for disc contacts at three separations, showing how the stiffness approaches the independent value at N below 14, but rises with $N^{1/2}$ at larger N

Therefore

$$\text{Adhesion force} = N\ (\pi D^3 WK)^{1/2}$$

showing proportionality to N. A similar calculation for the interacting case gives

$$\text{Adhesion force} = N^{3/4} (\pi D^3 WK)^{1/2}$$

demonstrating that adhesion force can rise with number of contact spots but not necessarily in proportion when the contact spots interact.

6.10 Conclusions

Adhesion forces have been utilised by biological systems in a variety of ways so that their study provides new insights into adhesion. In biological systems, contact spots are often split up into separated contact areas. This can either reduce adhesion of spiky spores and pollen grains which need to disperse, or increase adhesion of hairy insects, spiders and lizards which need to crawl up walls and adhere to ceilings. A general rule with climbing organisms is that more hairs and smaller contact spots are needed for larger creatures, with number of setai varying with organism mass to the power 0.7. A simple analysis based on JKR contact mechanics suggests that adhesion should rise with $N^{1/2}$, but more detailed investigation shows that force can rise with N for independent contacts or with $N^{3/4}$ for interacting contacts. These ideas have been applied to the design of 'gecko tape' which can adhere to smooth surfaces by a similar mechanism, and to the design of nano-patterned surfaces which can give interesting adhesion effects.

References

1. Gerber, P.J., Lehmann, C., Gehr, P., Schurch, S., Wetting and Spreading of a Surfactant Film on Solid Particles: Influence of Sharp Edges and Surface Irregularities, Langmuir 22 (2006) 5273–81.
2. Arzt, E., Gorb, S., Spolenak, R., From micro to nano contacts in biological attachment devices, PNAS 100 (2003) 10603–10606; Keane, P., Reid, I., Jarvis. S.P., Adhesion measurements on the nanoscale attachment structures of a range of spider species, Proc Adhesion Society, 33 Annual Meeting, Daytona Beach Fl. Feb 21 2010.
3. Autumn, K., Liang, Y.A., Hsieh, S.T., Zesch, W., Chan, W.P., Kenny, T.W., Fearing, R., Full, R.J., Adhesive force of a single Gecko foot hair, Nature 405 (2000) 681–85.
4. Breidbach, O., Mikrokosmos 69 (1980), 200–201.
5. Schliemann, H., Funkt. Biol. Med. 2 (1983), 169–177.
6. Liang, Y. C. A., Peattie, A. M., Hansen, W. R., Sponberg, S., Kenny, T. W., Fearing, R., Israelachvili, J. N. & Full, R. J., Proc. Natl. Acad. Sci. USA 99 (2002) 12252–12256. pmid:12198184
7. Arzt, E., Enders, S. & Gorb, S., Z. Metallkde. 93 (2002) 345–353.
8. Kendall, K., Sticky solids, Contemporary Physics 21 (1980) 277–297
9. Kendall, K., Molecular adhesion and its applications, Kluwer, New York 2001.
10. Autumn, K., Peattie, A.M., Mechanisms of adhesion in geckoes, Integrative comparative Biol 42 (2002) 1081–90.
11. Autumn, K., Dittmore, A., Santos, D., Spenko, M., Cutkosky, M., Frictional adhesion: a new angle on gecko attachment, J. Exp. Biol., 209 (2006) 3569–3579.
12. Irschick, D. J., C. C. Austin, K. Petren, R. Fisher, J. B. Losos, and O. Ellers. A comparative analysis of clinging ability among pad-bearing lizards. Biol. J. Linn. Soc, 59 (1996) 21–35
13. Dellit, W.-D. Zur anatomie und physiologie der Geckozehe., Jena. Z. Naturw, 68 (1934) 613–656
14. Ruibal, R., and V. Ernst., The structure of the digital setae of lizards. J. Morph, 117 (1965) 271–294
15. Kendall, K., Thin film peeling: the elastic term, J Phys D Appl Phys 8 (1975) 1449–52.
16. Kendall, K., Interfacial dislocations spontaneously created by peeling, J Phys D Appl Phys 11 (1978) 1519–27.
17. Kendall, K., Shrinkage and peel strength of adhesive joints, J Phys D: Appl Phys 6 (1973) 1782–7.
18. Gurney, C. and Hunt, J., Quasistatic crack propagation, Proc R Soc Lond A299 (1967) 508–524.
19. Williams, J.G., A review of the determination of energy release rates for strips in tension and bending, Strain Anal Engng Design 28 (1993) 247–256.
20. Drory, M.D., and Hutchinson, J.W., Measurement of the adhesion of a brittle film on a ductile substrate by indentation, Proc R Soc Lond A452 (1996) 2319–2341.
21. Kendall, K., Preparation and properties of rubber dislocations, Nature, 261 (1976) 35–6.
22. Schallamach, A., How does rubber slide? Wear 17 (1971) 301–12.
23. Kendall, K., Interfacial dislocations spontaneously created by peeling, J Phys D: Appl Phys 11 (1978) 1519–27.
24. Kendall, K., Strengthening of adhesive joints by dislocations, Phil Mag 36 (1977) 507–15.

25. Kendall, K., Control of cracks by interfaces in composites, Proc R Soc Lond A341 (1975) 409–428
26. Kendall, K., Transition between cohesive and interfacial failure in a laminate, Proc R Soc Lond A344 (1975) 287–302.
27. Cook, J., Gordon, J.E., A Mechanism for the Control of Crack Propagation in All- Brittle Systems, Proc. Roy. Soc., vol. A282 (1964), p. 508–520.
28. Chung, J.Y., Chaudhury, M.K., Roles of discontinuities in bio-inspired adhesive pads, J Roy Soc Interface 22 (2005) 55–61.
29. Lee, J., Majidi, C., Schubert, B., and Fearing, R., Sliding induced adhesion of stiff polymer microfiber arrays: 1. Macroscale behaviour, Journal of the Royal Society, Interface (10.1098/rsif.2007.1308)
30. Schubert, B., Lee, J., Majidi, C., and Fearing, R., Sliding induced adhesion of stiff polymer microfiber arrays: 2. Microscale behaviour, Journal of the Royal Society, Interface (10.1098/rsif.2007.1309)
31. Qu, L., Dai, L., Stone, M., Xia, Z., Wang, Z.L., Carbon nanotube arrays with strong shear binding-on and easy normal lifting-off, Science 322 (2008) 238–42
32. Chen, B., Wu, P.D., Gao, H., Hierarchical modelling of attachment and detachment mechanisms of gecko toe adhesion, Proc R Soc A464 (2008) 1639–1652
33. Glassmaker, N.J., Jagota, A., Hui, C-Y., Noderer, W.L., Chaudhury, M.K., Biologically inspired crack trapping for enhanced adhesion, PNAS 104 (2007) 10786–91.
34. Chen, B., Wu, P., Gao, H., Pre-tension generates strongly reversible adhesion of a spatula pad on substrate, J R Soc Interface 6 (2009) 529–537.
35. Oliver, A.E., Ngassam, V., Dang, P., Sanii, B., Wu, H., Yee, C.K., Parikh, A.N., Cell attachment behaviour on solid and fluid substrates exhibiting spatial patterns of physical properties, Langmuir (2009) 10.1021.
36. Dulcey, C. S., Georger, J. H., Krauthamer, V., Stenger, D. A., Fare, T. L. and Calvert, J. M. *Science* 252 (1991) 551–554
37. Chen CS, Mrksich M, Huang S, Whitesides GM, Ingber DE., Geometric control of cell life and death. Science 276 (1997) 1425–28.
38. Kendall, K., The stiffness of surfaces in static and sliding contact, PhD thesis, Cambridge University 1970.

Chapter 7
Measurement Methods

All of the nerves observed were growing upon the surface of the cover-slip

Harrison 1914

This chapter provides an overview of the different measurement methods for studying adhesion phenomena. Cells, viruses, and nanoparticles have a typical length scale ranging from tens of micrometres down to several nanometres and usually reside in an aqueous solution that significantly decreases adhesion forces. The small length and force scales as well as the liquid environment pose distinct requirements for the instruments used to observe adhesion processes.

In general we can divide the techniques into three subgroups illustrated in Table 7.1.

First, we discuss purely observational measurements where the adhesion process can be monitored while different parameters of the adhesion geometry, chemistry and physical properties are varied. These are mainly light microscopy techniques. The second group uses active forces to perturb the adhesion process thereby directly measuring the forces involved and elucidating certain feedback mechanisms. Here, atomic force microscopy, micro-pipette aspiration and optical tweezers are most prominent. Last but not least, there is the cell counting method, in which cells are attached to a surface or to themselves and counted after applying known detachment stresses, usually Brownian impacts or liquid shearing forces. For example, the rotating disc instrument allows direct measuring of the adhesive force between a large number of cells and their substrate. Additionally, counting aggregates gives a measure of adhesion.

The conclusion is that measurement methods play a large part in our understanding of adhesion processes. New methods are therefore required, especially to focus on the functions of adhesion molecules.

K. Kendall et al., *Adhesion of Cells, Viruses and Nanoparticles*,
DOI 10.1007/978-90-481-2585-2_7, © Springer Science+Business Media B.V. 2011

Table 7.1 Three measurement groups

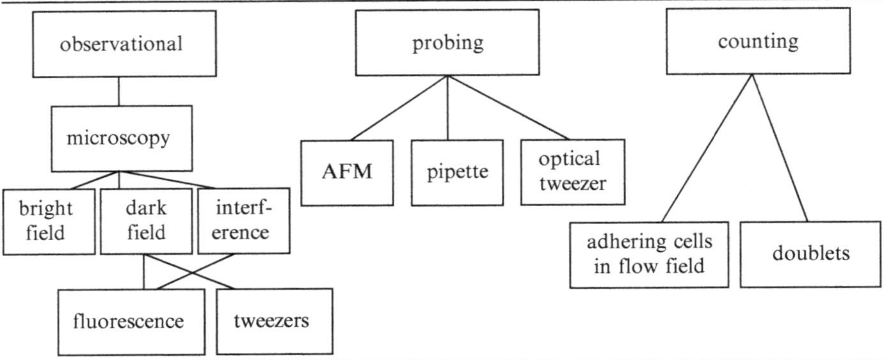

7.1 Light Microscopy

The typical length scale of cells adhering to glass substrates is up to 100 μm while bacteria, viruses, and nanoparticles are even smaller, from a micron down to several nanometres. They are therefore too small to be directly observed by our eyes and so magnifying instruments are required to investigate and measure their adhesion behaviour. Although the first accounts on magnifying lenses date back to mediaeval times, it was Robert Hooke[1] in 1665 who systematically described the microscope in his book 'Micrographia'. He coined the term 'cell' to describe the structures he saw magnified 50 times in a slice of cork. But he failed to see bacteria. Antoni van Leeuwenhoek was the first person to resolve such small cells with a microscope magnifying 250 times and he wrote to Hooke to describe his fantastic observations. 'I then most always saw with great wonder ...there were many very little animalcules very prettily a-moving'.[1] Since that time, light microscopy has become an essential tool for biologists. Continuous improvement and specialisation for different applications including adhesion measurement has been observed over the past decades.[2]

The simplest light microscopy technique is brightfield microscopy. Light from a light source (e.g. a halogen lamp) is focused with a condenser lens onto the specimen on a glass slide. The transmitted light then is focussed by the objective lens that magnifies the image, seen using an eyepiece or recorded using a camera. This transmission technique can only be used to distinguish regions of different transmittance, so called amplitude objects. It is suitable for observing cell walls in plant cells for example or to see the cell outline and some features of mammalian cells cultured on a substrate. By inserting a dark field patch stop and a direct illumination block into the light path, only scattered light from the sample becomes visible and the background is dark. This method is darkfield microscopy and yields a higher contrast between sample and background. To further elucidate details of a cell, especially when spread on a substrate, more specialised setups are required, for example phase contrast, a sophisticated improvement of the simple bright field technique. Frits Zernike[3] invented it in 1932 and received the Nobel prize in 1953.

Phase contrast microscopy makes use of the fact that the optical path length is the thickness multiplied by the refractive index of the object, so called phase objects. Light that travels through a region of higher refractive index yields a phase difference to background light that did not pass this object. To generate two different light beams two phase rings are inserted into the light path, one at the illumination aperture and a second one in the back focal plane of the objective. With this setup one can image cell compartments without staining them, making this an invaluable tool for non-invasive observations and especially live cell imaging.

7.2 Reflectance Interference Contrast Microscopy

Although all the different microscopy techniques above are able to magnify and monitor cells and bacteria, they do not allow for direct observation of the adhesion process, for example the contact area or height between cell and substrate. Here, reflectance interference contrast microscopy (RICM) comes into play. This special technique was first described by Curtis[4] in 1964 as interference reflection microscopy and was successively improved by several groups in the following decades. It is the direct microscopic analogue to the macroscopic observation methods for adhesion as discussed in Chapter 1 for rubber spheres and is based on the Newton fringes to determine the distance between a cover slip and a cell. The principle of the different light paths reflected at the glass and the cell is sketched in Fig. 7.1 to show how the contact spot can be readily seen and measured. Changes in cell contact spots are thus monitored, especially when chemical changes are made, for example by adding adhesion molecules.

Monochromatic light is polarised and directed into an antiflex objective (an objective with a $\lambda/4$ wave plate on top). Then the light is reflected at the substrate and the object above but will have a distinct phase correlation according to the distance to the substrate.[5] Then the light passes an additional polariser to discard all stray light and since the light is monochromatic Newton fringes will appear

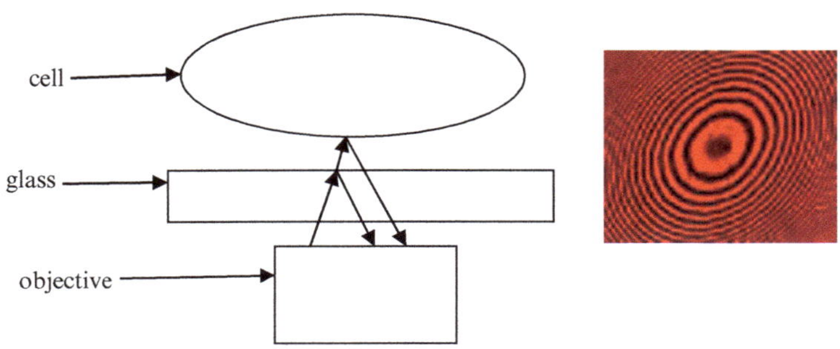

Fig. 7.1 Principle of RICM. Light is reflected at the glass and at the cell (*left*) and creates Newton fringes (*right*). From the interference pattern it is possible to calculate the distance between object and surface

for a spherical object as showed in Fig. 7.1. The black spot in the centre indicates the zone of adhesion (i.e. zero distance from the object to the surface). Using a fast image processing system, Zilker and Sackmann[6] optimised RICM, to monitor the flickering of a red cell membrane in contact with a substrate. This enabled the study of the dynamic equilibrium of adhesion that is a continuous process of binding and unbinding. Another sophisticated improvement was dual colour RICM[7] where light of two different wavelengths was used to determine the absolute object-substrate distance in a much larger range.

7.3 Fluorescence Microscopy, Interference Contrast, Total Internal Reflection

Seeing the contact spot is important; however it is also vital to observe changes in the cell structures such as the actin filaments. This is best achieved by Fluorescence Microscopy, a light microscopy technique in which dye molecules (fluorophores) are excited with ideally monochromatic light that matches the excitation maximum then using a filter system only the emitted light of these molecules is observed (Fig. 7.2). This has become one of the most valuable tools for cell biologists since it allows for selectively imaging certain molecules within cells and tissues.

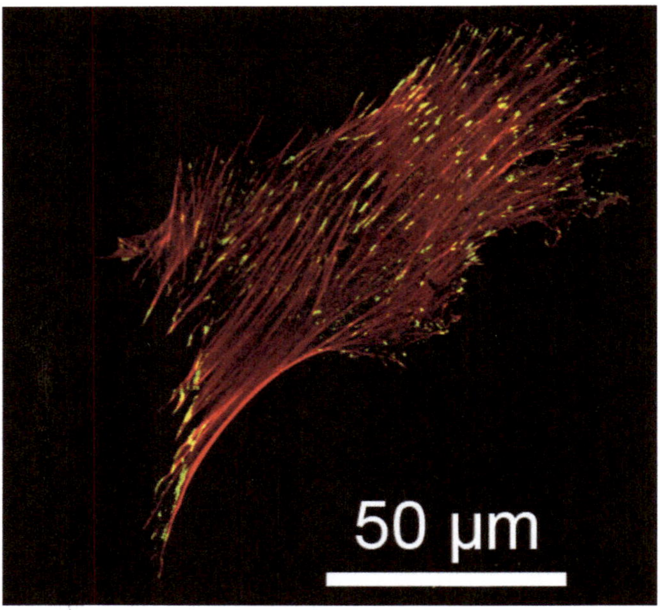

Fig. 7.2 Fluorescence microscope picture of a human mesenchymal stem cell adhering to glass; actin cytoskeleton in red, focal adhesion contacts (vinculin) are stained green and show up colocalized with actin in yellow

To visualise distinct molecules, proteins that bind selectively to the molecule of interest are then tagged with a fluorescent dye. Much progress followed from the discovery of the green fluorescent protein which can be added to a protein of interest by genetic engineering so that the protein of interest is visible even in living cells, an invention which was recognised in the 2008 Nobel Prize for chemistry.[8]

Another technique that emerged from fluorescence microscopy uses the fact that two fluorophores that are within a few nanometers of each other and have overlapping emission and excitation spectra can transfer energy without radiation between the donor and acceptor dye. This mechanism is called fluorescence (or Förster) resonant energy transfer[9] FRET and was discovered by Theodor Förster in 1948. It can be used to monitor the distance between the two dyes because, roughly speaking, the acceptor dye will only emit if the pair is together, up to 10 nm apart, while the donor dye will only be visible when separated. Therefore it can be used as a nanometre proximity sensor – ideal to look at adhesion events at the small length scale of proteins and viruses.

Fluorescence interference contrast microscopy is based on the same principle as Reflection Interference Contrast Microscopy. Fluorescent dyes are excited on a silicon substrate with a silicon oxide layer on top as transparent spacer. The reflective surface causes a sinusoidal modulation of the fluorescence intensity that depends on the distance between fluorophore and substrate. By using special wafers with a square step pattern of different silicon oxide thicknesses this technique allows for absolute height measurements in the nanometre range. The theory was first published by Lambacher and Fromherz, then improved and refined.[10] Although substrate preparation is somewhat demanding, this method became a valuable tool to measure the substrate spacing of lipid bilayers and cell membranes or to investigate fluorescently labelled proteins.

Total internal reflection fluorescence (TIRF) microscopy is a special derivative of fluorescence microscopy that uses the evanescent field of the incoming light totally reflected at the surface above its critical angle (Fig. 7.3). First described by Daniel Axelrod[11] in 1981, it allows one to illuminate only a thin slice above the substrate. The penetration depth of the evanescent field depends on the wavelength λ, the incident angle θ, on the refractive indices of the substrate n_0, and the surrounding media n_1, but is usually of the order of 100 nm. This makes the method very useful for observing adhesion events since only the fluorescent molecules in a thin region above the substrate will be excited and not the bulk sample. For example cells settling down from solution onto a surface can be imaged and their adhesive area quantitatively measured if their membrane is stained with a fluorescent dye.[12] This way, the spreading of the cell onto the substrate can be monitored in real time yielding insight of the cell-substrate interactions.

7.4 Atomic Force Microscopy

We will now switch towards the second group of instruments that actively manipulate the sample to investigate adhesion. The atomic force microscope (AFM) was inspired by the scanning tunnelling microscope (STM) and developed in 1986 by Binnig,

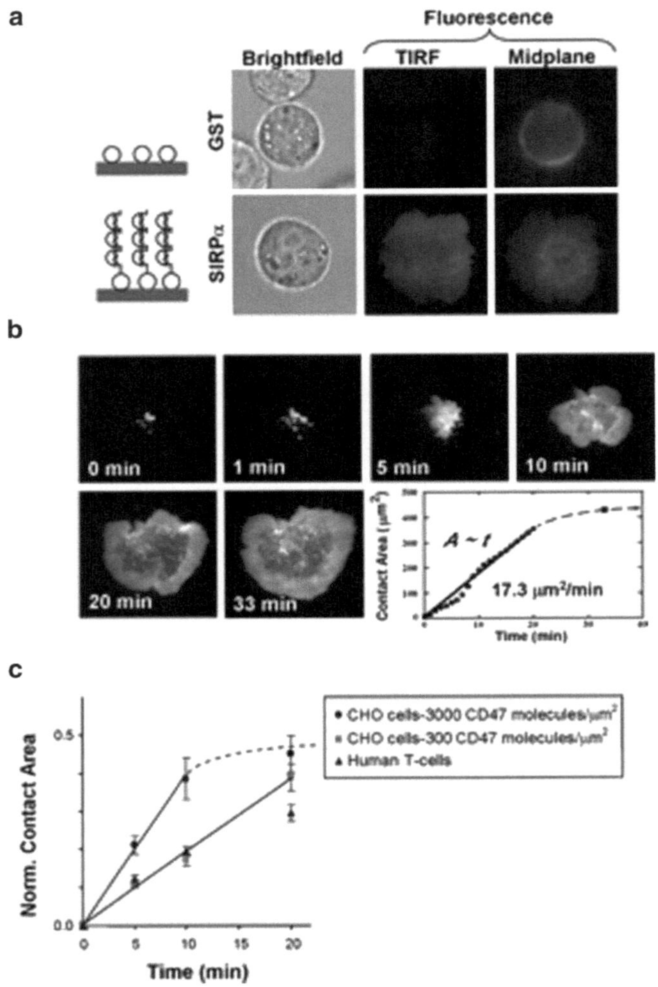

Fig. 7.3 (**a**) Two types of cells imaged by different methods, showing how the contact area can be plotted from the TIRF microscope pictures (**b**), to give contact area/cell area plot (**c**) for the two cell types[12] (with permission)

Quate and Gerber.[13] Originally invented to study solid surfaces at atomic resolution, the AFM has also become an invaluable tool to study soft condensed matter systems and biological specimens. Its ability to measure forces and image the topology of samples in an aqueous environment allowed the AFM to study adhesion phenomena in biological systems. Figure 7.4a sketches the general principle used in most AFMs for measurements in liquid. Laser light is reflected on the cantilever with a probe at its tip and the light hits a quadrant photodiode. If the cantilever is bent due to forces acting on the tip, the deflection d can be measured with the photo diode and the force is calculated according to Hooke's law by multiplying the deflection d by the spring

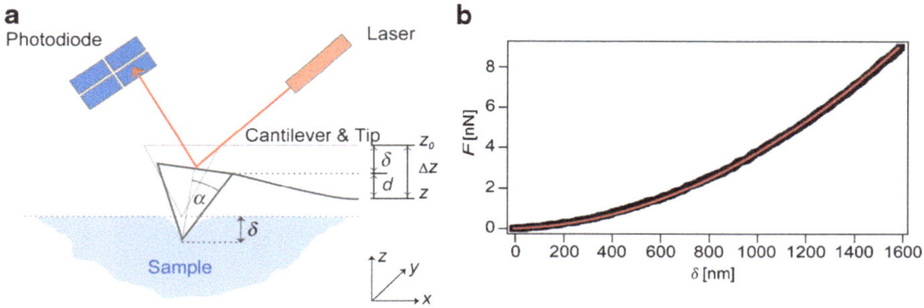

Fig. 7.4 (**a**) Schematic depiction of and AFM cantilever indenting a soft sample. The laser is reflected into a photodiode and thus the deflection *d* can be measured. Using Hooke's law the applied force can be calculated using the spring constant *k* of the cantilever. (**b**) Shows experimental data (black) and best fit (red) of a modified Hertz model to determine the Young's elastic modulus *E* with permission[15]

constant *k*. The cantilever or the sample is mounted on a stage that can be translated with piezo elements in all three directions with nanometre accuracy and on a length scale from nanometres up to hundreds of micrometres. That way the probe can scan across the sample in *x* and *y* and can measure the topography in *z*. The AFM can also be used to indent or pull samples and is now frequently used to measure the micro-elasticity of cells, tissues, or elastic substrates.[14]

Using a modified Hertz model,[15] theYoung's elastic modulus *E* of the samples can be measured as shown in Fig. 7.4b. Indenting over a larger grid can be used to create a force map – in essence an image of the sample's mechanical properties. The AFM can also be used to make direct adhesion force measurements by modifying the tip with certain molecules of interest or even 'gluing' cells on the cantilever.

7.5 Optical Tweezers

About two decades ago in 1986, the first proof of principle for the optical trap was demonstrated by Ashkin and his colleagues.[16] The optical trap uses the fact that refraction of light in a dielectric bead exerts a net force driving the bead towards the focus. This principle led to a Nobel prize in 1997 for Steven Chu who used it to cool down and trap single atoms which were used to form a Bose-Einstein condensate.

For biological systems, one can use the trap with beads in the micrometre range to probe like tweezers, creating and measuring forces in the range of ten to hundreds of picoNewtons, the ideal range for probing the mechanical and adhesive properties of cells.[17] More advanced setups have two or even more separately controllable beads using independent laser beams.[18] This enables one to apply and measure forces at multiple points with a high lateral and force resolution. Like AFM, optical tweezers are used to measure cell adhesion forces and single molecule binding forces using specific adhesion molecules because they have high force sensitivity and do not require the fixed cantilever, giving them additional degrees of freedom but a lower

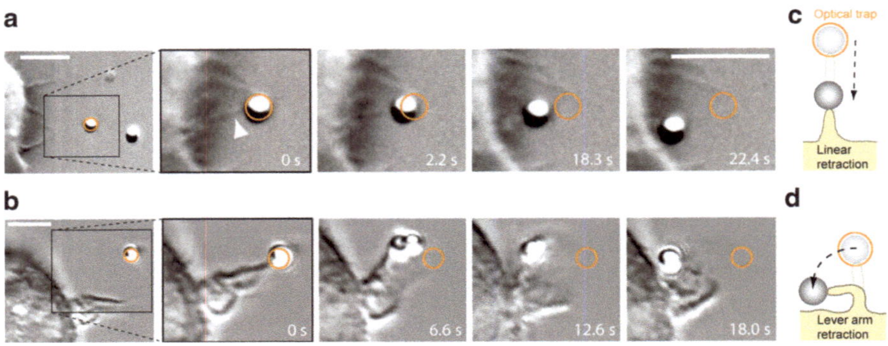

Fig. 7.5 (**a**) A coated polystyrene particle controlled by laser tweezers brought into contact with a living macrophage cell and pulled into contact by linear retraction of the filopodia; (**b**) the particle brought into contact by bending of the filopodia; **c** & **d**) schematic mechanisms Copyright Holger Kress, with permission[19]

maximum force compared to AFM. Figure 7.5 shows a coated 1 μm polystyrene particle manipulated by optical tweezers to adhere to a living macrophage cell.[19]

7.6 Micropipette Aspiration

The above techniques apply and measure forces in the pico and nanoNewton range, ideal to probe single molecules or small ensembles of molecules, but this force range is not enough to peel a fully spread cell off the substrate. For this higher force scale we need instruments with a much higher maximum force like micropipette aspiration as shown in Fig. 7.6. Micropipettes are extremely thin glass tubes with a diameter of only μm at the tip. They are operated with a small suction pressure (in the range of $1–10^4$ Pa) and can be used to aspirate small parts of a cell membrane or a lipid bilayer from a vesicle to get a grip on that object.[20] Then the whole pipette can be moved or the suction pressure slightly increased to exert forces on the object, for example to give a peeling force curve rather like that for peeling rubber from a glass surface. It can be seen that the kinetic effect is dominant here, with collagen adhering better to the cells than glass. In the same way also the mechanical properties of the membrane or bilayer itself can be probed and if using two pipettes with two vesicles, molecular interactions of proteins in these vesicles can be studied.[21]

7.7 Spinning Disc

All of the above mentioned techniques are essentially single sample measurement techniques – meaning one cell at a time. But especially for biological systems, good statistics are very important because of the intrinsic variability. To measure the average

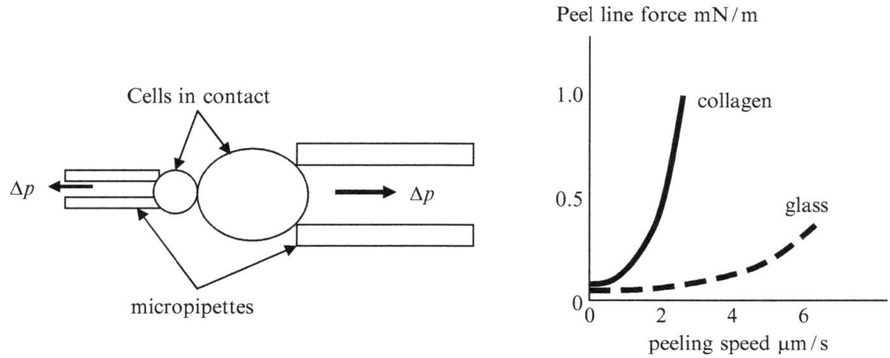

Fig. 7.6 *Left*; Schematic of double micropipette measurement[21]. *Right*; Peel curves for muscle cells showing force versus speed[20]

adhesion strength of cells to a certain substrate it is useful to probe a whole batch of cells simultaneously. A very useful technique for that purpose is the spinning disc device[22] which uses a disc rotating in a solution to exert a hydrodynamic force onto the cells adhering to it. The force can be well-controlled by geometry and rotation speed. Approximating the rotating disc with an infinite disc embedded in an infinite fluid and taking the no-slip boundary conditions between the solid disc and the liquid buffer into account a velocity gradient can be calculated that exerts a shear stress τ on the cells on the spinning substrate that depends on radial position r, the rotation frequency ω, the viscosity μ and the density ρ of the liquid:

$$\tau = 0.8\, r \sqrt{\rho\, \mu\, \omega^3} \tag{7.1}$$

After spinning for a certain time, the cells that are initially uniformly distributed on the disc are fixed and counted according to their radial position r. That way it is possible to screen a wide range of adhesive force for many cells in a single experiment.[22]

A typical result is shown in Fig. 7.7 for detachment of cells and the effects of molecules blocking the adhesion molecules i.e. integrins. Adhesion is seen to vary statistically across the cell population of around 6,000 cells, and the strong effect of the adhesion molecules is evident.

7.8 Cell Adhesion by Flow Methods

In practical cell culture applications, fluid flow methods of estimating cell adhesion have been most useful. In particular, the radial flow chamber shown in Fig. 7.8 was developed to study both deposition and removal of cells at surfaces.[23] The apparatus has been used to measure adhesion of many cell types to different surfaces, including diatoms[24] and *Pseudomonas* fluorescens.[25]

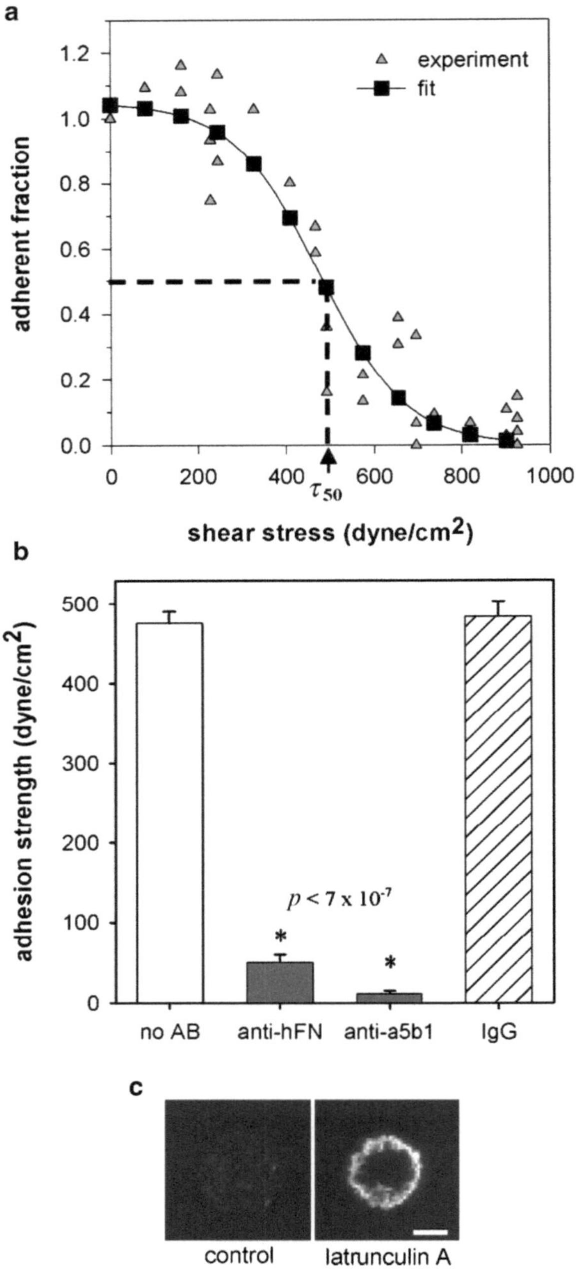

Fig. 7.7 Spinning disc adhesion measurements of cell adhesion strength; (**a**) Detachment profile showing fraction of adherent cells versus applied shear stress for cells adhering to 5-μm-diameter islands for 16 h. Experimental points were fitted to a sigmoid curve to obtain the shear stress for 50% detachment (τ_{50}). (**b**) Antibodies against human FN or $\alpha_5\ \beta_1$ integrin completely block adhesion to micropatterned islands. (**c**) Immunostaining for vinculin after the application of high detachment forces illustrating detachment mechanism. *Bar*, 5 μm[22] (with permission)

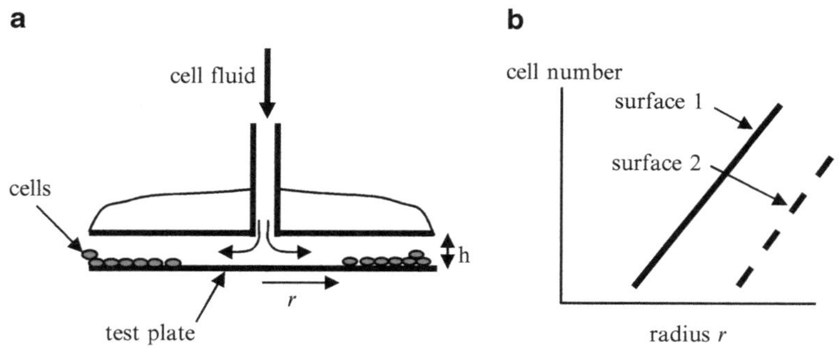

Fig. 7.8 (**a**) Radial flow chamber apparatus; (**b**) Results for cell number versus radius

The test apparatus is shown in Fig. 7.8a. The cell culture is grown to produce a fluid containing a concentration of cells. This fluid is pumped down a tube to impact onto the test plate, then to move radially along the plate
Shear stress τ is highest at the inlet in the middle of the plate, dropping off with radius r according to the equation

$$\tau = 3Q\mu / \pi r h^2 \tag{7.2}$$

where Q is the flow rate, μ the viscosity and h the gap. The results show that there is a critical radius for cell adhesion and another for cell detachment. Different test surfaces give different characteristic curves as shown in Fig. 7.8b. The interpretation of such curves in terms of the cell adhesion force or energy is difficult because it depends greatly on the assumptions about the cell geometry and the flow regime. However, this method is of practical utility for measurement of biofouling which is of enormous interest to industry.[26]

Another flow method for studying cell adhesion was developed by Bongrand and his associates.[27] The idea was to use a laminar i.e. Couette flow along a plane surface, such that the force on each cell was less than the strength of a single cell to surface bond, but sufficient to make the free cells move with a velocity of a few micrometres per second. By watching the movement of cells, bonding events and breaking events could be defined. Then the influence of adhesion molecules on these events could be observed. The basic scheme is shown in Fig. 7.9.

A rectangular cavity 1 mm deep was cut in a plastic block and a glass coverslip glued to it with silicone rubber to form a microscope chamber through which cell dispersions could be pumped with a motorised syringe. The shear rate G was around 3 s^{-1}. In this situation, the force experienced by a spherical cell bound to the substrate was about[28]

$$F = 8\mu D^2 G \tag{7.3}$$

where D was the cell diameter, G the shear rate and μ was the viscosity. Because the viscosity of dilute aqueous solutions is 0.0007 Pas and a typical cell diameter

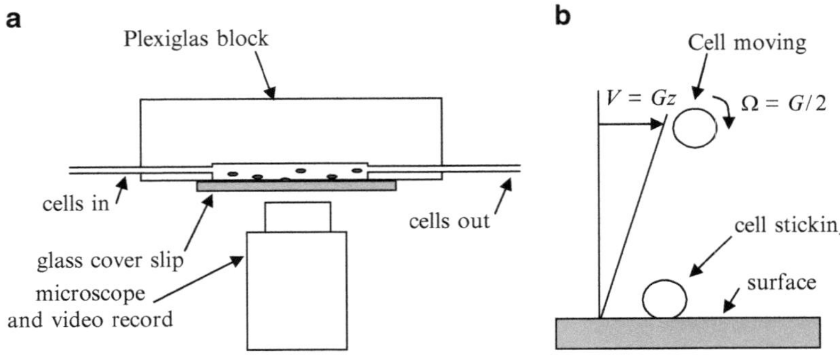

Fig. 7.9 (**a**) Apparatus for studying cell adhesion; (**b**) laminar flow near surface

is 8 μm, the force is $F = 0.35G \times 10^{-12}$ N, which is around 1 pN, much smaller than the force of a single bond. Of course, the cells some distance from the wall are moving and also rolling but these do not stick to the surface and are neglected.

The experiment consists in observing the cells and only observing those close to the wall. These move at constant slow speed until there is an adhesion event; then they stop, but start moving again later once the bond breaks due to Brownian collisions. A well defined arrest and continuation is shown in Fig. 7.10a.

A large population of T cell hybridoma B10BR cells coated with anti CD8 adhesion molecules was observed with the video microscope and cell arrests were measured as the shear rate was increased in the range 1–10 s^{-1}, to give the results shown in Fig. 7.10b. In the full line of this graph, the fraction of cells showing one stop, i.e. greater than 1 s, was plotted as the shear rate was increased. A significant fraction of cells showed binding events, which fell as the shear rate increased. Once the cells adhered, some could stick very strongly within a few seconds. In this case, the cells made molecular contact with the surface and required a large increase in flow rate, typically 1,000 times larger, to detach them. Thus there were two adhesion steps; an initial single bond; followed by a multiple strong bond.

The interesting feature of the experiments was the effect of adhesion molecules on the glass surface. The glass was treated with ethanol, then with glutaraldehyde in phosphate buffer, followed by incubation with 5 μg/mL adhesion molecules. In the control experiments, random antibodies were used, but in the broken curve of Fig. 6.8b anti-CD8 adhesion molecules were applied. These had a strong effect on increasing adhesion. Also, the adhesion events were more permanent. Such experiments were supported by further tests on adhesion molecule coated beads, to show that similar phenomena occurred.[29] Beads 1.4 μm diameter were coated with streptavidin and observed flowing across mica surfaces, either control or coated with biotin. The adhesion molecules caused five to 13 times more arrests.

Recently, a number of new studies have appeared using smaller optical chambers, typically 400 mm wide by 100 mm deep to mimic human capillaries with microfluidic pumps which can control flow rates to low values, while measuring the adhering cells by computer controlled movie camera.[30,31] A typical supplier is Cellix, whose

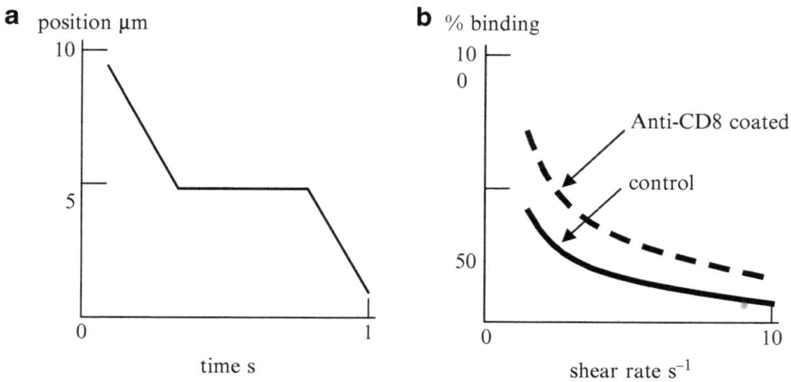

Fig. 7.10 (**a**) Observation of cell arrest and continuation; (**b**) Effect of flow rate and adhesion molecules on binding to surface of glass

instruments have been used to measure the effect of adhesion molecules on cell adhesion. T cells, platelets, tumour and other cells have been studied.

7.9 Cell Counting Methods

The third method widely used for measuring cell adhesion is cytometry or cell counting, first employed in 1969 by Orr and Roseman.[32] They passed a cell dispersion through a Coulter counter and measured the loss of single cells as aggregates formed. The counter in this case did not detect the aggregates themselves but could in principle have measured the whole size distribution of aggregated cells. The Coulter counter was invented by Wallace H Coulter in the late 1940s and its mode of action has found widespread use in the blood and cell industries.[33]

It is shown schematically in Fig. 7.11. Cells are dispersed in an isotonic solution which is stirred and pumped through a fine orifice. An electrode is positioned on each side of the orifice, passing a small current through the conducting fluid as it flows through the hole. As a cell goes through the hole, it causes an electrical blockage proportional to its diameter. Each pulse is counted and sized by the computer to give a distribution of cell sizes.

In more recent versions of this method, cells are labelled with fluorescent molecules and specific adhesive bonds between cells can then be measured using a laser and detector.[34] This has mainly been used to measure cell–cell interactions in the immune system, especially those involving toxic cells and target cells. The method can also be used to separate specific adhered cells. In a typical experiment,[34] mucosal epithelial cells were mixed with bacterial and candida organisms. By counting the aggregates in the electronic apparatus, the adhesion could be quantified. Alternatively, the engulfment of nano-particles by human phagocytes could be quantified by labelling the particles using fluorescent molecules, then counting the fluorescing cells.

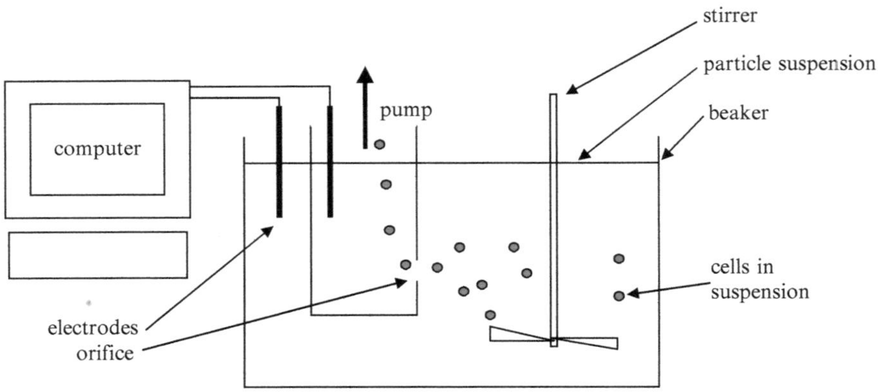

Fig. 7.11 Schematic of Coulter counter instrument for measuring cells

A 2008 study[35] developed a flow cytometric assay for quantitative determination of adhesive interactions of human endothelial cells (ECs) with tumour cells. EC lines established from human lymph node, appendix, lung, skin and intestine microvessels, labelled with PKH26-GL fluorescent dye, were grown to confluency in 24-well TC plates. Human colon adenocarcinoma cell suspension was overlaid onto labelled ECs, and allowed to adhere for 20 min at 4°C under static conditions. Non-adhering cells were collected first, and adhering tumour cells together with ECs were detached from the culture plate. Collected cell fractions were evaluated by flow cytometry. It was demonstrated that immortalised human microvascular ECs preserved their organ specificity. Colon carcinoma cells adhered preferentially to ECs of intestine origin. Similar experiments have been carried out on Streptococcus and Porphyromonas cells[36,37] suggesting that adhesion molecules play an important role in gum infections.

7.10 Adhesion by Counting Doublets

Recently, a more precise method for measuring and understanding cell adhesion has been devised.[38–40] The objective was to remove the need for probes because these damage the cells and change the conditions. Also, the idea was to produce an absolute measure of cell adhesion which did not demand new definitions of binding. It seemed logical to define adhesion of cells in terms of the two parameter model of adhesion interactions described in Chapter 3.

Consider a dilute dispersion of uniform spherical particles as shown in Fig. 7.12.[38]

These spheres experience Brownian motion and therefore diffuse in all directions, causing collisions between the particles. If there is adhesion between the particles, then a collision has a chance of creating a doublet, that is two particles adhering together at the point of contact. If the adhesive bond is weaker than kT,

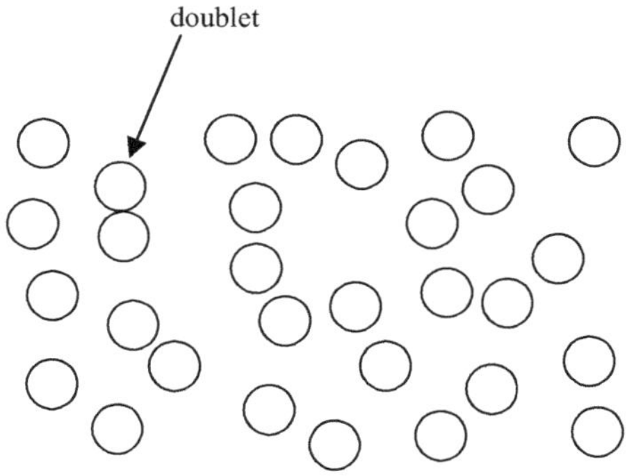

Fig. 7.12 Dispersion of spherical particles

then thermal collisions can break this bond in a period of time. The spheres will then separate and move apart. Thus there is a dynamic equilibrium between joining and separation, giving a certain number of doublets in the suspension at equilibrium, after a suitable time has elapsed for diffusion to take place. High adhesion should give a larger number of doublets and lower adhesion a smaller number. Hence there is a definite connection between sphere adhesion and the equilibrium number of doublets observed in a dilute suspension.

Of course, there are several assumptions in this argument. The main premise is that the spheres are all identical. This is not true of most cells which are known to have distributions of various molecular species on their surfaces. However, it is possible in principle to filter out any rogue doublets formed by unusually tacky cells. Equilibrium can then be re-established. Repeating this filtering and equilibriation procedure several times should lead to a point where the remaining cells are more nearly equal. A second assumption is that the cells are spherical and equal in diameter. In fact human red cells are dimpled and range in size between 6 and 8 μm which will cause a small error.

The most interesting consequence of the above idea, that cell adhesion may be measured by observing the number of doublets at equilibrium in a dilute suspension, is that an exact mathematical solution can be found under certain circumstances, depending on the interaction between spheres when they collide. The simplest situation is that shown in Fig. 7.13 where a particle approaches its neighbour at constant speed until, at a certain separation, the particles are attracted to each other with an energy ε. If this energy remains constant until the spheres touch rigidly at the point of contact, then the square well potential is revealed. The approaching sphere travels at constant speed, is accelerated into the potential well, reflects rigidly on contact, and then is decelerated as the particles move apart. This 'hard sphere square well' which was first used by Alder and Wainwright[41] can be solved exactly to predict the number of doublets in a suspension.

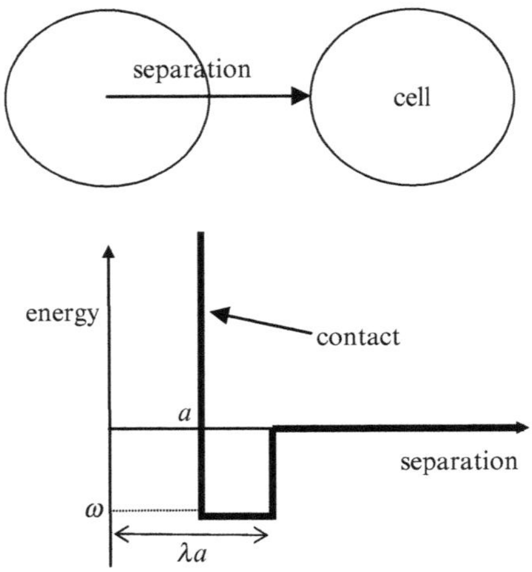

Fig. 7.13 Interaction energy between approaching spheres

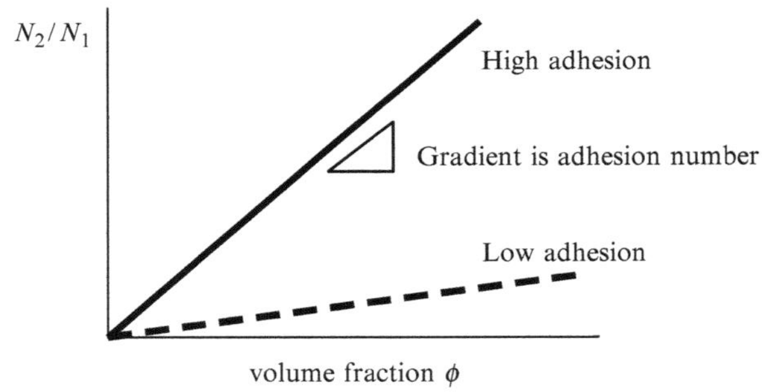

Fig. 7.14 Defining the adhesion number for cells

The mathematical result is that the ratio of doublets to singlets N_2/N_1 is proportional to the volume fraction ϕ of the cells and depends on the range λ and the energy ε of the well according to the equation below.

$$\frac{N.N_2}{N_1^{\,2}} = 4\,\phi\,(\lambda^3 - 1)\exp(\varepsilon/kT) \approx \frac{N_2}{N_1} \tag{7.4}$$

The conclusion of this argument is that a plot of doublet to singlet ratio versus particle volume fraction should yield a straight line passing through the origin. The gradient of the line is a measure of the adhesion which depends on range and energy of the interactions. Thus a high gradient signifies high adhesion and a low gradient low adhesion as shown below in Fig. 7.14. Thus an adhesion number can be defined as the gradient of this plot, to give a measure of the bonding of the cells. The experimental objective was to define this non-dimensional adhesion number for three different species of red cells, horse, rat and human.[40]

7.11 Experimental Results

The important conclusion from these arguments is that cell adhesion must depend on cell concentration. Cells will appear to stick more in proportion to their volume fraction. Of course this is a general law which applies to all reversibly adhering Brownian particles.

Red blood cells, erythrocytes, were used because of their low and reversible adhesion. Cells were prepared from three species, human blood from North Staffordshire Hospital, fresh horse blood in EDTA and fresh rat blood from Central Animal Pathology Ltd. Each blood sample was washed six to seven times in phosphate buffered saline to remove the non red cell components, before suspending in physiological saline solution, then examined by both optical and Coulter tests. Each species of cell was treated in three ways to judge the effect of surface adhesion molecules; by adding glutaraldehyde, fibronectin and papain.

The optical apparatus is shown in Fig. 7.15. The cells were placed in an accurately defined 10 μm space within a glass chamber which was imaged using a video microscope at 40× magnification. Each cell could then be clearly seen moving around with Brownian movement, while not overheating as occurred at 100× magnification. Pictures of the cells were taken at random locations in the chamber and the numbers of doublets and singlets were counted by image analysis software. Taking the ratio of doublets to singlets, the adhesion number was obtained.

The collision and adhesion events could be observed in experiments as shown in Fig. 7.15b which shows one field of view. There were several doublets which could be counted.

The second set of experiments to measure the doublet numbers used the Coulter Counter, which was set up in standard mode to count the individual red cells, as shown by the results of Fig. 7.16a. The strong peak showed a symmetrical distribution of single cells at a volume fraction near 10^{-5}.

At higher concentration, a shoulder appeared at a 13% higher diameter, 7.16b, and this was interpreted as a doublet peak. At still higher concentration of the red cells, the shoulder increased in size (7.16c), indicating that more doublets formed as the blood cells became more numerous. The number of doublets was measured and

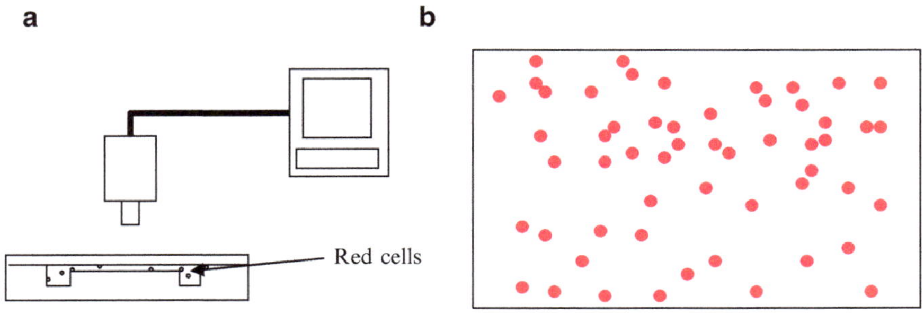

Fig. 7.15 (**a**) Video camera apparatus for observing red cells; (**b**) Field showing red cell doublets

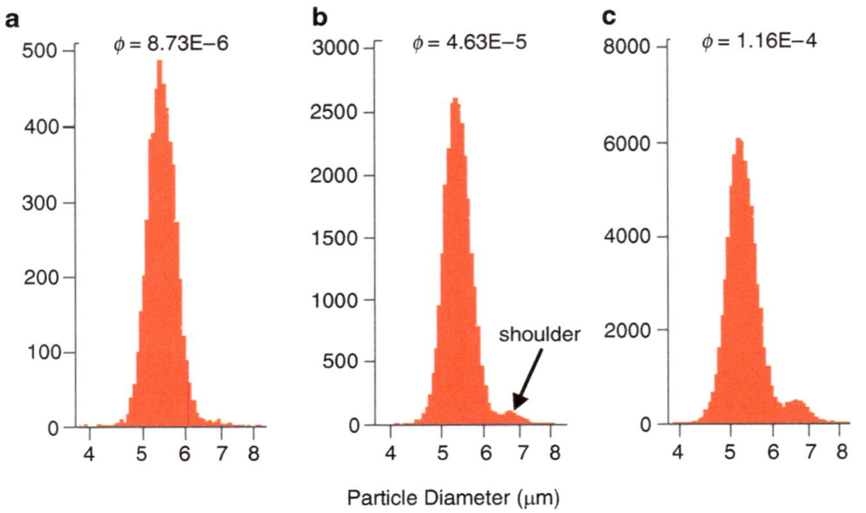

Fig. 7.16 (**a**) Coulter Counter result for human cells; (**b**) result at higher concentration showing shoulder; (**c**) larger shoulder at higher concentration

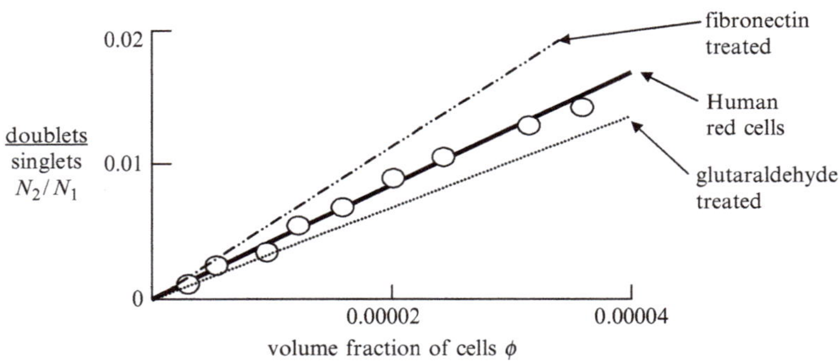

Fig. 7.17 Increase in doublets with concentration of red cells

divided by the singlet peak to obtain the ratio N_2/N_1. This was then plotted as a function of cell volume fraction to give the curve shown in Fig. 7.17. The results showed the doublets increasing in proportion to concentration and allowing the adhesion number to be found by determining the gradient. For human cells this was 420.

Horse and rat erythrocytes were then tested in the same way and shown to give significantly higher adhesion. Baskurt et al.[42] have shown that the aggregation of such cells is increased over human cells, but volume fraction effects were not taken into account. Popel et al.[43] recognised that horse cells stick better and this was attributed to the athletic nature of the animal. Table 7.2 quantifies the difference of adhesion in terms of the adhesion number $N_2/N_1\phi$.[40]

Table 7.2 Comparison between adhesion of various red cells

Animal	Adhesion number $N_2/N_1\phi$
Horse	1,488 ± 200
Rat	750 ± 4
Human	420 ± 5

Table 7.3 Effect of surfactants on horse red cell adhesion

Horse cell treatment	Adhesion number $N_2/N_1\phi$
Isoton	1,279 ± 203
Isoton + glutaraldehyde	1,020 ± 162
Isoton + fibronectin	1,399 ± 184
Isoton + papain	1,513 ± 295

These results show conclusively that rat cells are almost twice as sticky as human red cells, while horse erythrocytes are almost twice as adhesive as rat cells. Whether this can be explained in terms of the higher energy of the bonds, as defined by Eq. 7.4, or in larger range of bonds remains to be determined.

Addition of surface active molecules to the cell suspension was also studied. The results for human cells are illustrated in Fig. 7.17 which shows that fibronectin increased the adhesion whereas glutaraldehyde reduced it. The effect of additives on horse erythrocytes is shown in Table 7.3.

The control sample of horse cells in isoton showed somewhat weaker adhesion than the sample shown in Table 7.2. Such variation was found to be common in different samples of horse blood. Differences between animals in type, age, etc. and also in blood cell conditioning had a distinct influence which will be described in separate papers. It is evident from the results that glutaraldehyde reduced the adhesion by about 25% whereas fibronectin increased the adhesion by 10% and papain by 20%, changes which were comparable with the effects seen on human red cells but disappointingly small compared with the effects anticipated.[40]

Further extension of this doublet counting method to measuring adhesion of smaller spheres such as viruses and polystyrene nanoparticles has been achieved using laser tracking as shown in Section 9.9.

7.12 Conclusions

Measurement techniques are essential to the understanding of adhesion processes. There are numerous measurements for investigating adhesion of cells, viruses and nanoparticles, while new methods are constantly emerging. This chapter has focussed on some of the microscopy methods, some of the probe techniques and some cell counting measurements. Microscopy is vital to obtain qualitative ideas about adhesion mechanisms. Probes such as AFM and optical tweezers can be extremely useful in measuring the small forces involved, especially when adhesion molecules are introduced. Counting methods have the advantage that large numbers of cells can be involved to give reliable statistics. Obtaining the interaction potentials from these measurements is possible and needs much further study. Now we need to apply these measurements in the following chapters.

References

1. Hooke, R., Micrographia, Royal Society, London 1665. Schierbeek, A., The collected letters of Antoni van Leeuwenhoek, Springer, Berlin 1953.
2. Bradbury, S., Bracegirdle, B., Introduction to Light microscopy, Vol 42 Microscopy handbook series, Springer Verlag, Berlin 1998.
3. Zernike, F., How I discovered phase contrast, Science 121 (1955) 345–349.
4. Curtis, A.S.G., Mechanism of adhesion of cells to glass- study by interference reflection microscopy, J Cell Biol 20 (1964) 199–215.
5. Radler, J., Sackman, E., On the measurement of weak repulsive and frictional colloidal forces by reflection interference contrast microscopy, Langmuir 8 (1992) 848–53; Versucheren, H., Interference reflection microscopy in cell biology-methodology and applications, J Cell Sci 75 (1985) 279–301.
6. Zilker, A., Engelhardt, H., Sackmann, E., Dynamic reflection interference contrast microscopy- a new method to study surface excitations of cells and to measure membrane bending elastic moduli, J de Physique 48 (1987) 2139–51.
7. Schilling, J. Absolute interfacial distance measurement by dual wavelength reflection interference contrast microscopy, Phys Rev E69 (2004) 021901.
8. Tsien, R.Y., The green fluorescent protein, Ann Rev Biochem 67 (1998) 509–44.
9. Forster, T., Zwischenmolekulare energiewanderung und fluoreszenz, Ann der Physik 2 (1948) 55–75.
10. Lambacher, A., Fromherz, P., Fluorescence interference contrast microscopy on oxidised silicon using a monomolecular dye layer, Applied Phys A63 (1996) 207–16.
11. Axelrod, D., Cell-substrate contacts illuminated by total internal reflection fluorescence, J Cell Biol 89 (1981) 141–45.
12. Subramanian, S., Boder, E.T., Discher, D.E., Phylogenetic divergence of Cd47 interactions with human signal regulatory protein alpha reveals locus of species specificity- implications for the binding site, J Biol Chem 282 (2007) 1805–18.
13. Binnig, G., Quate, C.F., Gerber, C., Atomic Force Microscopy, Phys Rev Lett 56 (1986) 930–33.
14. Engler, A.J., Rehfeldt F., Sen, S., Discher, D.E., Microtissue Elasticity: measurements by atomic force microscopy and its influence on cell differentiation, Methods in Cell Biology, 83 (2007) 521–45.
15. Radmacher, M., Measuring the elastic properties of living cells by the AFM, Methods in cell biology 68 (2002) 67–90.
16. Ashkin, A., Bjorkholm, J.E., Chu, S., Caught in a trap, Nature 323 (1986) 585.
17. Sheetz, M.P., Laser tweezers in cell biology, Academic Press, San Diego 1998.
18. Grier, D.G., A revolution in optical manipulation, Nature 424 (2003) 810–816.
19. Kress, H., PhD thesis, Cell mechanics during phagocytosis studied by optical tweezers based microscopy, Cuvillier, Gottingen, 2006.
20. Ra, H.J., Picart C, Feng H, Sweeney HL, Discher DE, Muscle cell peeling from micropatterned collagen: direct probing of focal and molecular properties of matrix adhesion, J Cell Sci 112 (1999) 1425–36.
21. Leckband D., Beyond structure: mechanism and dynamics of intercellular adhesion. Biochem Soc Trans. 36 (2008) 213–20.
22. Gallant, N.G., Garcia, A.J., Quantitative analysis of cell adhesion strength, in Methods in molecular biology, Vol. 370, Adhesion protein protocols, ed Coutts, A.S., Humana Press, NJ,

2008. ch.7; Gallant,N.D., Michael,K.E., and Andrés J. García, A.J., Cell adhesion strengthening Contributions of Adhesive Area, Integrin Binding, and Focal Adhesion Assembly MBC 16, (2005) 4329–4340.
23. Fowler, H.W. and McKay, A.J. in 'Microbial adhesion to surfaces' (eds. Berkeley, Lynch, Melling and Rutter) Academic, London, 1980 pp143–61.
24. Woods, D.C. and Fletcher, R.L., Biofouling 3 (1991) 287–303.
25. Duddridge, J.E., Kent, C.A. and Laws, J.F., Effect of surface shear stress on the attachment of Pseudomonas fluorescens to stainless steel under defined flow conditions, Biotechnol and Bioeng 24 (1982) 153–64.
26. French, M.S. and Evans, L.V. in 'Algal Biofouling' (eds. LV Evans and KD Hoagland) Elsevier, Amsterdam, 1986 pp.79–100.
27. Pierres, A., Tissot, O., Malissen, B. and Bongrand, P., Dynamic adhesion of CD8-positive cells to antibody-coated surfaces: the initial step is independent of microfilaments and intracellular domains of cell-binding molecules. J Cell Biology 125 (1994) 945–53.
28. Segal, D.M. in 'Physical Basis of Cell Adhesion' (ed. P Bongrand) CPC, Boca Raton, 1988, pp157–72, ; Gorman, S.P., McCafferty, D.F., and Anderson, L., Letters in Applied Microbiol. 2 (1996) 97–100, Rolland, A., Merdrignac, G., Gouranton, J., Bourel, D., LeVerge, R. and Genetet, B., J Immunol Methods, 96 (1987) 185–93.
29. Pierres, A., Benoliel, A. and Bongrand, P., Use of a laminar flow chamber to study the rate of bond formation and dissociation between surface-bound adhesion molecules: effect of applied force and distance between surfaces. Faraday Discuss, 111, (1999) paper 24.
30. Fitzgerald et al, Cell based microfluidics, FASEB 22 (2008) 676.4.
31. Remuzzi, A., Giavazzi, R., Adhesion of tumor cells under flow, Methods in Molec Biol 96 (2008) 153.
32. Orr, C.W. and Roseman, S., Intercellular adhesion. I. A quantitative assay for measuring the rate of adhesion, J Membr Biol 1 (1969) 109–124.
33. Coulter Electronics Inc, Luton, UK.
34. Segal, D.M. in 'Physical Basis of Cell Adhesion' (ed. P Bongrand) CPC, Boca Raton, 1988, pp157–72, ; Gorman, S.P., McCafferty, D.F., and Anderson, L., Letters in Applied Microbiol. 2, 97–100 (1996) Rolland, A., Merdrignac, G., Gouranton, J., Bourel, D., LeVerge, R., and Genetet, B., J Immunol Methods, 96 (1987) 185–93.
35. Paprocka, M., Dus, D., Mitterrand, M., Lamerant-Fayel, N., Kieda, C., Flow cytometric assay for quantitative and qualitative evaluation of adhesive interactions of tumor cells with endothelial cells, Microvascular Research 72 (2008) 134–138.
36. Sethman CR, Doyle RJ, Cowan MM., Flow cytometric evaluation of adhesion of Streptococcus pyogenes to epithelial cells, J Microbiol Methods 51 (2002) 35–42.
37. Pathirana RD, O'Brien-Simpson NM, Visvanathan K, Hamilton JA, Reynolds EC., Flow cytometric analysis of adherence of Porphyromonas gingivalis to oral epithelial cells. Infect Immun 75 (2007) 2484–2492.
38. Kendall, K., Liang, W. and Stainton, C., New theory and observations of cell adhesion, Proc R Soc Lond A454 (1998) 2529–33.
39. Kendall, K. and Liang, W., Aggregate formation in colloid dispersions, Colloids Surf 131 (1998) 193–201.
40. Kendall, K. and Attenborough, F., Cell-cell adhesion of erythrocytes, J Adhesion 71 (2000) 41–51.
41. Alder B.J. and Wainwright, T.E., Studies in molecular dynamics.1. general method, J Chem Phys 31 (1959) 459–66.
42. Baskurt, O.K., Farley, R.A. and Meiselman, M.J., Am J Physiol:Heart & Circulatory Physiology 42 (1997) H2604–12.
43. Popel, A.S., Johnson, P.C., Kavenesa, M.V. and Wild, M.A., J Appl Physiol 77 (1994). 1790–94.

Part III
Detailed Research

Chapter 8
Adhesion of Nanoparticles

characteristic wandering of the cells with definite relations to the web fibers

Harrison 1914

Nanoparticles are ubiquitous: in the vacuum of space where they are visible through their spectral signatures,[1] and also on earth where they are present in the atmosphere as aerosols, in fresh waters where they occur as humic substances causing the brown colour in bog water and as clay particles from erosion of rocks, and in the sea where they can be precipitated from silicate and calcium-based solutions. Nanoparticles have found numerous industrial applications; in pigments, cements and coatings; they can also be used as functional additives giving optical or electronic effects, as reinforcing agents and also to control the stiffness of soft solids.[2] There is much known about material at the nanoscale.

Nanoparticle adhesion is important because it allows capture and aggregation when contact is made with other materials. Also, the toxicity of nanoparticles is related to their adhesion to molecules, viruses and cells in the body. By understanding such adhesion, it may be possible to understand the huge death toll, approximately 2 million per year, caused by humans contacting 'inert' fine particles, and another 100 million caused by ingesting viruses.

This chapter seeks to define nanoparticles and describe their occurrence, manufacture and applications. Then it describes a new method for measuring nanoparticle adhesion based on laser tracking to determine the number of aggregates in a dispersion at equilibrium. Theory shows that the more aggregates are found, the more must be the adhesion between the particles.

8.1 Nanoparticles in Space

The idea that the cosmos is full of nanoparticles became widely considered after Harry Kroto and his colleagues received the Nobel prize for fullerenes and particularly C_{60}, the new football shaped carbon molecule[3] about 1 nm in diameter which they had detected in the spectra received from stars. Now it has been possible for

K. Kendall et al., *Adhesion of Cells, Viruses and Nanoparticles*,

DOI 10.1007/978-90-481-2585-2_8, © Springer Science+Business Media B.V. 2011

various space missions, Ulysses, Galileo, Cassini and Helios, to collect dust samples, enabling us to discuss the range of nanoparticle compositions and sizes from 1 to 1,000 nm to find out how the particles stick together and eventually form planets which make up our solar system, ultimately leading to life.

About 1% of the cosmos is in the form of fine particles, mainly less than 1 μm in diameter, most of them formed from asymptotic giant branch stars as a steady outflow. However, some particles are thought to be formed in star explosions and it is interesting to learn that a supernova can produce around one solar mass of particles.[4] These particles have been detected from earth by observations at infrared, radio, millimetre, and submillimetre frequencies, demonstrating that particles of silicate, carbon and organic materials are dispersed in the interstellar gas which makes up most of the intergalactic material.

The Helios mission[5] collected dust data for interstellar particles which were analysed by impact ionisation detectors on-board to show the deep penetration of a nanoparticle stream into the Solar System. It was demonstrated that gravitation focusing facilitated the entry of micrometre-size material around the earth, while radiation pressure prevented 75% of the nano-material from penetrating into the innermost regions of the Solar System. A flux value of about $2.6 \times 10^{-6}\ m^{-2}\ s^{-1}$ was derived for micrometre-size grains. The density of particles was about 3×10^{-25} kg m^{-3} in the local interstellar cloud, contrasting with the 10^{-8} kg m^{-3} typically found in our atmosphere. Compositions of particles were assessed by the time-of-flight mass spectrometer subsystem of the Helios instrument, showing that a varying mixture of various minerals and carbonaceous compounds was present. A typical dust analysis rig is shown in Fig. 8.1.

A dust trajectory module consists of four sensor wire grids mounted between two electrical shielding grids 40 mm apart. Each sensor grid consists of 15 parallel wire electrodes and particle trajectories are determined by the measurement of the electric signals that are induced when a charged grain passes through the position sensitive electrode system. The objective of the trajectory sensor is to measure dust charges in the range 10^{-16}–10^{-13} C and dust speeds in the range 6–100 km s^{-1}.

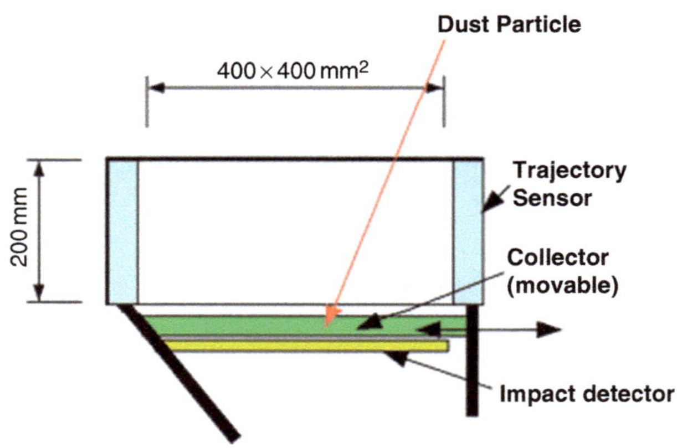

Fig. 8.1 Schematic cross-section of a space vehicle dust analysis unit showing the trajectory sensor, the collector material and the impact detector[4]

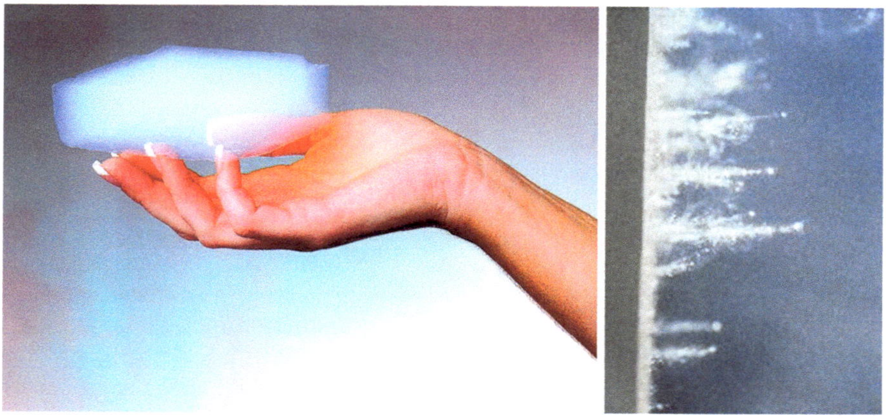

Fig. 8.2 *Left*: Aerogel in hand; *right*: Particle tracks in aerogel[7]

The particle collector material can be silica aerogel, polymer or metal foil. Silica gels are very low density (Fig. 8.2), as low as 0.002 g cm^{-3}, and the impacting particles bury themselves in it but survive to allow subsequent chemical analysis. Non-silica aerogels, e.g. carbon gels, are better for detecting silicates.

Polymer foils are made in multilayer composites with gaps in between to collect and identify inorganic particles.[6] The capture medium is a stack of very thin (8 and 40 μm) polyimide foils, supported on poly-tetrafluoroethylene sheet frames, surrounded by a protective aluminium casing. The uppermost foil has a very thin metallic coating for thermal protection and resistance to atomic oxygen and ultra-violet exposure. Analysis of impact residue is performed using energy dispersive X-ray spectrometers. Impact may cause disruption and melting, but some residue retains sufficient crystallographic structure to show clear Raman lines, diagnostic of the original materials.

Typical samples returned to Earth by the Stardust spacecraft[7] appeared to be weakly aggregated mixtures of nanometer-scale grains, with occasional much larger (over 1 μm) ferromagnesian silicates, Fe–Ni sulphides, Fe–Ni metal, and other phases. The variety of olivine and low-Ca pyroxene compositions requires a wide range of formation conditions, probably reflecting very different formation locations in the protoplanetary disc. The restricted compositional ranges of Fe–Ni sulphides, the wide range for silicates, and the absence of hydrous phases indicated that there was little or no aqueous alteration.

8.2 Nanoparticles in the Atmosphere

Nanoparticles have been identified in all atmospheric layers, and increasing evidence from epidemiology indicates their potential to damage human health.[8,9] While this relationship is not fully understood, and the mechanistic pathway between exposure and health outcome remains obscure, it has prompted increasing interest in nanoparticle measurement in cities where hundreds of millions of people now live.

Atmospheric nanoparticles occur as solids or liquids, formed by a number of natural processes, for example drying of sea-spray to form NaCl nano-crystals, sulphates formed from plankton excretions, solidification of volcanic emissions, or nucleation of smoke particles from forest fires. In atmospheric science, nanoparticles are known as adventitious or ultrafine particles (UFPs), or particulate matter less than 0.1 μm ($PM_{0.1}$). UFPs <0.1 μm may also be confusingly described as nucleation or Aitken[10] mode aerosols, after the Scottish scientist who devised instruments for measuring and counting nanoparticles in clouds. Particle fractions that include large and nanoparticles are termed PM (particulate matter) below a certain aerodynamic size; for example, $PM_{2.5}$ is below 2.5 microns and PM_{10} measures particles with an aerodynamic diameter below 10 microns. The background concentration of natural nanoparticles is in the lower μg m^{-3} range, but PM concentrations in cities where air pollution is poorly regulated can reach several mg m^{-3} (in the developing world, or in the London smogs of the 1950s). Nanoparticles tend to dominate the particle number and surface area (see Fig. 8.5), while particles exceeding 1 μm contribute most to the mass.

Emissions of nano-particles in cities are associated with human activity[11] and there is increasing legislation to reduce their emissions. Urban particles may arise from (1) photochemical production of nanoparticles from organic precursors or (2) emissions of nanoparticles from high temperature combustion. If molecules emitted as gases do not deposit on existing particle surfaces, then new particles are nucleated. For example, Dunn et al.[12] reported that nucleation events near Mexico City only occurred during daylight hours at low particle concentrations, when ambient sulphur dioxide concentrations significantly increased above background levels.

Nanoparticles are affected by local atmospheric conditions such as gas concentrations, temperature and humidity, and they may grow quickly through water accretion, gas condensation and/or coagulation processes. Nano articles are removed by growth, coagulation and diffusion to surfaces, whereas larger particles disappear by sedimentation and deposition processes. While airborne, nanoparticles participate in important physical atmospheric processes such as cloud formation, precipitation, light scatter and absorption. In addition, they play a crucial role in atmospheric heterogeneous chemistry.

In urban atmospheres, airborne nanoparticles are well characterised. Several authors report that 15–20% of $PM_{2.5}$ mass is made up of nanoparticles,[12] especially in urban street canyons. Vehicles emit tiny soot particles directly to the atmosphere. Roadside measurements show that freshly emitted soot particles are 30–200 nm in diameter and this size distribution rapidly shifts upwards with downwind distance from roadsides.[13–14] These directly emitted chain aggregates of 25 nm sub-particles (Fig. 8.3 top) quickly adhere to form larger agglomerates (Fig. 8.3 bottom). The soot contains mainly elemental carbon, with organic carbonaceous material such as polyaromatics adsorbed to the expansive particle surfaces.[15] Ultrafine soot particles from a modern diesel car are approximately 50–70 nm in size, while the mean mass diameter is around 300 nm.

Spherical particles of fly-ash or metal oxides deposited from lubricating oil additives can also be widely recognised in urban atmospheres as shown in Fig. 8.4. Large amounts of fly-ash used to be emitted from coal fired power stations during the last century, but these are now reduced by collection on electrostatic precipitators in the chimneys.

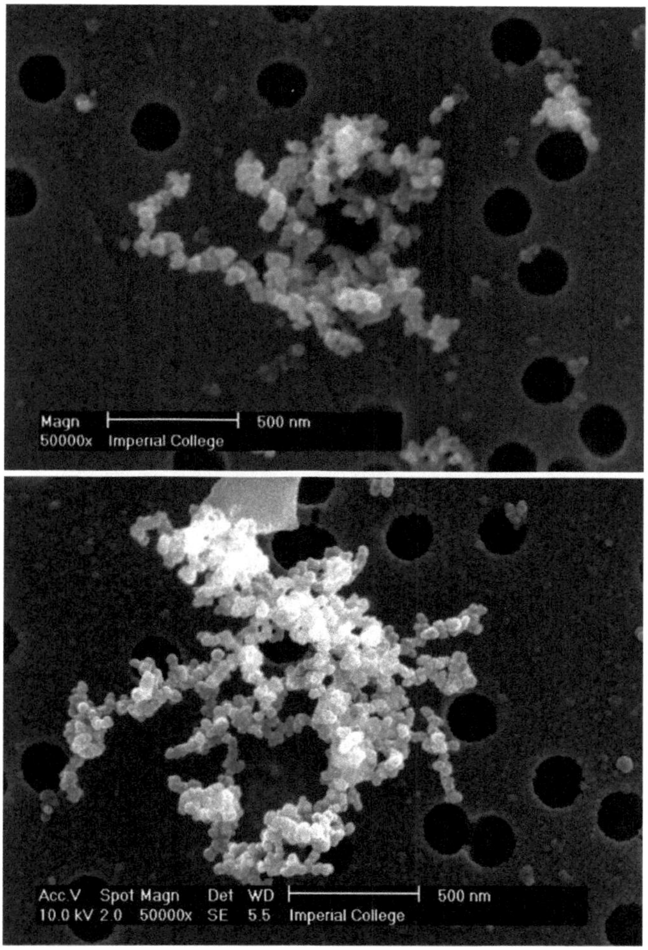

Fig. 8.3 Chain aggregates *top*; larger agglomerates *bottom*. Copyright M. Kendall and E. Wigzell

8.3 Characterisation of Atmospheric Nanoparticle and Their Health Effects

One of the problems of atmospheric nanoparticles is defining the nomenclature, since widely different names have been used including aerosols, UFPs, Aitken mode, nucleation mode etc. as shown in Fig. 8.5. Under 7 nm, nanoparticles are measured using condensation particle counters (CPC, e.g. TSI) and ultrafine condensation particle counters (UCPC, e.g. TSI). Newer UCPC instruments can now measure nanoparticles under 3 nm. Number concentrations are measured by Scanning Mobility Particle Sizer (SMPS). Newly developed techniques such as novel chemical ionisation mass spectrometers – the cluster chemical ionisation mass spectrometer (the Cluster-CIMS) or

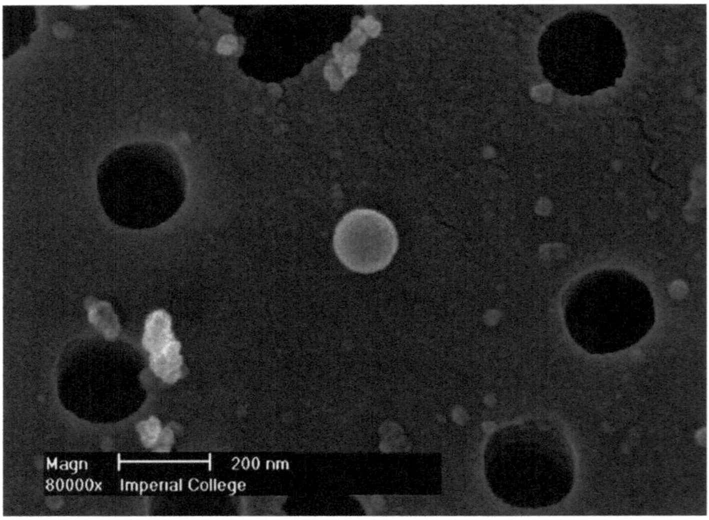

Fig. 8.4 Spherical fly-ash particle on collection filter. copyright M. Kendall and E. Wigzell

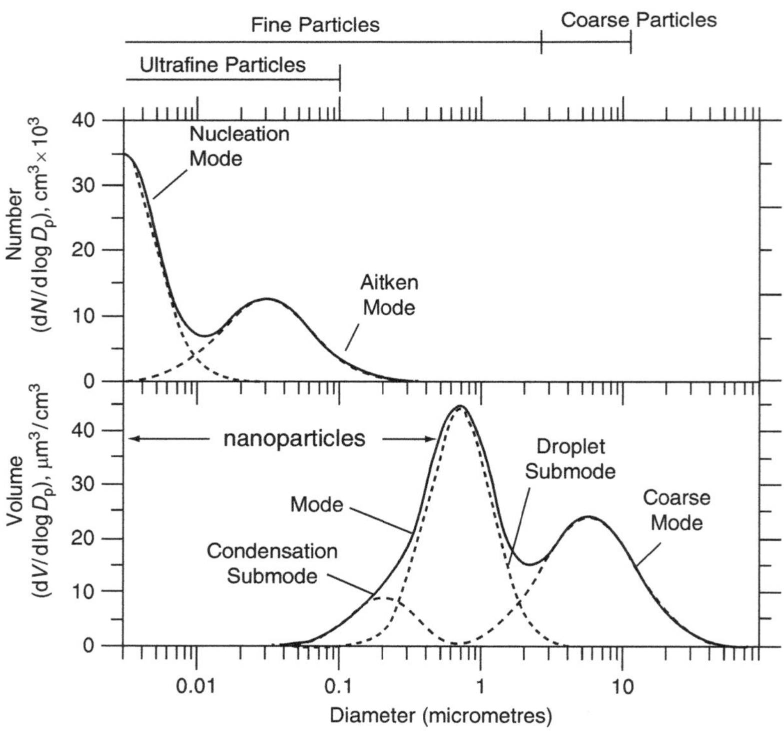

Fig. 8.5 Size distribution of atmospheric particles showing different confusing nomenclature and the dominance of nanoparticles by number[13]

Neutral Cluster and Air Ion Spectrometer (NAIS) – measure the pre-nucleation clusters below 3 nm in the laboratory and in the atmosphere. Studies of UFP formation and tropospheric characterisation have been conducted in remote locations such as forests, the marine boundary layer and Antarctic together with urban atmospheres. These studies have covered some of the least and most polluted atmospheres in the world.

The typical number and volume distributions of atmospheric particles with the different modes[13] are shown in Fig. 8.5. Large particles with aerodynamic diameters above 10 μm have short atmospheric lifetimes of a few hours, because of sedimentation. By contrast, nanoparticles have atmospheric lifetimes up to days or weeks, but are prone to aggregation. UK Government measurements of nanoparticle concentrations on Marylebone Road in Central London show that submicron particle number concentrations can reach the hundreds of thousands level during peak traffic volumes, and that traffic is the major source of these peak concentrations.

Some characteristics of particles and aerosols in ambient atmospheric and industrial settings can be seen below in Table 8.1.[13]

Epidemiological studies showed an increased health risk with increasing airborne nanoparticle mass exposures.[16] These exposures, especially of susceptible groups such as children and the elderly have life-shortening and life-threatening consequences[17] as discussed later in Chapter 11. Both the European Union (EU) and the World Health Organisation (WHO) have introduced increasingly stringent values for aerosol mass concentrations (WHO 2006; EC Directive 2008/50/EC11[18]). But since nanoparticles account for a small fraction (<5%) of the total mass, they are not well represented or controlled by a mass-based measurement. There are currently no regulations for particle number or surface area of nanoparticles causing health effects, but reasonable relationships with mass measurements exist. Insoluble particles such as soot, which are not efficiently removed from the lung, are believed to pass through cell walls, are surface reactive (damage cells), carry adsorbed toxic materials e.g. carcinogenic polycyclic aromatic hydrocarbons and have high mobility within the human organism so that exposure may result in widespread distribution around the body. The health risk of 1–100 nm particles is not well known yet, but remains the focus of significant research programmes around the world[18] as described in Chapter 11.

Table 8.1 Classification of particles at different sizes

nanoparticles	microparticles	macroparticles	
1nm	1μm	1mm	1m
fume	mist		
	dust spray		
smog	cloud fog drizzle	rain	
tobacco smoke	fly ash		
carbon black			
silica fume	sulphuric acid mist		
	coal dust cement		
	milled flour		
	clay silt fines and coarse sand	gravel	
virus	bacterium red cell hair		

8.4 Nanoparticles in Water

Nanoparticles are also prevalent in water and can be classified[19] into the several types shown in Fig. 8.6, which shows how engineered nanoparticles, e.g. titanium dioxide pigments[20,21] manufactured at the five million ton scale and polymer latex particles which are produced at the 20 Mte/a level, have added to the natural burden of siliceous and humic nanoparticles normally present in water, plus the human detritus of soot from combustion and wear products e.g. rubber from car tyres. The cut-off point in particle size is generally set arbitrarily at the 450 nm filter scale.

In some cases these nanoparticles are dispersed and can pass through filters, but more commonly, the nanoparticles adhere to each other and to surrounding surfaces to form aggregates which are bigger in size but contain significant internal porosity and surface area. Adhesion is therefore significant in determining the bioavailability of the nanoparticles to organisms.

Generally, the nanoparticles are described by their core molecular structure e.g. TiO_2 for white pigment, but this can be misleading because the detailed crystallinity of the material is modified by dopants such as Al_2O_3 in the plasma process for producing rutile to give enhanced light scattering rather than anatase crystals, and the individual nanoparticles are often coated with silica and zirconia to prevent yellowing of the pigment after exposure to sunlight. Thus, the precise crystalline nature of the nanoparticle is important. Finally, the reaction of the particle at the surface with water to form hydrated/ionised material, and with surfactant molecules to provide organic layers has enormous impact on the adhesive forces between particles. Therefore, the nanoparticle must be viewed as a complex structure with a solid core, often a nanocrystal with facets, surrounded by a variable, hydrated "corona" – material containing ions and surfactant molecules which could be large adhesion molecules such as glycoproteins. A typical structure is described schematically in Fig. 8.7 for an oxide.

When such nanoparticles mix with natural waters containing humic acids and other surface active materials, they become coated with the natural surfactant molecules and this changes their behaviour substantially. A scheme for the aggregation of the nanoparticles as they adhere together is shown in Fig. 8.8. The nanoparticles are first precipitated by hydrolysis. For example hematite can grow from ferric

Natural nanoparticles		**Breakdown particles**	**Engineered NPs**
inorganic	organic	tire particles	pigments
silicates	polymers	brake debris	polymer latex
clays	fulvic acids	catalyst	surfactants
mica	humic acids	metal wear	metal oxides
kaolinite	viruses	soot	metals Au, Pt, Ag
oxides	funguses	fly ash	quantum dots
silica	bacteria		fullerenes
carbonates	peptides		carbon
phosphates			nanotubes

Fig. 8.6 Classification of aqueous nanoparticles[19]

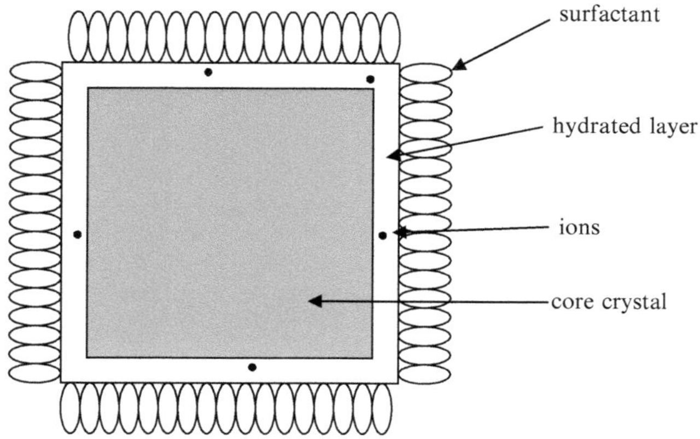

Fig. 8.7 Schematic diagram of a nanoparticle structure in water dispersion

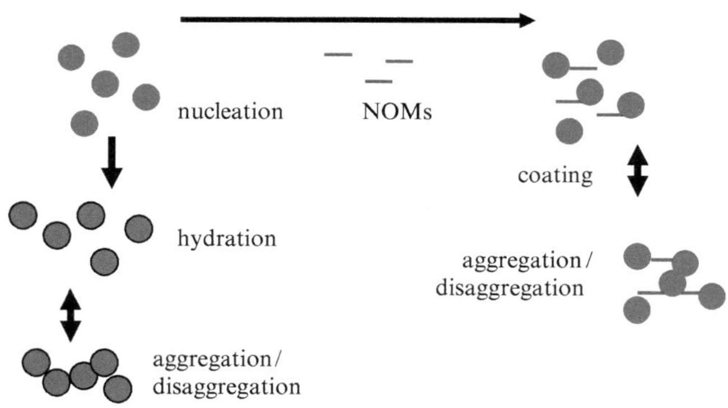

Fig. 8.8 Mechanisms by which aqueous nanoparticles aggregate or interact with natural organic materials (NOMs) which are generally present as humic substances, etc

chloride solution. Below pH 6 these particles are sufficiently charged by hydrogen ions that they repel each other electrostatically and remain dispersed.[22,23] However, slight adhesion exists between the particles and they can aggregate to form reversible doublets and triplets which are weakly bonded. Natural organic materials (NOMs) can coat the surfaces to create a surfactant layer which strongly influences the particle adhesion, altering the aggregate size distribution and affecting the bioavailability and toxicity of the nanoparticles. Other contaminants such as metal ions like calcium, copper and lead can be strongly adsorbed on hematite and titania, especially on the finer particles.[24,25]

These processes have been studied using hematite nanoparticles[24] and the different shapes of aggregates determined in the natural situation are shown in Fig. 8.9. The fractal dimension of iron-oxide NPs was found to vary in the presence of humic acid (HA) molecules. For instance, the electron micrograph shows the variation of aggregate structure due to the addition of Suwannee river humic acid molecules at

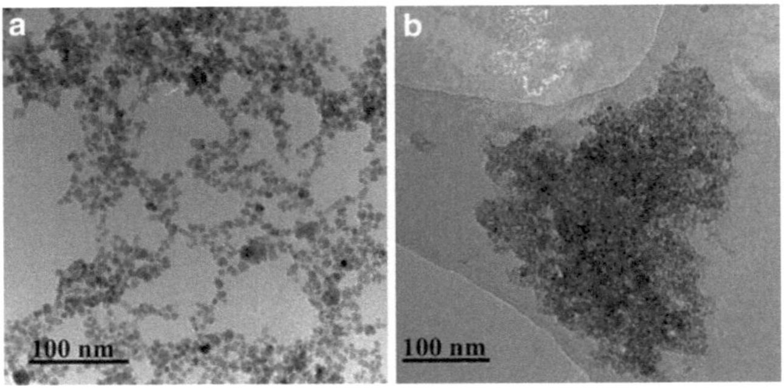

Fig. 8.9 *Left*: Transmission electron micrograph (TEM) of hematite nanoparticle aggregate; *right*: Aggregate formed in presence of humic substances[24]

pH 6. In the absence of HA, iron-oxide nanoparticles form open porous aggregates with a fractal dimension (D_1 1.16 ± 0.06, D_2 1.78 ± 0.06 and D_3 1.87 ± 0.06), whereas in the presence of HA, they form compact aggregates with a fractal dimension (D_1 1.74 ± 0.10, D_2 1.95 ± 0.01 and D_3 2.06 ± 0.02).

The humic acid has clearly had two major effects; first restricting the growth of individual nanoparticles so that the grain size is significantly smaller, second reducing the adhesion between the grains such that the aggregate structure can be more compact as the particles can adjust their positions to gain a more dense packing.

Nanoparticle may impact the environment in three possible ways: (1) direct effect on biota, i.e., toxicity, (2) changes in the bioavailability of toxins or nutrients, (3) indirect effects on ecosystem, i.e., break-up of refractory natural organic substances and (4) changes of the environmental microstructures. Understanding the interactions between the particles and their coatings is important to ensure environmentally sustainable production and use. It is evident that the exact behaviour will depend in a complex manner on nanoparticle properties, organic matter type and concentration and on solution conditions such as pH and ionic strength.

Surface coating, aggregation and disaggregation will largely determine the bioavailability of the nanoparticles through controlling (1) transport in surface and ground waters and (2) sedimentation in surface waters or deposition and filtration in soils and groundwaters. Stabilisation by surface coating may maintain suspension within the water column increasing their transportation distances/rates and bioavilability for fish and waterborne species. Aggregation leads to NP settling to the sediments where benthic organisms become the key receptor for NPs.

8.5 Synthetic Nanoparticle Polymers; Latex Coalescence

An adhesive nanoparticle technology which came to fruition during the twentieth century is that of synthetic polymer latex. When Columbus travelled for the first time to the New World more than 500 years ago, he found that the natives played games

Fig. 8.10 TEM of synthetic polystyrene latex particles

with a rubber ball which they had made by gathering the natural latex from certain trees. This milky fluid exuding from the tree-bark could be dried and used in several interesting ways; as a glue to stick things together; as a waterproof coating for fabrics; or as an elastic material for ball games. The natural latex was somewhat unstable and putrescible, but once the trick of adding ammonia as a stabiliser was discovered, the latex could be stored, transported and used for many applications.

Above their glass transition temperature, the spheres are rubbery, but on cooling turn glassy like the polystyrene shown in Fig. 8.10. These spheres are not compliant enough to deform on drying and so give the powdery ordered rafts which can be viewed in the electron microscope. Rubbery latex by contrast is compliant and dries to form a tenacious film which can be used as adhesive, waterproofing agent or paint.

The secret of artificially making such milky dispersions was found by Hofmann[26] and coworkers at Bayer in 1913. There was a need at that time to find substitutes for natural rubber for the manufacture of tyres. By taking a synthetic rubber precursor, for example butadiene, which is an organic liquid, adding it to water in the presence of a dispersing agent such as blood serum, then shaking, a milky fluid like the natural polymer latex could be produced. This has been enormously successful for producing the polymer materials which lie at the heart of our modern civilisation. Typical applications are adhesives, paints, condoms, tyres, window frames, clothes and shoes.

The way in which rubbery latex particles stick together on drying depends very much on the prevention of premature adhesion of the spheres during the drying process. Electrostatic charge repulsion and steric hindrance of adsorbed molecules prevent the spheres sticking prematurely, so they dance around to adjust their positions until they are in almost close-packed structure before the adhesion kicks in. On drying further, the work of adhesion increases to produce adhesive interactions between the spheres, allowing JKR theory to be used to calculate the contact spot size, as shown in Fig. 8.11.

The JKR explanation of latex coalescence was proposed in 1982.[27] Padget had observed the hexagonal structure of coalesced rubber latex (Fig. 8.11a) and Kendall had measured the contact spot sizes between latex particles using electron microscopy (Fig. 8.11b). When the results were plotted in Fig. 8.11c, they fitted both the

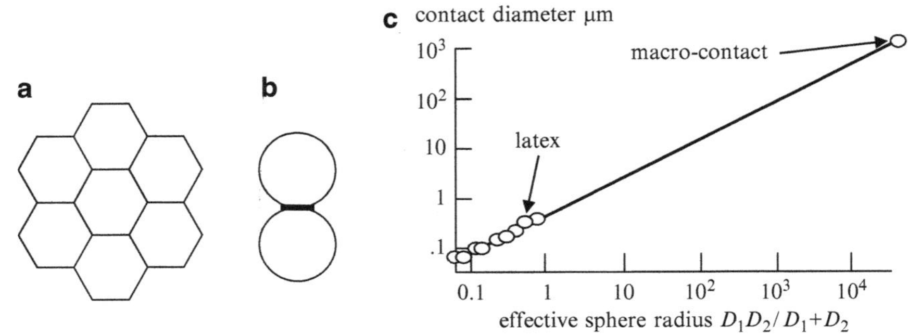

Fig. 8.11 (**a**) Hexagonal structure of coalesced film; (**b**) contact size measurement of latex particles; (**c**) electron microscope and optical measurements confirming JKR theory for latex[27]

JKR equation and the macroscopic observations, taking the elastic modulus to be 5.64 MPa and the work of adhesion to be 26.5 mJ m^{-2}.

The theory of coalescence required two stages:[27] the first was an agglomeration step which was driven by drying of the latex film, allowing the particles to be pushed together into close-packed adhesive contact, but with small adhesion because of the presence of water; the second was an adhesion step in which the work of adhesion increased as the last water was removed and the particles pulled each other together elastically, causing coalescence and shrinkage of the polymer film.

The theory applies properly to elastic particles. In general, polymer spheres are viscoelastic, depending on molecular weight and cross-link density. Consequently, this theory cannot explain the kinetic and irreversible nature of latex contacts which often show pronounced sluggishness and hysteresis in the adhesion process.

8.6 Synthetic Inorganic Nanoparticles

One of the earliest synthetic nanoparticle process inventions was that of Indian Ink, a suspension of carbon black which after drying became waterproof. Such inks came originally from China some 5,000 years ago.[28] Lampblack was prepared by burning pine wood and collecting the soot in a furnace container. The fine black powder from the top of the furnace was mixed with glue made by boiling animal skins, together with other additives like crushed pearl, egg white and musk, pounded 30,000 times to break up aggregates, then strained through cloth to produce a fine ink. This was a superb example of the interaction of the collagen polymer solution with the carbon particle surface, to give the irreversible adhesion of the coated nanoparticles on drying.

Another significant early invention, about 2,000 years old, was volcanic ash as a cement which would harden and remain strong in the presence of water. The Romans wished to build durable sea-walls and found that a certain ash from the region of Pozzuoli, so-called pozzolanic material, would harden after mixing with water and would not be weakened by further immersion in sea water. Some of the sea defences built with this material survive to this day. Interestingly, the pozzolanic cement has lasted better than the rocks in the original walls. It was not until 1824 that Joseph Aspdin managed to make such a cement synthetically by heating clay and limestone together in a furnace, then grinding the product to a fine powder which he called Portland cement. This has now been developed to be the largest synthetic material on the planet[29] with about three billion tons produced annually. By mixing the cement with polymers to improve compaction and moulding, a product called Macro-Defect-Free (MDF) cement was produced, giving strengths 20 times that of conventional products, so that objects like the cement guitar[30] (Fig. 8.12a) can be produced. Strength, stiffness and toughness were much improved over ordinary cast cement products as shown by the stress/strain curves plotted in Fig. 8.12b.

Even more ancient, about 5,000 years old, was the discovery of the sintering process to make weather resistant clay bricks. Mud and clay, which both contain nanoparticles, have been used as building materials for millennia, and continue to be useful in dry countries,[31] mainly as adobe (Fig. 8.13) which houses three billion

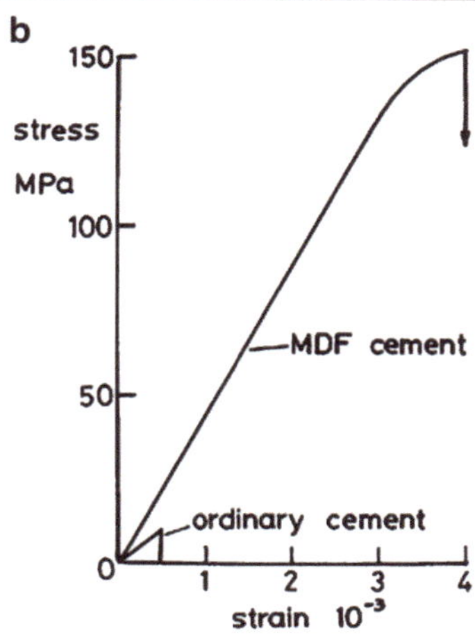

Fig. 8.12 (**a**) Cement guitar made from rolled sheets of alumina cement[30] (**b**) Stress/strain curves showing improved properties of Macro-Dedect-Free (MDF) over ordinary cement

Fig. 8.13 Adobe building from Albuquerque

people worldwide. Clay makes up about 30% of adobe and is important because of its mouldability when wet; it can be plastically formed into elegant and large shapes, for bricks or concrete. After drying, the clay is very strong but suffers from the disadvantage that it becomes weak again on rewetting. This problem was cured by heating the clay particles to high temperatures such that the particles adhered more strongly together. Temperatures of 700°C were sufficient to prevent rapid degradation by water. But above 1,100°C, the material became totally resistant to moisture.[32] Essentially, the nanoparticles were sintering together by solid-state diffusion at elevated temperatures to reduce the surface energy, causing enlarged contact areas. Such sintering technology has been extended over the past century to adhere nanoparticles together, producing ceramic and metal materials of all kinds, from electronic packaging, to magnets for electric motors, to nuclear fuel pellets.

Summarising these well-known synthetic routes, it is clear that there are three methods for producing strongly adhering nanoparticles: first the smoke or gas-phase process which is widely used for carbon black, silica fume and titania pigment manufacture, followed by polymer coating and drying; second the liquid phase precipitation route in which nanoparticles deposit from a liquid and subsequently make irreversible contact as in the Portland cement process; and thirdly the solid-state sintering route in which nanoparticles are compacted, then heated such that diffusive shrinkage occurs to extend the adhering contacts. Let us consider these in more detail.

8.7 Gas-Phase Nanoparticle Production

There is massive world production of gas phase nanoparticles for products which include carbon black reinforcing agents for rubber tyres, titania pigments for whitening paint, silica for optical fibres and specialised metal particles for catalysts.[33] The oldest process is pyrolysis of petroleum or natural gas, heating the material to 1,100–1,500°C by partial combustion with air such that fine smoke particles are produced which can be collected by cyclone and bag filters. The most recent is the production of sunscreen nanoparticles of titania by oxidation of $TiCl_4$ to make 50 nm particles to absorb UV from sunlight while giving a translucent cream for application to the skin of sunbathers. Such particles are shown in Fig. 8.14. Titanium tetrachloride, like silicon tetrachloride, is a volatile liquid which fumes in air or water vapour. In other words, the liquid forms a smoke at the surface as the $TiCl_4$ evaporates and reacts rapidly to form nuclei of nanoparticles in the gas phase.

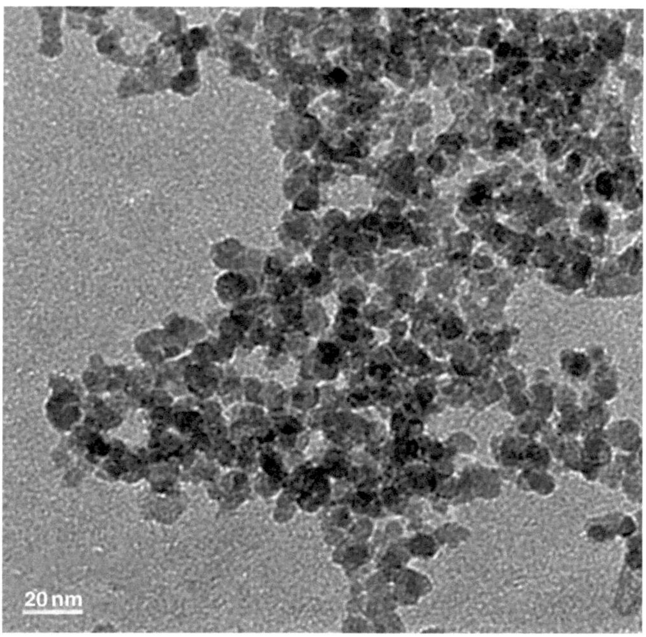

Fig. 8.14 TEM picture of TiO_2 nanoparticles for sunscreen

These particles aggregate to form fractal clusters which can be filtered out and ground into oily pastes for mixing into sunscreen formulations.

More specialised metal particles for use as catalysts may be made by evaporation of the metal followed by condensation and liquid capture at low pressures with inert gas such as argon. These products have the benefit of better purity because there are no chemical reactions producing by-product contaminants.

8.8 Liquid Phase Preparation of Nanoparticles

Nanoparticles can grow in liquids from dissolved species such as chloride or nitrate salts, forming a range of beautiful particle shapes, as described in many papers by Matijevic and his colleagues.[34] An interesting example is iron oxide which is found in most waters at the 100 ppb level, and which can deposit on bathroom surfaces to cause adherent brown stains which are very difficult to wipe off. Many different states of iron oxide and hydroxide exist but hematite is perhaps the most common and can readily be grown to controlled sizes in the following preparation.[35]

A typical synthesis of hematite is by forced hydrolysis of homogeneous $FeCl_3$ solutions under controlled conditions. 2.43 g of $FeCl_3.6H_2O$ was diluted in 12.5 ml of 3.75×10^{-3} M HCl and mixed with 487.5 ml of 3.75×10^{-3} M HCl preheated to 100°C. Vigorous stirring was applied during addition to ensure a homogeneous mixture, such that uniform nuclei were formed in the suspension. Growth of the monosize hematite particles to 60 nm diameter was achieved by incubating the mixture for 24 h at 100°C, then cooling to room temperature before carrying out the washing step. Centrifugation was used to separate the particles after adding KCl to flocculate the particles. The supernatant liquid was discarded and the sediment was redispersed in 10^{-3} M HCl by ultrasonication. This washing procedure was repeated five more times. The suspension was stored at pH3 and diluted with $HClO_4$ solutions to give various concentrations around 100 ppm by volume and pH values to vary the adhesion between particles. The particle size distributions were measured directly on the samples by transmission electron microscopy (TEM, Hitachi H7000, Japan) after evaporating a droplet of suspension on a carbon coated grid. Adhesion could be increased by raising the pH to 6 as shown in Fig. 8.15 where 60 nm diameter single particles can still be seen but where most particles are adhering in large aggregates.

Another way in which nanoparticles can grow is directly on the solid surface by reaction and deposition. This is the well-known rusted bolt effect where steel nuts and bolts corrode, depositing iron hydroxide particles between the iron surfaces such that the bolts cannot be unscrewed. The most well-developed example of this phenomenon is Portland cement made from tricalcium silicate which reacts in water to form calcium silicate hydrate nanoparticles that deposit directly on the cement crystal surface. This effect is shown in Fig. 8.16a. C3S is the shorthand nomenclature for a tricalcium silicate crystal which has reacted at its right hand surface with water to produce a deposited gel of hydrate. Close inspection reveals that the gel is made up of 10 nm diameter particles which have stuck firmly to the

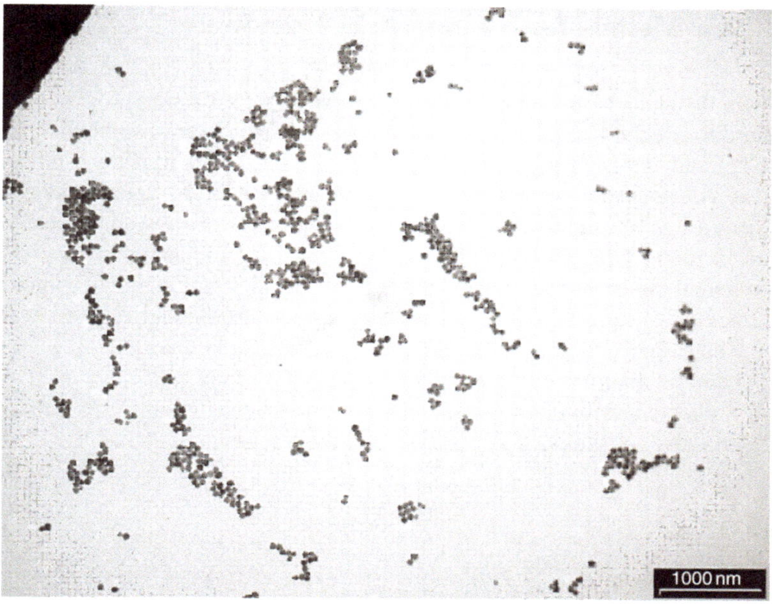

Fig. 8.15 TEM picture of hematite nanoparticles showing individual grains and also aggregates formed by adhesion in suspension

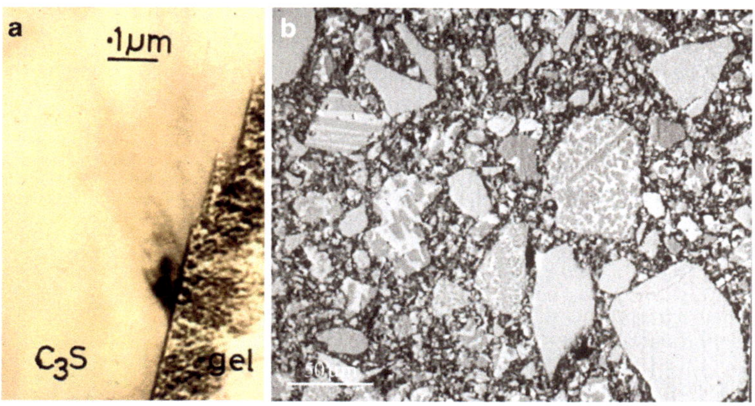

Fig. 8.16 (**a**) TEM showing nanoparticles of calcium silicate hydrate deposited on a crystal of tricalcium silicate after contact with water (Picture by Dr G W Groves). (**b**) SEM of hydrate particles sticking grains of cement together in a well-packed MDF cement composition

C3S surface and built up into a layer about 1 μm thick. Such a thickness of adherent gel is sufficient to bind the residual 20 μm cement particles together as shown in Fig. 8.16b. This is the mechanism of setting of hydraulic cements which is now used at the three billion ton scale for construction purposes.

8.9 New Method for Nanoparticle Tracking

The main difficulty with nanoparticles is following the adhesion process because it is so hard to see the fine particles in the microscope. In order to solve this problem, Bob Carr and his colleagues, when working at the germ warfare establishment based at Aldermaston in England, invented the Nanosight apparatus[36–39] which comprises a window containing the nanoparticle dispersion viewed vertically using an optical microscope, a horizontal laser beam causing the nanoparticles to scatter light to reveal the Brownian movement of nanoparticles, and a movie camera/computer tracking system which measures the particle diffusional random walk and plots the size distribution calculated from Stokes-Einstein theory.

A schematic diagram of their apparatus is shown in Fig. 8.17.

The Stokes Einstein equation gives the diffusion coefficient D in terms of the particle diameter d, the fluid viscosity η and the temperature T,

$$D = \frac{kT}{3\pi\eta d} \tag{8.1}$$

allowing the particle size to be calculated from the tracks.

A typical single frame for 62 nm diameter (TEM result) monosize hematite particles is shown in Fig. 8.18, with the computed particle size distribution shown superimposed on it. The result shows that the particles have mainly a hydrodynamic diameter about 90 nm but there is also a second peak of doublet aggregates and a third peak of triplets plus a smaller number of larger aggregates.

This result was markedly different from dynamic light scattering (DLS) data from instruments such as the Brookhaven or Malvern Nanosizer which showed a single monosize peak of nanoparticles almost 80 nm in diameter as shown in Fig. 8.19, falsely suggesting that the nanoparticles were fully dispersed. This dynamic light scattering method cannot distinguish aggregates from the primary particles because the peaks are too close together to be resolved. Also it gives a

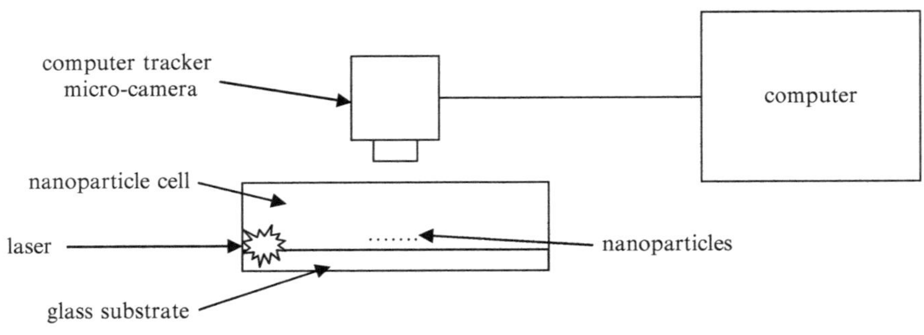

Fig. 8.17 Schematic diagram of the Nanosight equipment showing the laser illuminating the nanoparticles near the coated glass surface, the movie camera capturing the scattered light to give random walk tracks which are analysed by the computer

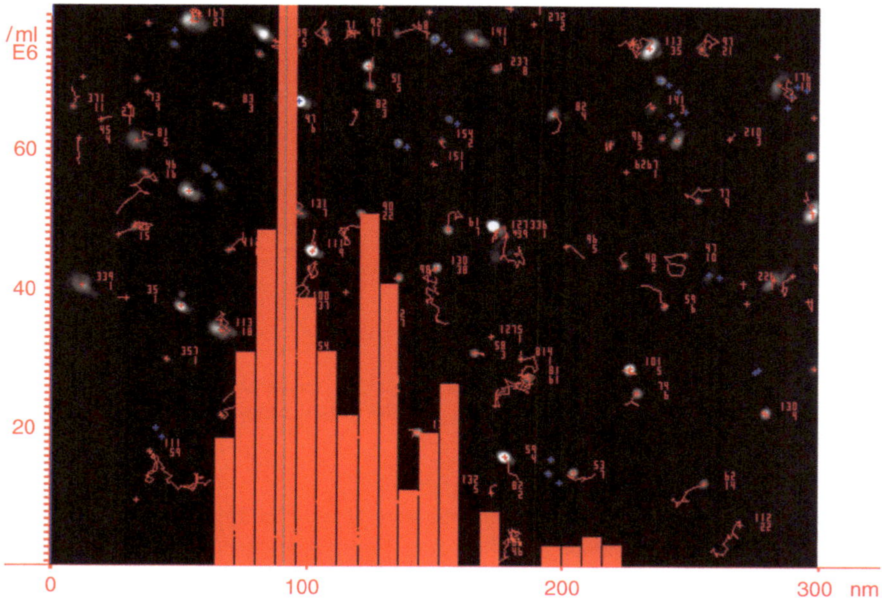

Fig. 8.18 Still from Nanosight movie of 62 nm hematite pH2, $\phi = 2$ ppm showing nanoparticle tracks with histogram of particle sizes superimposed

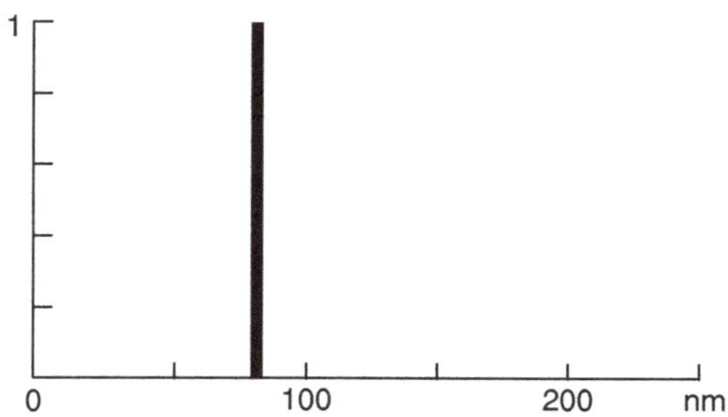

Fig. 8.19 Dynamic light scattering results for 62 nm diameter hematite

slightly larger particle diameter than that indicated by TEM because it is averaging the results of aggregated nanoparticles.

The real nature of the dispersion was shown by depositing the dispersed nanoparticles in fine aerosol form onto carbon grids and examining them in the transmission electron microscope (TEM) as shown in Fig. 8.20a. TEM showed clearly that, although most of the particles were singlets, there was a range of doublets, triplets, etc. up to larger aggregates of 16 particles. By counting the distribution of aggregates in the micrographs and comparing with the numbers measured in the Nanosight instrument, Fig. 8.20b showed that the TEM and Nanosight results were in reasonable agreement, but slightly lower than the off-lattice model prediction based on a square well interaction

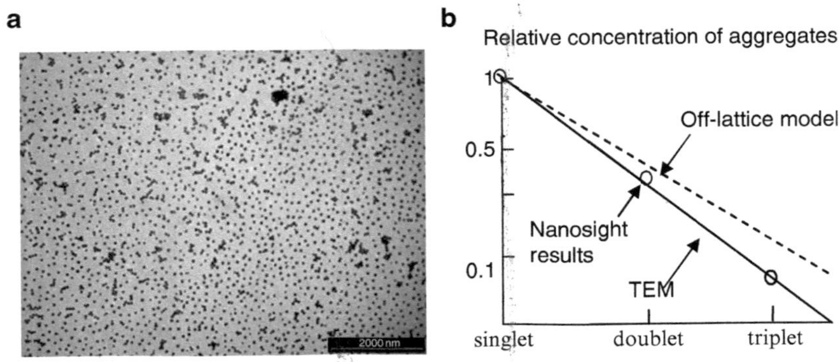

Fig. 8.20 (**a**) Transmission Electron Microscope (TEM) image showing singlets, doublets, triplets and larger aggregates. (**b**) Comparison of TEM results, nanosight results and off-lattice model

potential.[40] These results show that the number of doublet aggregates can be measured by this new laser scanning method in real time in real environments in contrast to DLS which cannot distinguish closely spaced aggregate peaks.[41]

8.10 Adhesion from Doublet Numbers

If the number of doublets is measurable, then the particle adhesion energy may be calculated from an aggregation theory. To test this idea, the Nanosight method was applied to dispersions of 100 nm diameter polystyrene (PS) latex particles as shown in the SEM pictures of Fig. 8.21. Again, these dispersions were found to contain singlets and doublets indicating that small but significant attractive forces were causing aggregation.[46]

The counts of the aggregates obtained from the SEM pictures corresponded reasonably to the measurements made in the Nanosight instrument as shown by the plots in Fig. 8.22 from singlets to quadruplets.

These results allowed an estimate to be made of the attractive forces holding the nanoparticle aggregates together. The ratio of doublet to singlet numbers was measured for PS and plotted as a function of particle concentration in Fig. 8.23. The results for doublet numbers gave nearly a straight-line dependence on particle concentration (volume fraction) as expected from the statistical mechanics theory originally described in 1998[42–45] i.e.

$$\frac{N_2}{N_1} = 4\phi\{[\frac{(z+r)}{r}]^3 - 1\}\exp(\frac{\varepsilon}{kT}) \tag{8.2}$$

based on a square well model of the particle interaction potential where N_2 is the number of doublets, N_1 the number of singlets, ϕ is the volume fraction of particles, ε is the adhesion energy, and z is the width of the square well added to particle radius r.

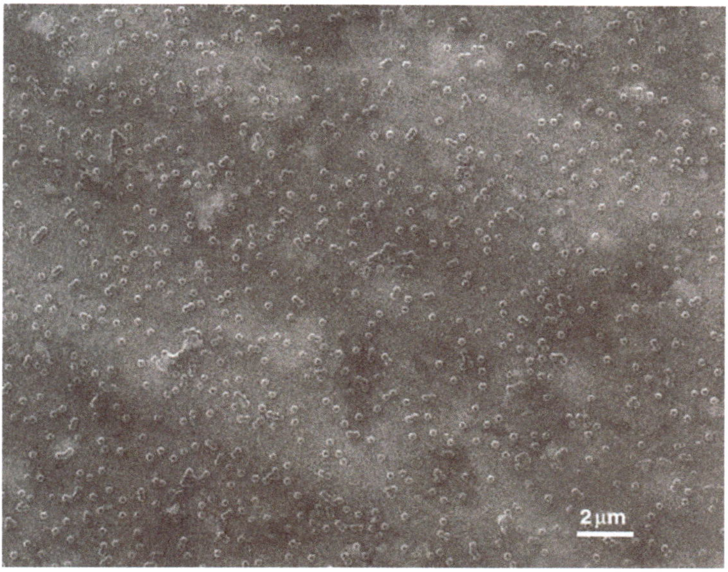

Fig. 8.21 Scanning electron micrograph of 100 nm diameter polystyrene particles[46]

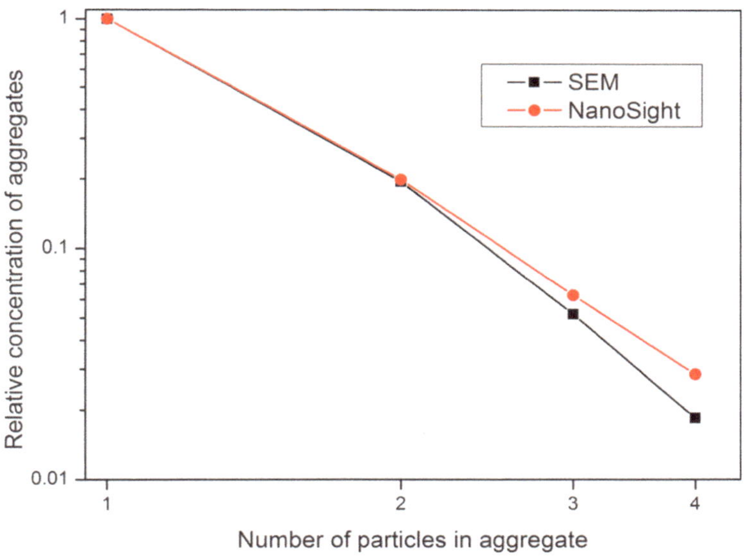

Fig. 8.22 Comparison of SEM and nanosight results for 100 nm diameter polystyrene particles (pH = 5.6, $T = 25°C$)

The experiment was repeated on four different sizes of the polystyrene latex to measure the adhesion energy ε and the extent z of the attraction. The results shown in Fig. 8.23 suggest an adhesion energy $\varepsilon = 2\ kT$ and $z = 600$ nm, assumed constant for all four particle diameters.[46] The experiments gave fair agreement with the

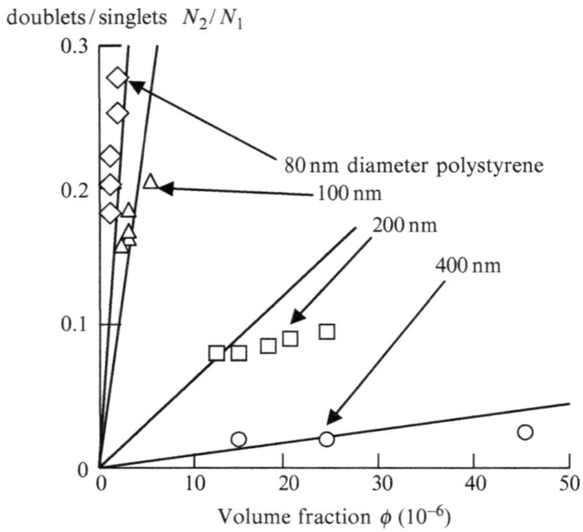

Fig. 8.23 The ratio of doublet to singlet numbers as a function of particle concentration for polystyrene spheres with various particle sizes for comparison with the lines plotted from Eq. 9.2 with $z = 600$ nm and $\varepsilon = 2\ kT$[46]

theoretical model, giving an approximate description of the forces holding polystyrene aggregates together in well dispersed colloids, although the detailed shape of the attractive potential curve is not known. It is interesting that the adhesion gap z is so large, suggesting that the long range electrical forces are dominant.

8.11 Detection of Adhesion with FemtoNewton Resolution

It is also clear from the results that very small forces ε/z are being detected directly in this experiment, since 2 kT/600 nm leads to a force estimate of 10 fN. This is very much smaller than has ever been detected previously with STM and AFM.

Previous measurements of the interactions between spherical particles have been of two kinds: optical measurements of the movement of free well-separated μm scale spheres moving with Brownian dynamics in an optical trap,[47–49] which allows the definition of the potential versus separation curve with pN resolution; and AFM which requires fixed spheres attached to surfaces that can be moved together controllably while monitoring the interaction force with a sensor,[50–52] typically with 100 pN resolution. Both methods require large spheres and neither can observe the particles moving from attached to free. The Nanosight method allows measurements of nanoparticles both in contact and freely diffusing and so covers the whole potential curve, but requires more than one particle diameter to extract the two parameters of energy and range. This could have significant application in a range of applications of nanoparticles from diagnostics to electronics to virus studies as described in the next chapter.

8.12 Conclusions

Nanoparticles have been detected in many situations; in space, the atmosphere and the oceans. The two main sources are nucleation from the gas phase and precipitation in liquids. Aggregation of the particles occurs even when adhesion is weak to form doublets, triplets, etc. in a gas-like dispersed state. However, stronger adhesion causes a phase change to agglomerates which can be fluffy and open for high adhesion greater than 10 kT or compact and dense when adhesion is lower, around 1 kT. These agglomerates sediment quickly and are removed.

Man-made nanoparticles have been used for millennia in construction of adobe and concrete, in making Indian ink, and in fashioning films from latex, utilising the natural adhesion of the nanoparticles to each other and to neighbouring surfaces. But now that epidemiology has shown diseases caused by nanoparticles, especially in cities where vehicle emissions are prevalent, regulations have been imposed to limit their concentrations. The mechanisms of toxicity will be described in Chapter 11.

Light scattering has been used to detect nanoparticles but is deficient in measuring small aggregates such as doublets and triplets caused by adhesion and readily seen by electron microscopy. These aggregates have now been detected online by a new method of laser tracking. This new technique has now been used to measure extremely small adhesion between nanoparticles.

References

1. Grün, E., Baguhl, M., Svedhem, H., Zook, H.A.: In situ measurements of cosmic dust. In: Grün E. (ed) Interplanetary Dust, pp. 295–346. Springer, London and Berlin (2001).
2. Kendall, K., Molecular Adhesion and its applications, Kluwer, New York 2001.
3. Kroto, H. W., Heath, J. R., Obrien, S. C., Curl, R. F., Smalley, R. E., C60: Buckminsterfullerene, Nature 318 (1985) 162–163.
4. Srama, R. et al, Sample return of interstellar matter, Experimental astronomy, (2008) 10.1007/s10686–008–9088–7.
5. Altobelli, N., E. Grün, E. and Landgraf, M., A new look into the Helios dust experiment data: presence of interstellar dust inside the Earth's orbit, Astronomy and Astrophysics 448 (2006) 243–252.
6. Kearsley, A.T., Graham, G.A., Burchell, M.J., Taylor, E.A., Drolshagen, G., Chater, R.J., McPhail, D., Mulpex: A compact multi-layered polymer foil collector for micrometeoroids and orbital debris, Advances in Space Research 35 (2005) 1270–81.
7. Zolensky, M.E. et al Mineralogy and Petrology of Comet 81P/Wild 2 Nucleus Samples, Science 314 (2006) 1735–1739; http://stardust1.jpl.nasa.gov/photo/aerogel
8. Peters A. et al., Respiratory effects are associated with the number of ultrafine particles. Am J Resp Crit Care Med. 155 (1997) 1376–1383.
9. Peters, A., Dockery, D.W., Muller, J.E., Mittleman, M.A., Increased particulate air pollution and the triggering of myocardial infarction. *Circulation* 103 (2001) 2810–2815.
10. Aitken, J., On dust, fogs, and clouds. Transactions of the Royal Society of Edinburgh 30 (1883) 337–368.

11. Morawska, L., Ristovski, Z., Jayaratne, E.R., Keogh, D.U., Ling, X., Ambient nano and ultrafine particles from motor vehicle emissions: Characteristics, ambient processing and implications on human exposure Atmospheric Environment 42 (2008) 8113–8138.
12. Dunn, M.J., Jimenez, J.L., Baumgardner, D., Castro, T., McMurry, P.H., Smith, J.N., Measurements of Mexico City nanoparticle size distributions: Observations of new particle formation and growth. Geophysical Research Letters, 31 (2004) L10102.
13. Seinfeld, J.H., and Pandis, S.N., Atmospheric chemistry and physics. From air pollution to climate change, John Wiley and Sons, inc., New York 1998.
14. Shi, J.P., Evans, D.E., Khan, A.A., Harrison, R.M., Sources and concentration of nanoparticles in the urban atmosphere, Atmospheric Environment 35 (2001) 1193–1202.
15. Wahlin, P., Measured reduction of kerbside ultrafine particle number concentrations in Copenhagen, Atmospheric Environment, 43 (2009) 3645–3647.
16. Ayres, J.G., (chairman) Long-term exposure to air pollution: effect on mortality, COMEAP report (2009) ISBN 978–0–85951–640–2.
17. Dockery, D.W., Pope, C.A., Xu X, Spengler, J.D., Ware, J.H., Fay, M.E., Ferris, B.G., Speizer, F.E., An Association between Air Pollution and Mortality in Six US Cities, New England Journal of Medicine, 329 (1993) 1753.
18. World Health Organisation (WHO). 2006. Air quality Guidelines for particulate matter, ozone, nitrogen dioxide and sulphur dioxide. Global update 2005. Summary of risk assessment. WHO; Oberdorster, G. et al. 2005. Nanotoxicology: An emerging discipline evolving from studies of ultrafine particles. Environ Health Perspec, 113 (2005) 823–839.
19. Christian, P., Von der Kammer, F., Baalousha, M., Hofmann, T., Nanoparticles: structure, properties, preparation and behaviour in environmental media, Ecotoxicology (2008) 10.1007/s10646–008–0213.
20. Baan, R., et al. Carcinogenicity of carbon black, titanium dioxide, and talc. The Lancet Oncology. Vol. 7 (Apr. 2006). P. 295–296.
21. Park, E., Yoon, J., Choi, K., Yi, J., Park, K., Induction of chronic inflammation in mice treated with titanium dioxide nanoparticles by intratracheal instillation, Toxicology 260 (2009) 37–46.
22. Kendall, K., Kosseva, M.R., Adhesion of nanoparticles fouling glass surfaces, J Adhesion 81 (2005) 1–14.
23. Kendall, K., Kosseva, M.R., Nanoparticle Aggregation influenced by Magnetic Fields. Colloids & Surfaces A: Physiochemical & Engineering Aspects, 286 (2006) 112–116.
24. Baalousha M, Manciulea A, Cumberland S, Kendall K, Lead JR., Aggregation and surface properties of iron oxide nanoparticles: influence of pH and natural organic matter. Environ Toxicol Chem 27 (2008) 1875–1882.
25. Giammar DE, Maus CJ, Xie L., Effects of particle size and crystalline phase on lead adsorption to titanium dioxide nanoparticles. Environ Eng Sci 24 (2007) 85–95.
26. Hofmann, German patent 250690 to Bayer Farbenfabriken (1913).
27. Kendall, K. and Padget, J.C., Int J Adhesion and Adhesives 2 (1982) 149–54.
28. Needham, J., Science and Civilisation in China, Cambridge University Press, Cambridge, 1985, pp.239–47.
29. Bye, G.C., Portland Cement, 2nd Edition, Thomas Telford, 1999, ch. 1
30. Kendall, K., Howard, A.J., Birchall, J.D., The relation between porosity, microstructure and strength, and the approach to advanced cement-based materials, Phil Trans R Soc Lond A310 (1983) 139–153.
31. Witynski, K and Carr, J.P., Adobe details, Gibbs Smith, PO Box 667, Layton UT, 2002.
32. Kang, S-J, Sintering: Densification, grain growth and microstructure, Elsevier Oxford 2005.
33. Friedlander, S.K., Smoke, dust and haze: Fundamentals of aerosol behaviour, 2nd edition, Oxford University Press, 2000.
34. Matijevic, E., Scheiner, P., J Colloid Interface Science 63 (1978) 509–524.
35. Kendall, K., Amal, R., Jiang, X. and Yu, A., Effect of adhesion on aggregation in nanoparticle dispersions, J Adhesion 83 (2007) 573–585.

36. Carr, R., Hole, P., Malloy, A., Weld, A., Smith, J. & Warren, J. The real-time, simultaneous measurement of size, surface charge and fluorescence of populations of nanoparticles in liquids. Particle Systems Analysis, *2008*, Stratford on Avon UK, pp1–5.
37. Carr, R., Weld, A. & Smith, J. Seeing and Sizing Nanoparticles in Liquids: Multi-particle Tracking of Brownian Motion. Development and Applications of Nanotechnology and Microscopy, Pittsburg Conference(PITTCON) (3rd April 2008).
38. Malloy, A. & Carr, R. Nanoparticle tracking analysis - The Halo (TM) system. Part. Syst. Charact. 23 (2006) 197–204.
39. Carr, R. Patrick Hole and Andrew Malloy. 8th International Congress on Optical Particle Characterisation, Karl-Franzens University Graz, Austria (9–13 July 2007).
40. Babu, S., Gimel, J. C. & Nicolai, T. Phase separation and percolation of reversibly aggregating spheres with a square-well attraction potential. J. Chem. Phys. 125, (2006) 184512 1–10.
41. Dyuzheva, M. S. & Klyubin, V. V. Measurement of continuous particle size distributions of finely dispersed powders by the dynamic light scattering. Colloid J. 65, (2003) 567–570.
42. Kendall, K., Liang, W. & Stainton, C. New theory and observations of cell adhesion. Proc. R. Soc. Lond. A 454, (1998) 2529–2533.
43. Kendall, K. & Stainton, C. Adhesion and aggregation of fine particles. Powder Techno. 121, (2001) 223–229.
44. Stainton, C., Liang, W. & Kendall, K. Formation and fracture of adhesive bonds between colloidal spheres. Eng. Fract. Mech. 61, (1998) 83–91.
45. Liang, W. & Kendall, K. Aggregate formation in colloidal dispersions. Colloids Surfaces A 131 (1998) 193–201.
46. Kendall, K., Dhir, A., Du, S., A new measure of molecular attractions between nanoparticles near kT adhesion energy, Nanotechnology 20 (2009) 0275701.
47. Polin, M., Roichman, Y. and Grier, D.G., Autocalibrated colloidal interaction measurements with extended optical traps, Phys Rev E77 (2008) 051401–17.
48. Roberts, G.S., Wood, T.A., Frith, W.J. and Bartlett, P., Direct Measurement of the effective charge in nonpolar suspensions by optical tracking of single particles, J Chem Phys 126 (2007) 194503–1–12.
49. Sainis, S.K., Germain, V. and Dufresne, E.R., Statistics of particle trajectories at short time intervals reveal fN-scale colloidal forces, Phys Rev Lett 99 (2007), 018303–1–4.
50. Gunning, A.P., Chambers, S., Pin, C., Man, A.L., Morris, J. and Nicoletti, C., Mapping specific adhesive interactions on living human intestinal epithelial cells with AFM, FASEB 22 (2008) 2331–9.
51. Goncalves, R.P. and Scheuring, S., Manipulating and imaging individual membrane proteins by AFM, Surface Interface Anal, 38 (2008) 1413–8.
52. Thornton, J.T., Microscopy Microanalysis, Applications of AFM in the pharmaceutical sciences, 8 (2002) 742–743.

Chapter 9
Adhesion of Viruses

surface of structures...serve as a base upon which the cells might creep

Harrison (1914)

Viruses exist in a wide spread of varieties with interesting mechanisms for adhering to each other, to neighbouring particles and particularly to living cells which they target by specific means. If this adhesive targeting mechanism changes slightly, then different cells, for example human rather than bird in the case of avian influenza, may be singled out for virus attachment and new infections can result, causing dangerous disease pandemics.

Each virus particle, called a virion, may be viewed as a polymer nanoparticle, rather like polystyrene but containing more complex molecules which can assemble and break up depending on the environment. These polymeric virus nanoparticles are normally not very sticky in water and can be observed oscillating in Brownian motion, forming small numbers of doublets and triplets which are a measure of the low adhesion between the virions. This self-adhesion can be varied by altering the chemical conditions of the suspension, for example pH, salt and polymer concentration.

Adding polymer molecules to the suspension can increase the adhesion by the depletion effect and crystals of virus particles may solidify out of the liquid. The way in which the free-floating viruses solidify into crystals is another measure of the weak van der Waals bonding between the particles.

By adding other particles or cells to the virus suspension, the virus adhesion to other materials process can be investigated to give a precise measure of the adhesion process, especially the adhesion energy and range of attraction. This adhesion depends greatly on the surfactant molecules in solution within the dispersion. The presence of other particles may influence the virus-cell adhesion process and give complex behaviour. More subtle adhesion effects, such as release of budding viruses from infected cell surfaces, have been used to devise new drugs such as Tamiflu, while other phenomena like assembly of virions within cells require further elucidation.

K. Kendall et al., *Adhesion of Cells, Viruses and Nanoparticles*,
DOI 10.1007/978-90-481-2585-2_9,

9.1 Variety of Virus Particles

There is an enormous variety of viruses affecting humans. More than half of US adults are infected with cytomegalovirus (CMV) which enlarges cells without much damage, whereas other dangerous viruses e.g. HIV AIDS need to be kept to a low level, with 3.1million deaths worldwide in 2004.[1] The question is 'how does each of these particles adhere to and infect cells?'

Virus particles can be found in a number of geometries including spherical, rod-like and head-tail as shown in Fig. 9.1. Typically they are about 100 nm in size but can range from 10 to 200 nm, depending on the species.[1, 2] Here we focus mainly on spherical viruses because these most resemble the inanimate particles described in Chapter 8.

A virus is viewed as a complex and unique type of organism which may be classified by a taxonomy system overseen by the 'International Committee on Taxonomy of Viruses' (ICTV) that identifies them into more than 5,000 different particle types, at present comprising 3 orders, 73 families, 9 subfamilies, 287 genera and 1,938 species.[1] Current information about the database (ICTVdb) is available on websites.[3–5]

The function of viruses is to transfer their reproductive molecules into cells where they can replicate. Consequently virus particles are much smaller than cells which have to contain them. Viral essential components are nucleic acid (NA) which comes in two forms, deoxyribonucleic acid (DNA) and ribonucleic acid (RNA), each of which can be double strand (DS) or single strand (SS). Additionally, the RNA can be positive sense (same as messenger RNA, mRNA) or negative (complementary to mRNA). These molecules are encased in a capsid, an assembly of protein molecules or capsomeres, which usually give a regular geometric structure. In addition, some virus particles are enclosed in a bilayer lipid envelope which is obtained from the host cell when the virus is released, making the particle larger. On the membrane surface are glycoprotein molecule spikes as shown below for a typical virus particle. Some of these are the adhesion molecules which appear to influence the targeting of specific cells.

One of the larger spherical virus particles is Herpes, which was known to Hippocrates because of the creeping skin lesions (Greek, herpein; to creep) and was first distinguished in transmission electron microscope pictures in 1953.[6] Its structure[7] obtained from cryo-electron microscopy is shown schematically in Fig. 9.2.

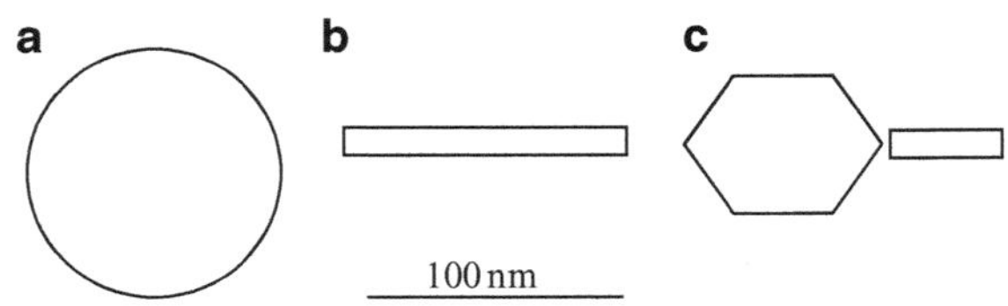

Fig. 9.1 Geometries of virus particles; (**a**) sphere e.g. herpes; (**b**) rod e.g. tobacco mosaic; (**c**) head-tail e.g. phage

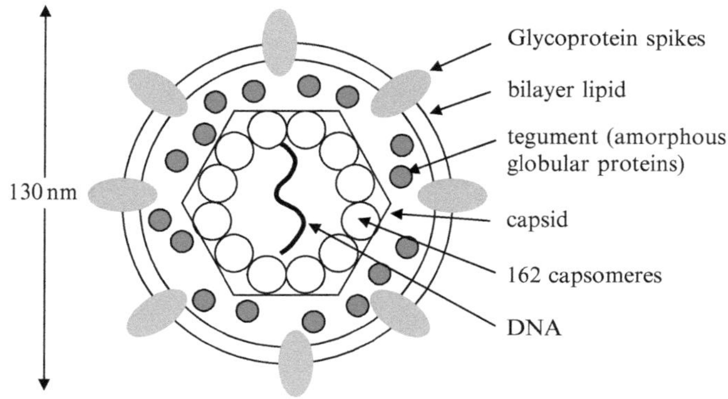

Fig. 9.2 Schematic structure of Herpes virus particle[7]

At the centre of the particle is a single strand of DNA. This is surrounded by 162 capsomeres, approximately spherical protein particles arranged in an icosahedral structure. Outside this is the tegument of amorphous globular proteins, enveloped in the bilayer membrane, studded with a number of glycoproteins which stick out like spikes from the particle.

9.2 Observing the Adhesion of Viruses by TEM & X-Ray

Electron microscopy has been the major technique for observing virus particles in contact with cells, ever since the first transmission electron microscope (TEM) images from 1939[8] showed tobacco mosaic virus particles. TEM gave a significant indication of virus particles adhering both to cell walls and to themselves to form crystalline arrays as in Fig. 9.3.

Stanley[10] in 1935 had originally observed X-ray patterns showing crystalline tobacco mosaic virus (TMV) and received the Nobel Prize in 1946. He had interpreted the patterns as protein but they also contained nucleic acid and were actually virus crystals. The rod-like virus particles could readily be seen like log-rafts in the TEM.

The crystallisation of satellite tobacco mosaic virus (sTMV) particles, which are spherical, can now be studied using atomic force microscopy and the growth of the ordered layers measured under various conditions of supersaturation and temperature.[11] Clearly, the self adhesion of the virus particles is very small, around 3 $kT/2$ the thermal energy, otherwise the particles would aggregate into random structures as the particles adhered instantly. It was estimated from the crystal growth behaviour of the sTMVs that the energy of the crystal edge was 0.26 mJm^{-2}, twenty times less than typical hydrocarbon spheres adhering in water.

Many types of virus particles have been crystallised since the 1950s enabling much information to be gathered about the structures and the chemistry of crystalline layers. An interesting study was by Casselyn et al. on Brome mosaic virus.[12] A dispersion of brome mosaic virus (BMV) particles was purified from infected barley leaves by

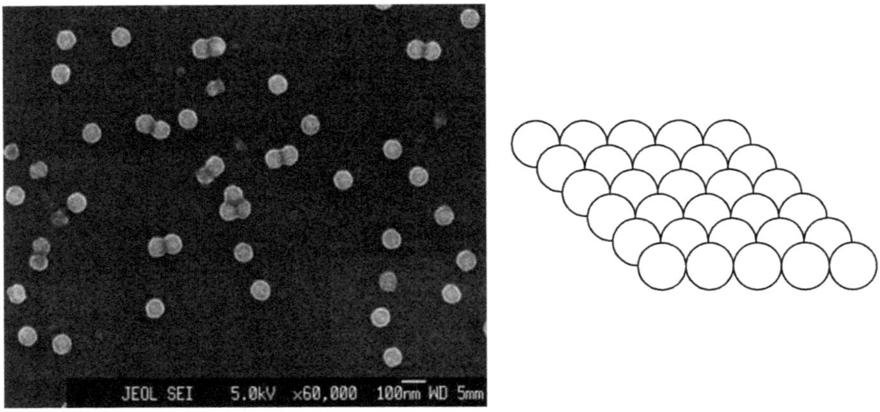

Fig. 9.3 *left*; Scanning electron micrograph of adenovirus showing particles adhering to form doublets and triplets[9] (with permission). *Right*; crystalline array of spherical viruses forming an adherent raft on a surface

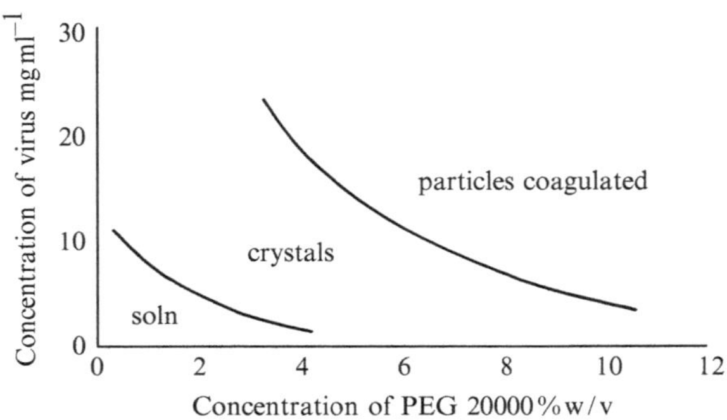

Fig. 9.4 Phase diagram [13] of virus structure in buffer as a function of BMV concentration and PEG 20,000

centrifugation and concentrated at 40 mg/ml in 20 mM sodium acetate buffer at pH 5.9. Three polyethylene glycols (PEGs) of molecular weights 3,350, 8,000 and 20,000 Da were prepared in solutions at 5%, 10% and 20% in the same buffer. The virus sample was then rapidly mixed with PEG and buffer in about 10 ms using a stopped flow device to provide a range of virus/PEG concentrations. Crystallisation was observed by small angle X-ray scattering (SAXS) to see the nucleation and growth of structure. Low concentrations of virus and PEG did not nucleate crystals, but crystals appeared at higher PEG, with high molecular weight being more effective. The virus crystals were face centred cubic (FCC) with a unit cell size of 39.1 nm. There was a sharp transition in the phase diagram from solution phase virus to crystal phase with no amorphous material (Fig. 9.4 left) but at high concentrations, there were no crystals but only amorphous solid deposits, showing that the attractive virus forces were high.

The critical nucleus size was 36 virus particles and the critical activation energy 65 kT. It is evident from these experiments that the virus particles were all exactly the same diameter and grew into crystals like polystyrene latex which also gives face centred cubic (FCC) crystals at the phase boundary. The PEG molecules were acting to increase the adhesion attraction between the virus particles by causing a depletion effect between the viruses.

A more complex phase diagram[13] was discovered for tomato bushy stunt virus (TBSV) which was encouraged to crystallise by adding either ammonium sulphate or PEG8000. Three morphologies of crystal were seen, the most common being the body centred (BCC) structure with 38.3 nm unit cell parameter. TBSV was the first icosahedral virus to be crystallised[14] and those early X-ray investigations had led to the understanding of the virus architecture by 1970.[15]

The problem with both X-ray and electron microscope results is that, although the virus particles can be seen adhering to themselves and to cells, there is no information on the adhesion forces. However, some structural information about the surface adhesive interactions and mechanisms can be gleaned from TEM as described next.

9.3 TEM Investigations of Adhesion Mechanism

Transmission Electron Microscopy (TEM) of viruses has improved to such a degree that our knowledge of the surface binding structures increased substantially.[16] Together with realistic docking models based on molecular structures, this endeavour suggests that better understanding of virus adhesion is appearing. Analysis of the pictures reveals the nature of the virus particle and the structure of the surface active molecules which cover the virus surface as shown below.

Figure 9.5a shows a stereo image of a feline calicivirus (FCV) particle reconstructed from 90 TEM defocus paired images and 1117 particle images by computer analysis described by Bhella et al.[16] The pictures show the classic structure of FCV with rhomboid arch shaped dimeric capsomeres at the icosahedral twofold symmetry axes (C-C dimers) and about the 5-fold axes (A-B dimers).

Several papers[17, 18] have described the feline junctional adhesion molecule 1 (fjAM-1) as being important to calicivirus infection. These were produced from RNA of Crandell Reese feline kidney cells and amplified by reverse transcription PCR. The product was used to generate a eukaryotic expression plasmid containing the fjAM-1 molecules which were then separated and purified. After that, the adhesion molecules were attached to the virus particles by incubating overnight at 4°C. To produce vitrified samples for electron microscopy (Fig. 9.5b) the samples were plunged into liquid ethane cooled by liquid nitrogen, then viewed at low temperatures on a cryo-stage giving a pixel size of 0.22 nm. The larger structure of the coated virus was evident with two FjAM-1 molecules adhering to each capsomere in a head to tail arrangement, one lying flat on the virus and the other sticking out. The two molecules could be fully distinguished at this resolution, possibly due to steric overlap.

Figure 9.5c is interesting because it is a difference map between A and B which reveals movements in the capsids beneath the adhesion molecules. This suggests that the adhesion molecules have changed the conformation of the virus significantly. By considering higher resolution analysis of the molecules in contact,

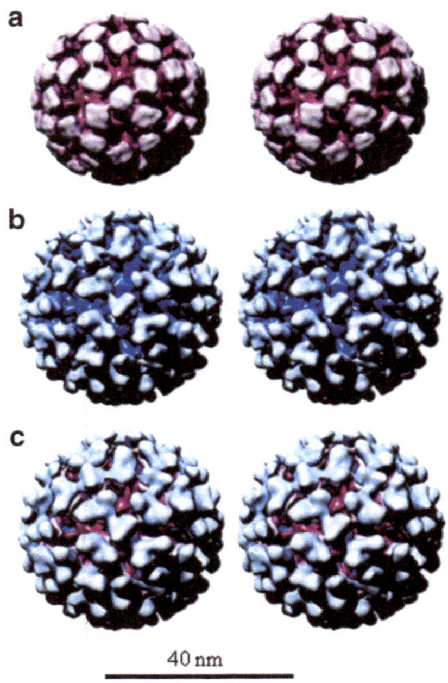

Fig. 9.5 Stereo pair images of three-dimensional reconstructions of (**a**) unlabelled FCV virions 1.6 nm resolution. (**b**) fJAM-1-labelled FCV virions viewed along the icosahedral twofold symmetry axes. (**c**) A difference map highlights additional density attached to the P2 domain of the labelled reconstruction. Difference density is also visible in the P1 domain, indicating changes in capsid conformation induced by fJAM-1 binding (with permission)

Bhella et al. were able to propose atomic structures for the fJAM-1 adhesion molecule and the underlying virus protein 1 (VP1) capsomere molecule (Fig. 9.6).

By constructing these models for the contact between the cell surface molecules i.e. receptors and the virus capsid protein VP1, it is possible to speculate on the specific amino acids which are important to the binding process and to the infectivity of the virus.[18] Understanding these structures is important in defining the molecular basis of infection. However, the problem is that the forces and the energetics cannot be measured by these static methods. Other methods are now emerging to measure the adhesive forces and energetics, as described next.

9.4 Atomic Force Microscope (AFM) Studies

A neat way to study the adhesion of virus crystals is the AFM which probes the surface with a sharp tip to give nanometre resolution. The benefits of this technique are twofold: first the viruses are imaged in realistic wet situations, avoiding the problems of dried out, high vacuum TEM; second, the AFM can measure forces on the virus to understand how it infects and releases the genetic material from inside the capsid.

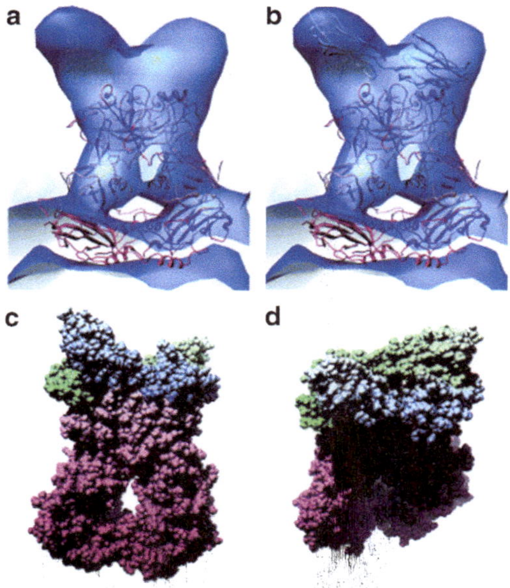

Fig. 9.6 Docked homology models for [16] (**a**) FCV VP1 (*pink ribbon*). (**b**) fJAM-1 (*blue ribbon*) in the 18-Å resolution reconstruction for FCV labelled with fJAM-1 (*transparent blue surface*). (**c**, **d**) Atomic-resolution representation of the modelled structure for a FCV A-B dimer of VP1 (*pink spheres*) labelled with soluble fJAM-1 (*blue and green spheres*) (with permission)

A typical picture is shown in Fig. 9.7, illustrating the comparison between TEM and AFM results on Herpes Simplex virus (HSV), one of the most common human pathogens.[19]

Figure 9.7a shows a TEM picture of HSV-1 capsids after treatment with 0.2% Triton X-100 detergent which removes the outer membrane to reveal the underlying protein structures. Figure 9.7b shows the same capsids adhering to poly-L-lysine coated mica in buffer solution revealing the uniform, spherical form of the viruses. The hexagonal stacking of the capsomeres is demonstrated in Fig. 9.7c and the diameter determined from the height at 123 nm in Fig. 9.7d, agreeing with cryo-TEM observations.[20] At higher magnification and slower scanning speed, the hexagonal structure, the triangular faces, individual capsomeres, pentons and hexons were revealed (E,F,G,H).

These results have proved that the original invention of the AFM by Binnig et al.,[21] leading the way to his Nobel prize in 1986, has opened the way to see viruses in realistic environments[22] while probing their structures, mechanical properties and also the changes which occur when other surfaces and proteins such as adhesion molecules are present. A typical finding is that certain surfaces attract and adsorb the virus particles whereas others cause the viruses to break up.[19] For example, graphite (highly oriented pyrolytic graphite) allows the viruses to adhere intact in the buffer solution, as does poly-lysine-coated mica, whereas poly-lysine-coated glass breaks up the virus and leaves 40 nm spheres, presumably the capsomeres, plus DNA coils attached to the surface (Fig. 9.8). This result is very significant

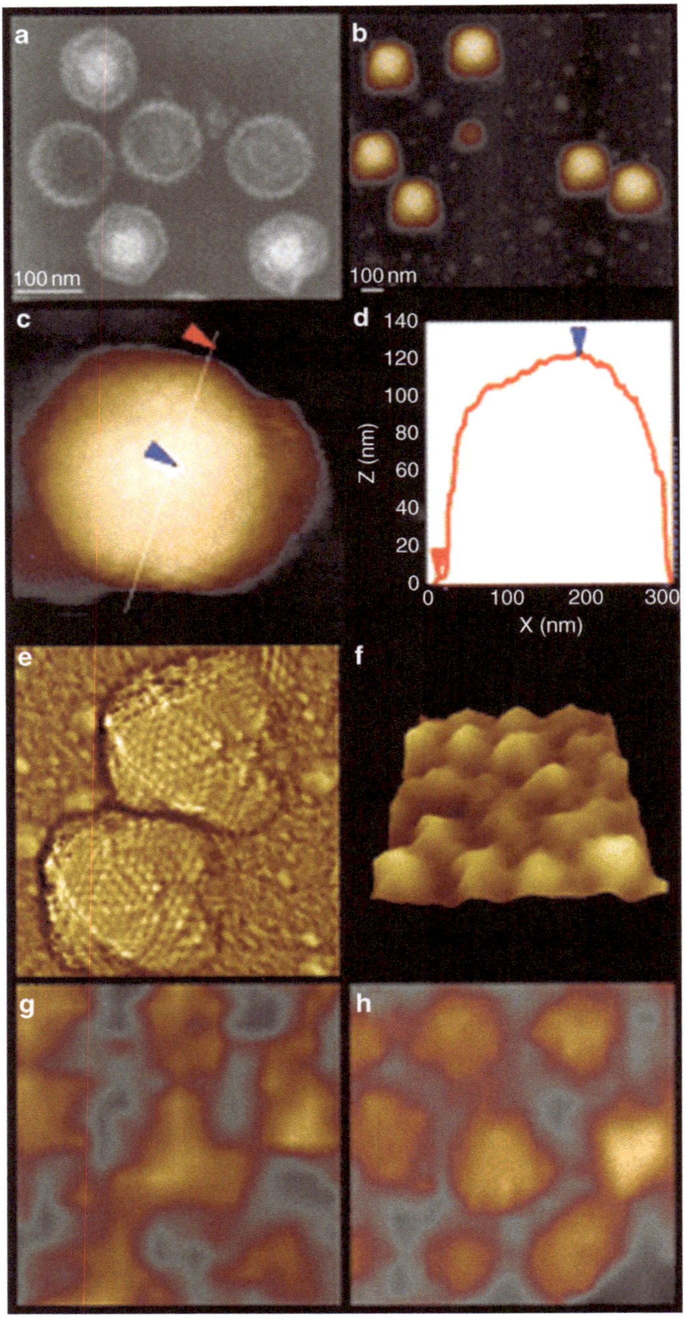

Fig. 9.7 Shape and surface topography of the HSV-1 capsids studied by EM and AFM.[19] (**a**) EM image of the capsid preparation. (**b**) HSV-1 capsids adsorbed to the surface of poly-L-lysine-coated mica (1.60×1.38 µm). (**c**) A single capsid with the typical hexagonal outline. (**d**) Cross-section of a capsid showing its height of 123 nm. (**e**) Two adhering capsids with apparent icosahedral symmetry and clearly distinguishable triangular faces (510×560 nm). (**f**) A fragment of the HSV-1 surface with clearly distinguishable capsomeres (100×100 nm). (**g**) A penton surrounded by five nearest neighboring hexons. (**h**) A hexon. Pentons and hexons images are 40×40 nm (with permission)

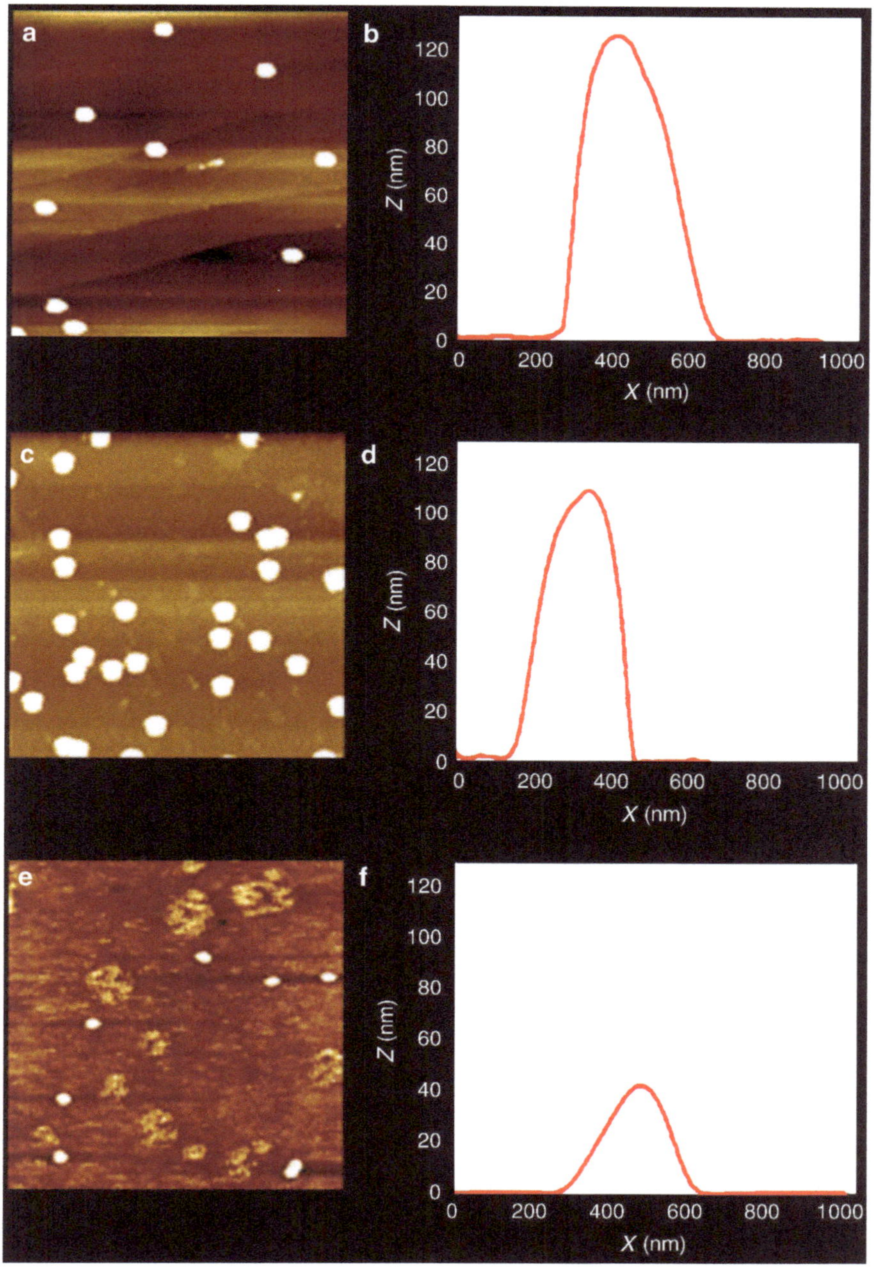

Fig. 9.8 AFM analysis of HSV-1 capsids on three surfaces,[19] (**a**) graphite; (**c**) poly-lysine coated mica; (**e**) poly-lysine-coated glass; (**b**, **d**, **f**) cross-sections of the particles (with permission)

because it shows the adhesion molecule (in this case poly-L-lysine) is not the only determinant of virus adhesion; the underlying substrate is crucial.

The AFM probe was also used to test the mechanical compression of the virus as Tan et al. had done on polymer spheres in 2004.[23] By squeezing 40 separate virions and plotting the force deflexion curves, the stiffness was calculated to be 523 pN/nm, which decreased to 356 pN/nm once the DNA was released by adding 1 M guanidine hydrochloride solution. The capsids were found to be exceptionally strong, recovering elastically from forces up to 7 nN. Above this limit, the capsids suffered irreversible mechanical failure, leaving individual capsomeres stuck to the substrate. The significance of these properties to delivery mechanisms has been discussed by Roos et al.[24] It appears that the virus is strong because the DNA is packed inside to crystalline density and this genetic material exerts an outward pressure on the protein capsomeres which self-assemble around it. Presumably, this high strength allows the capsid to be handled by molecular motors within cells until the release and delivery of the DNA is chemically stimulated.

The AFM could also be used in principle to measure the adhesion of the viruses by scraping the surfaces to find the detachment conditions but this has not yet been achieved. However, sensor methods have been applied to this problem.

9.5 Sensor Methods for Detecting Adhesion

A number of physical measurement techniques have been applied to detect virus and other particles adhering specifically to surfaces, including quartz crystal microbalance[25,26] (QCM), surface plasmon resonance,[27] and capacitative methods.[28]

Figure 9.9 shows the mode of operation of the QCM instrument.[25] The transducer surface is coated with receptor adhesion molecules to which the target virus binds. Virus particles accumulate on the surface but there is also some non-specific binding by other particles, for example leucocytes, which are only weakly held. The surface is oscillated at increasing amplitude until bond rupture occurs and the falls in the adsorbed mass together with bond-rupture noise peaks are measured.

Because the weakly bonded material falls off at lower amplitude, the specifically bonded viruses can be measured, with the peak heights giving the concentration of the virus particles. A typical adhesion molecule couple is biotin-streptavidin, much studied and widely available, giving a quantifiable strong bond which is easy to store and safe in application.[29] Experiments using polymer latex and microbeads showed that the specific adhesion could be detected by the quartz microbalance method.

Herpes Simplex virus (HSV) was tested using this method by Cooper et al.,[30] where he termed it Rupture EVent Scanning (REVS). The experiment is depicted in Fig. 9.10 which shows an AT cut quartz crystal of diameter 8.25 mm to which an electric field was applied using gold electrodes adhering via chromium interlayers. On the surface of one gold electrode, a Self Assembled Monolayer (SAM) of mercapto alkane was deposited, then the desired adhesion molecules were attached

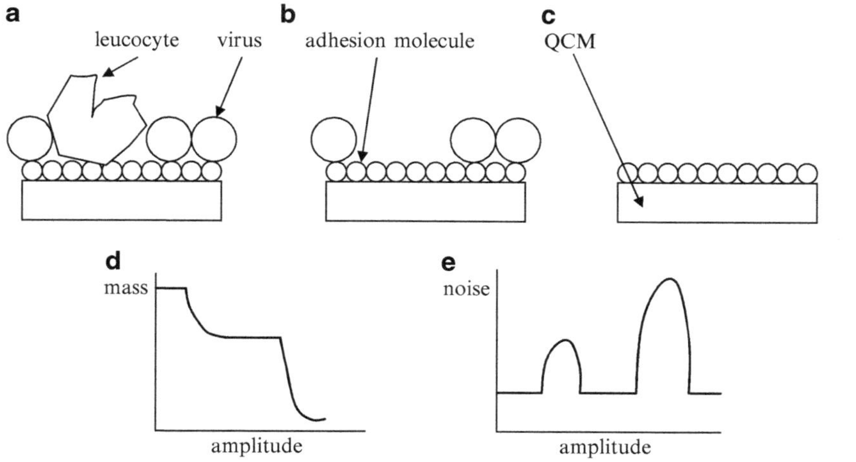

Fig. 9.9 Mode of operation of the quartz crystal microbalance virus sensor. (**a**) QCM coated with receptor molecules, with adhering viruses and other particles; (**b**) removal of non-specific bonded material at low amplitude; (**c**) removal of viruses; (**d**) mass change; (**e**) noise results from bond breakage

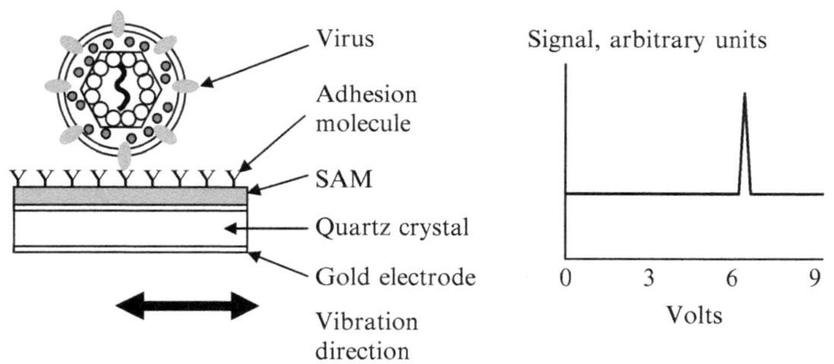

Fig. 9.10 (**a**)Virus particle adhering to a quartz crystal through an adhesion molecule, then being vibrated to cause detachment, which emits a sound. (**b**) Signal at 7.4 V showing detachment of 500,000 virions as amplitude was increased

giving HSV particle adhesion. After adhesion, the surfaces were washed with phosphate buffered solution (PBS) to remove loosely adhering material.

The quartz crystal was excited by a signal generator and the amplitude increased by gradually raising the voltage. At 7.4 V a sharp peak was seen which varied with the virus concentration over a range of six orders of dilution, indicating that the virus particles had detached. No signal was found after detachment, nor in the absence of virus, nor when the adhesion molecules were blocked. The results suggested that the method could detect the 500,000 virions adsorbed on the surface very sensitively. Cooper estimated the force applied to each virus particle was about 3 nN which corresponded to the breaking of many adhesion molecule bonds of 10–200 pN each.

9.6 Single Particle Fluorescence

In their 'Introduction to modern Virology', Dimmock, [1] Easton and Leppard stated that "viruses are too small to be seen except by electron microscopy." This is no longer true because direct observations can be made both by fluorescence microscopy and by laser light scattering, enabling the dynamic processes of virus adhesion and infection to be seen, even though the virus particle itself is too small to be imaged by optical microscopy.

Single particle fluorescence is a technique where a fluorescent label is attached to a virus protein, giving off coloured light when illuminated with an exciting beam. Shimomura in the 1960s had investigated the Green Fluorescent Protein (GFP) after extracting it from the Pacific Ocean jellyfish Aequorea victoria and showed that the intense green light was emitted from a particular protein sequence in the presence of oxygen on illumination with blue light. Chalfie and his colleagues[31] later showed that the protein could be expressed in several bacteria and therefore used it to investigate bacterial processes. Tsien then improved the brightness of the green light and also developed further fluorescent systems with other colours such as red and cyan. For this breakthrough in observing cell structures, Shimomura, Chalfie and Tsien shared the Nobel prize for chemistry in 2008.

By incorporating the GFP into a virus protein, the mechanism of virus adhesion and infection of cells could then be followed in the optical microscope.[32] Desai and Person generated a new 12 kDa capsid protein which they called VP26-GFP by fusing the fluorescent protein with the VP26 capsid protein. The new protein was fluorescent and could be seen growing in cell cultures which had been infected by the virus, as shown in Fig. 9.11.

At early times, the virus was located in distinct regions of the nucleus of the infected cell, suggesting that there were certain compartments where the virus capsids were assembled. Later, the cytoplasm was fluorescing and then the membrane, showing that the virus components had been transported to the outer parts of the cells. The study showed that the fluorescent protein sequence could be incorporated into the virus structure without changing the function and infectivity.

Seisenberger and his colleagues[33] then showed in real time the infective process of a 100 nm adenovirus adhering to a HeLa cell, using one dye molecule in each virus to follow the movement of the virion to the cell wall and into the cytoplasm and nucleus as shown in Fig. 9.12. This method had also been used on adeno virus,[34] influenza, SV40, polyoma and leukaemia.[35]

The adeno-associated virus (AAV) particles diffused outside the cell as in track 1. Also they were observed bumping into the cell membrane in track 2. Typically a particle approached the cell and decelerated, then touched the cell about five times with a mean touching time of 62 ms interrupted by short diffusion paths near the cell surface. Track 3 shows several touching events followed by entry through the membrane and diffusion in the cytoplasm, as in track 4. A penetration efficiency of 13% was measured. Surprisingly, the time of contact at the entry stage was no different from the bumping contacts, typically 64 ms. This suggests that the time required for engulfment of the virion by the cell membrane, i.e. endocytosis, is short.

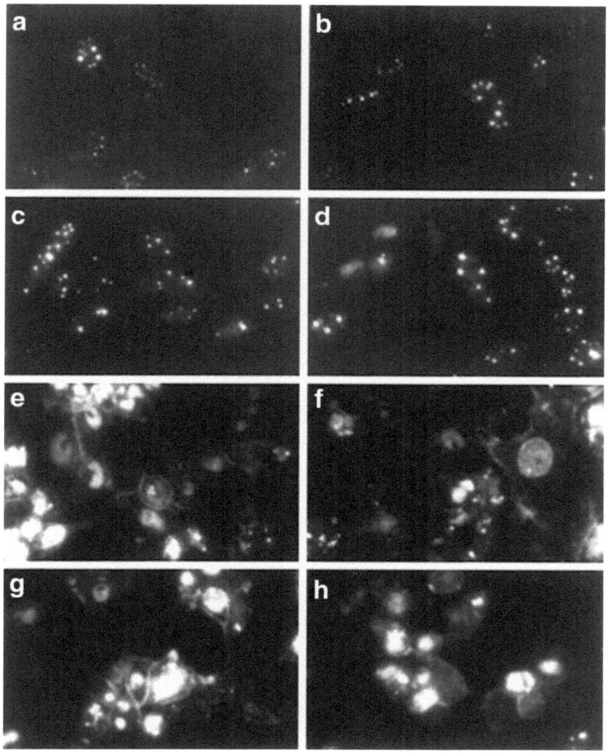

Fig. 9.11 Localisation of VP26-GFP in infected cells. Vero cells seeded at 3.5×10^4 cells/cm^2 in culture dishes were infected at an MOI of 10 PFU/cell. Cells were visualised live at various times after infection with an Olympus BH-2 fluorescence microscope (×40 objective). Photographs were taken at 6 (**a**), 8 (**b**), 10 (**c**), 12 (**d**), 16 (**e** and **f**), 20 (**g**), and 24 (**h**) h postinfection[32]

Once inside the cytoplasm, the virion could diffuse (53 trajectories) or occasionally show directed motion towards the cell nucleus (nine trajectories) indicating microtubule dependent transport. After 15 min, virions were detected in the nucleus and about one third of these showed directed motion along well defined pathways.

A key question relates to the adhesion of the virus particle to the cell membrane, and how surface active molecules can promote or inhibit entry. Cheshenko and his colleagues[36] studied Herpes Simplex in contact with cells and showed a complex interaction with surface molecules, suggesting interference with the calcium signalling pathway. Confocal microscopy and small interfering RNA (siRNA) were used to identify the source of the calcium and to dissect the requisite viral–cell interactions. Binding of HSV to human epithelial cells induced no calcium response, but shifting the cells to temperatures permissive for penetration triggered increases in plasma membrane calcium followed by a global release of intracellular calcium. Transfection with siRNA targeting the proteoglycan syndecan-2

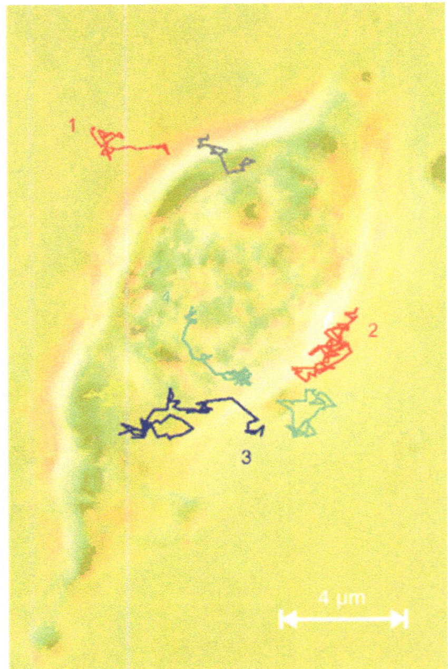

Fig. 9.12 Trajectories of single AAV-Cy5 particles indicating infectious entry pathways of AAVs into a living HeLa cell. The traces show diffusion in solution (1), touching at the cell membrane (2), penetration of the cell membrane (3), diffusion in the cytoplasm (3 and 4), penetration of the nuclear envelope (4), and diffusion in the nucleoplasm[33] (with permission)

blocked viral binding and abrogated any calcium response. Transfection with siRNA targeting nectin-1, a glycoprotein D receptor, also prevented both membrane and intracellular calcium responses. In contrast, the membrane response was preserved after transfection with siRNA targeting integrin αv, a novel glycoprotein H receptor. The membrane response, however, was not sufficient for viral entry, which required interactions with integrin αv and release of inositol-triphosphate receptor-dependent intracellular calcium stores. Thus, calcium seemed to play a critical, complex role in HSV entry.

9.7 Self Adhesion and Aggregation by Light Scattering

Perhaps the most direct and new way to visualise virus particles is by laser light scattering using the Nanosight[37] instrument, in which the virions appear as bright specks undergoing Brownian motion, described in detail in Chapter 8. Originally developed by Bob Carr and his colleagues to measure virus particle sizes, the instrument is now being used to detect many different nanoparticle species,[38] and also to measure very small adhesion forces.[39] It differs from the dynamic light scattering (DLS) instruments such as the Malvern Nanosizer by tracking individual bright spots and calculating their sizes from the Stokes–Einstein equation, to deliver a movie of virion tracks such as those shown in the single frame of Fig. 9.13a.

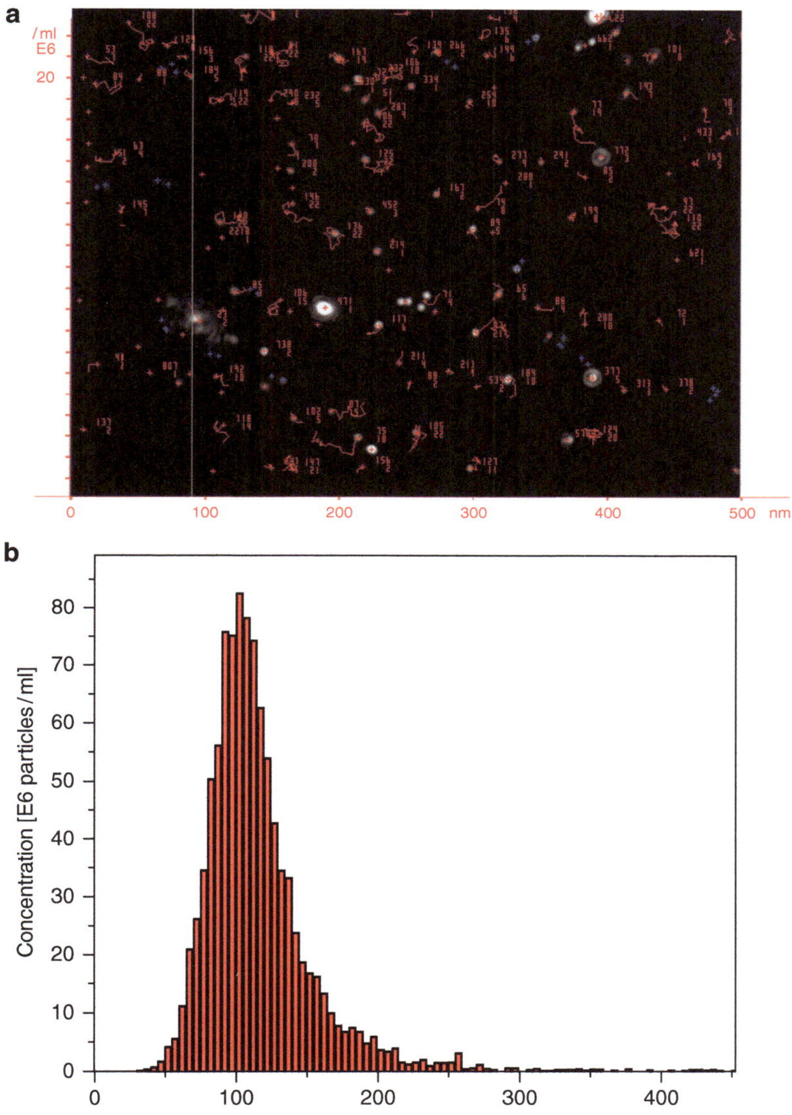

Fig. 9.13 (**a**) AdLz (E1A deleted) Adenovirus tracks from a single frame of the movie delivered by the Nanosight instrument. (**b**) Histogram plot of measured virion sizes for sample diluted 200 times [40], showing the concentration in 5 nm bins

Each track in the beam was then analysed to obtain the particle diameter, and the concentration of measured particles in the laser beam volume was plotted as a histogram of 5 nm bins in Fig. 9.13b to show the diameter distribution of the virus particles. It was clear that the virus particles had a tight distribution with a mean diameter of 100 nm and few large aggregates, indicating low self-adhesion.

The problem with this result was that the Nanosight was clearly not registering all the particles in the field of view. The virus,[40] AdLz (E1A deleted) Adenovirus of TEM particle size 90 nm, was grown by Dr Susan Morris at Warwick University in collaboration with Dr Clive Sweet of Birmingham University and was purified then concentrated to a phosphate buffer solution (PBS) suspension containing 40.10^6 pfu (plaque forming units) per ml, as measured by the standard infection test. The main reason for this problem was the laser tracking problem with overlapping tracks. The tracking system could not distinguish overlapping tracks and so could not count the large number of virions present.

This problem was confirmed when the sample was diluted 400 and 700 times for comparison with the other diluted specimens, 50, 100, 200 as shown in Fig. 9.14. The concentration measured by the Nanosight instrument seemed to rise as the concentration was reduced, because more independent tracks were being detected as the particles became separated on dilution. However, at dilutions of 400 and 700, the instrument seemed to be giving sensible concentration results, showing that the Nanosight concentration decreased as dilution proceeded. However, below 10^5 particles per ml, the number of tracks was limited and statistics were insufficient. The conclusion was that the Nanosight could only be used to give reliable results in the range of 1–10 million particles per ml, corresponding to volume fractions around 1 ppb to 10 ppb, i.e. 10^{-9}–10^{-8}, for adenovirus of 100 nm diameter. In order to verify the results shown in Fig. 9.14, calibration experiments were conducted using well characterised polystyrene particles.

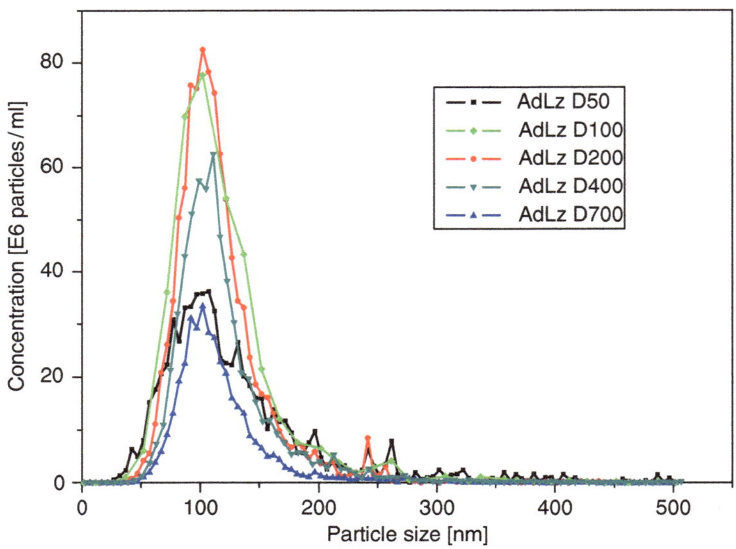

Fig. 9.14 Nanosight tracking results for five different dilutions of adenovirus showing that concentrations are only accurate for dilutions higher than 400, i.e. less than 10^5 pfu/ml, corresponding to about 2 million particles/ml

9.8 Virus Calibration Using Standard Polystyrene Latex

The sample of adenovirus described above, diluted 400 times was compared directly with a standard polystyrene nanoparticle dispersion measured separately after diluting 8,000 times, shown in Fig. 9.15. The standard polystyrene

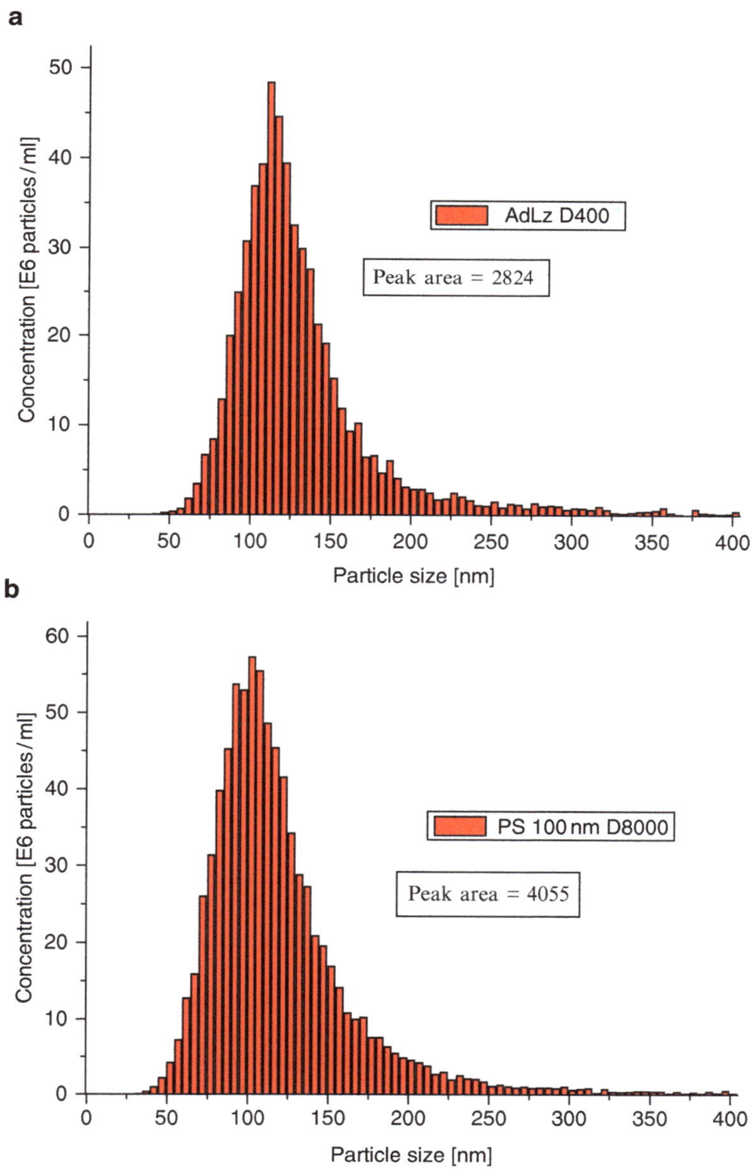

Fig. 9.15 (**a**) Histogram of AdLz sample diluted by PBS in 400 times. (**b**) Histogram of 95.6 nm PS aqueous solution with a concentration of 2.5×10^9 particles/mL

sample concentration was originally 2.08×10^{13} particles/mL, and after dilution was 2.6×10^{9} particles/mL. The two samples behaved comparably in the separate experiments, indicating that the concentrations, the particle size and the states of aggregation (i.e. self-adhesion) were similar in the two cases. Comparing the peak areas for the single particle Gaussian fits, we calculated the concentration of the original AdLacZ sample was 6.97×10^{11} particles/mL (viral particles (VP) per ml).

The second issue was that the number of virus particles measured by the Nanosight was found to be considerably higher than the infective particle numbers measured by plaque forming assay. Usually the explanation for this is the presence of viral particles which are not infective because of damage or lack of reproductive material. The particle infectivity ratio, that is the total number of particles divided by the infective particles (VP/pfu) for adenovirus was calculated to be 17.9 from this experiment.

9.9 Self-Adhesion of Viruses

Having measured the particle concentrations with good precision, it was then possible to use the Nanosight to count the number of doublets as a measure of self-adhesion. To estimate self-adhesion of the nanoparticles, the number of doublet aggregates was obtained from the histogram by fitting two Gaussian curves to the results, one for single particles assumed to be uniformly distributed and a smaller curve for doublets and large aggregates. Figure 9.16a shows the computer fit for 95.6 nm polystyrene nanoparticles coated with HAS in PBS, the doublet peak at 1.49 times the primary particle diameter and large aggregates to give the measured result for particle size distribution. The ratio of doublets to singlets was then calculated and plotted against virus concentration to give the adhesion characteristic curve fitting equation 8.2.

Figure 9.16b shows the ratios for 95.6 nm polystyrene in aqueous solution and Adenovirus at different concentrations. It was evident that the results gave an approximate fit to equation 8.2 but with substantial scatter. The adhesion curve for the virus was a little lower than that for the polystyrene indicating slightly lower adhesion. Using this method, it was not possible to distinguish the energy ε and the range z to describe the potential curve of equation 8.2. However, for polystyrene, it was possible to use several particle sizes to obtain the two parameter fit as shown in Fig. 9.17.

An interesting feature of this experiment was the possibility of measuring the adhesive interaction between the virus particles (VP) and the polystyrene (PS) particles. If the virions were identifying the PS as infectable units by recognising molecules on their surfaces, then a third peak of 'attacked particles' (AP) would have been observed at 500 nm diameter as indicated in the schematic of Fig. 9.18. This experiment is being carried out now.

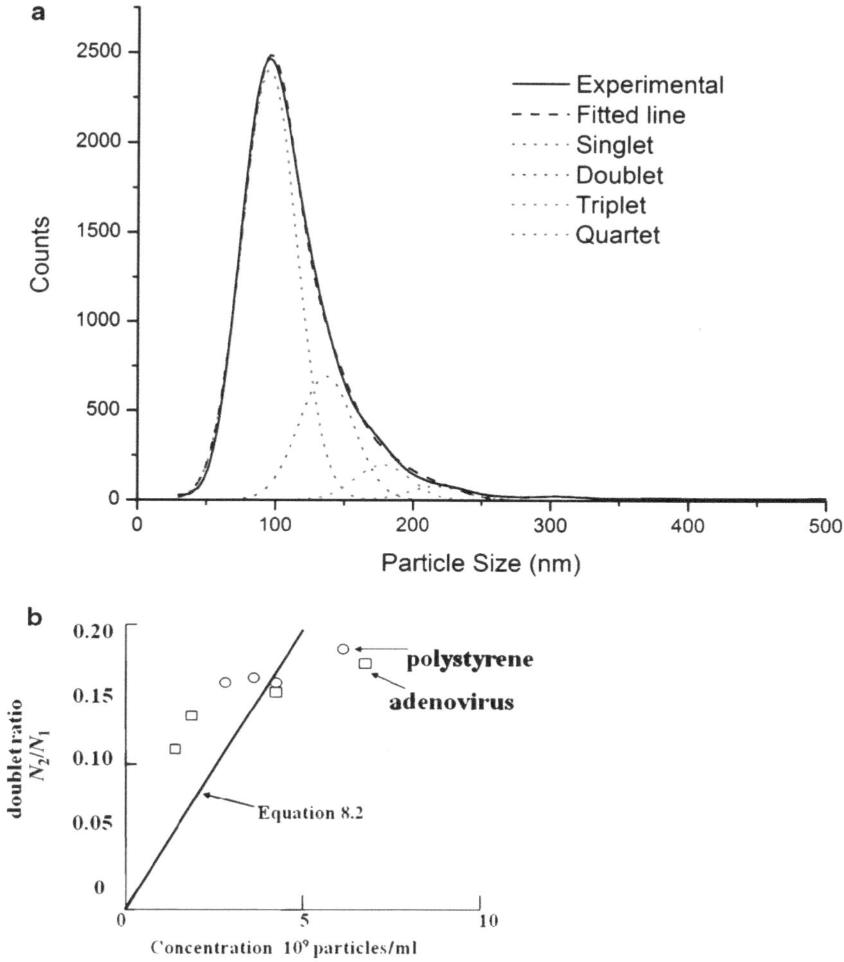

Fig. 9.16 (**a**) Singlet and aggregate peaks fitted to results for 95.6 nm polystyrene coated with HAS in PBS at concentration of 2.5×10^{-7} g/ml; (**b**) plots of doublet ratio N_2/N_1 as function of concentration for virus and polystyrene, showing that the adhesion characteristic curves were similar

9.10 Drug Treatments Associated with Adhesion

Around forty antiviral drugs have been approved to combat viral diseases, most of them against AIDS.[1] Adhesion is involved in at least two of the major viral infection processes: 1) attachment of the virus to cell surface molecules (prevented by e.g. pleconaril, synagis and respigam) and 2) detachment of the budding virus particles from the infected cell (controllable by e.g. zanamivir and oseltamivir).

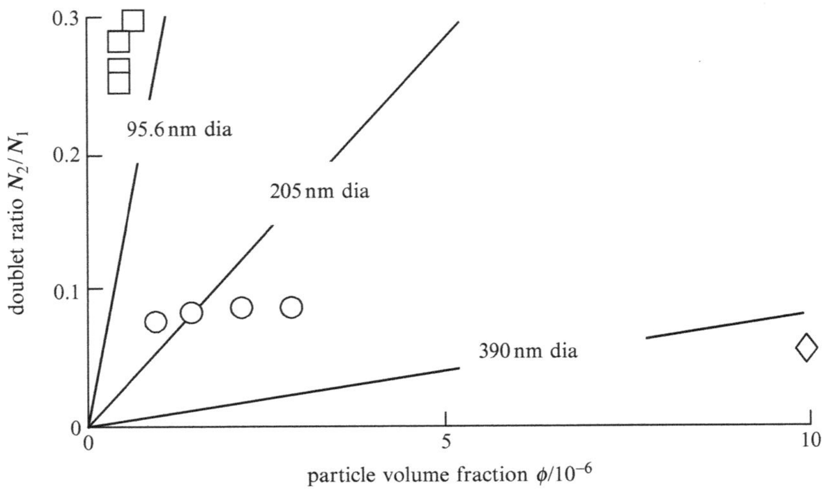

Fig. 9.17 Plot of adhesion characteristic curves for polystyrene latexes of several diameters in PBS, showing a fit to the adhesion energy 3.5 *kT* and range parameter $z=600$ nm

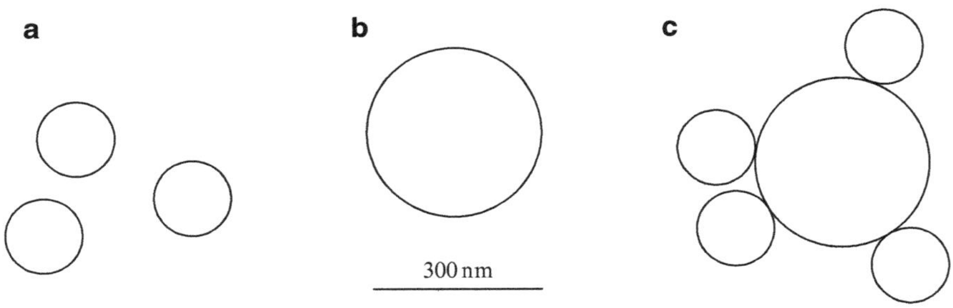

Fig. 9.18 (**a**) Virus particles 100 nm dia; (**b**) polystyrene particle 300 nm dia; (**c**) virus particles adhering to PS particles to give a 500 nm aggregate[41]

The overall cycle of viral infection and release is shown in Fig. 9.19.[1] Virus particles diffuse near the cell surface and eventually touch, though the first contacts may be fleeting and quickly followed by release. Eventually a location is reached where the virus particle gains entry through the cell membrane. The viral components then invade the nucleus and reproduce before reassembling near the cell wall. Virions can then escape from the cell by crossing and detaching from the cell membrane to continue the infection cascade process. The problem is that the infection proceeds rapidly, typically 1 h to move from one cell to the next, and so drug treatment must be given before the infection or within 6 h after infection to be effective.

The roles of drug molecules in reducing adhesion is not always clear, complicated by commercial conflicts and over-simplification of complex or multiple molecular interactions. One drug claiming to reduce the adhesion of viral

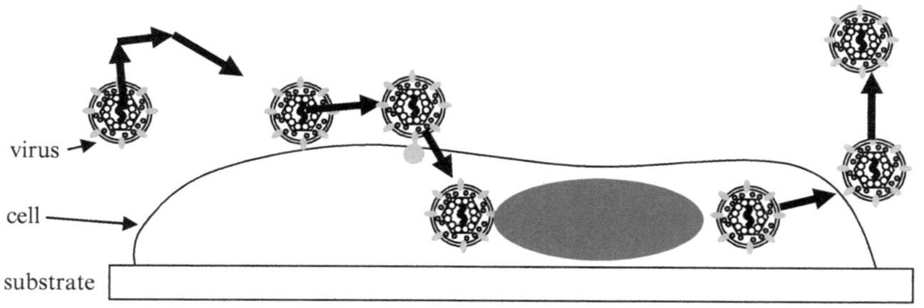

Fig. 9.19 Process of viral infection of cell showing the diffusion, contact, travel along cell membrane, entry and release after reproduction

particles is pleconaril[42] (VP-63843) 3-[3,5-dimethyl-4[3-(3-methyl-5-isoxazolyl) propyloly] phenyl-5-(trifluoromethyl)-1,2,4-oxadiazole, which is a novel, broad spectrum antipicornaviral agent that supposedly acts by altering receptor binding and viral uncoating. Pleconaril is orally bioavailable and achieves serum concentrations in excess of those required to inhibit 90% of clinical rhino- and enteroviral isolates in vitro. It claims the additional advantage of achieving several fold higher concentrations within the central nervous system and nasal secretions than in serum, a characteristic that is highly desirable for an antiviral targeted towards viruses known to cause central nervous system and upper respiratory tract infections. Approximately 80% of an orally administered dose is excreted in the faeces within 48 h. Urine excretion accounts for the remainder of the drug. Pleconaril is said to have demonstrated an excellent safety profile in dose escalation and clinical studies. Clinical studies have reported a reduction in the duration and intensity of symptoms in children and adults with enteroviral meningitis and in adults with rhinoviral respiratory tract infections treated with pleconaril. Lastly, pleconaril has demonstrated efficacy in the treatment of severe life-threatening enteroviral infections of the newborn and in immunosuppressed individuals. Pleconaril has shown some side effects and was therefore not approved in 2002 by the FDA.[43] Pleconaril shortened the viral meningitis illness compared to placebo recipients but only modest benefit was gained in severe diseased patients.

A review by Chen et al.[44] considers how molecules can attach to the dangerous enterovirus 1 (EV1) and how slight changes in the molecular structure can lead to anti-viral activity. Numerous synthetic compounds exhibit in vitro antiviral activity by binding in a hydrophobic pocket beneath the canyon floor in the centre of the viral protein 1 (VP1), preventing viral attachment or uncoating. Pleconaril, one of the WIN compounds (a class of anti-EV and anti-rhinovirus compounds targeting the event of uncoating during viral replication), is a successful clinical candidate – it acts as a small-molecule inhibitor of EVs and rhinoviruses. It was developed to treat diseases associated with picornavirus infections. In a virus-induced cytopathic effect assay, pleconaril was found to act against not only prototypic EV strains but also 215 clinical isolates that represent the most commonly isolated EV serotypes. Pleconaril inhibited the 50% replication of all clinical isolates at a concentration of ≤ 0.03 μM. The results of a Phase III clinical trial that involved nearly 2,100 participants showed that treatment with pleconaril reduced the duration and severity of a cold in picornavirus-infected patients. In 2003, pleconaril was licensed to Schering-Plough by Viropharma. In 2007, a Phase II clinical trial to investigate the effects of pleconaril nasal spray on common cold symptoms and asthma exacerbations following exposure to rhinovirus was completed.

9.11 Preventing Virus Detachment

At the detachment part of the virus lifecycle, zanamivir, the first of the neuraminidase inhibitors, (trade name Relenza from GlaxoSmithKline) and oseltamivir (trade name Tamiflu from Roche) have been useful molecules to prevent influenza virus release from infected cells. Zanamivir was invented in 1989 by a group working in Australia and the company Biota then sold the patent to Glaxo. It was approved in the USA in 1999 and began to be applied but had the disadvantage that it was delivered by inhaling the powder. Oseltamivir was produced as an oral drug in 1999 by Gilead Sciences and sold to Roche who have disseminated it widely. Figure 9.20 shows the molecular structures.

The way in which these molecules were discovered was described by Graeme Laver.[46] Their invention was based on the model derived from transmission electron microscope pictures which showed flu virus particles to be covered with two kinds of glycoprotein spikes projecting about 13 nm from the membrane: the two virus surface molecules hemagglutinin (HA) which is rod-like and neuraminidase (NA) an enzyme which is mushroom shaped. The flu virus shown schematically below is not spherical and of regular size as some other viruses are. Its surface contains typically 2,000 HA molecules and 500 NA spikes, virus surface molecules which are thought to cause adhesion to cell receptor molecules, thereby promoting recognition and infection.

George Hirst, working at the Rockefeller Institute New York in the 1940s, showed that flu virus particles could adhere to red blood cells at room temperature, causing the cells to clump together, the so-called hemagglutinin test. After warming to body temperature of 37°C the cells dispersed and the virus came off. Alfred

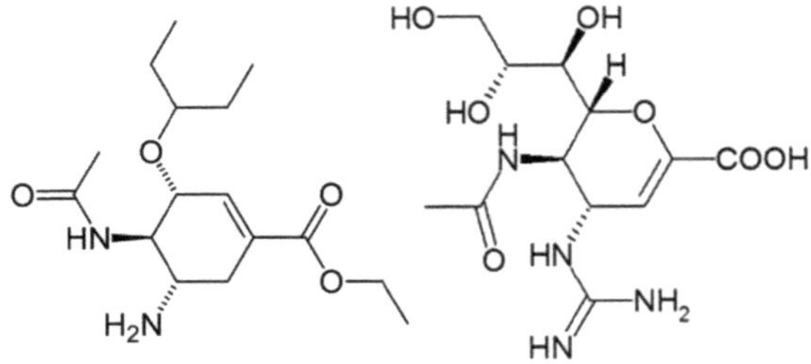

Fig. 9.20 *left*: Oseltamivir (Tamiflu); *right*: Zanamivir (Relenza)[46]

Gottschalk at the Walter and Eliza Hall Institute, Melbourne, Australia believed this was due to the neuraminidase cleaving the adhesive bonds. Scientists quickly realised that if a blocking molecule, a 'plug-drug' could be developed that stopped this enzyme working, it might provide a cure for the flu by preventing the virions from detaching from their victim cells. It took more than 50 years for this idea to develop fully.

Robin Valentine and Graeme Laver, working at the National Institute for Medical Research in London, proved by using electron microscopy that there were two distinct molecules with different functions on the surface on the influenza virus. One was the glycoprotein, hemagglutinin, which attaches the virus to adhesion molecules on the cells, and the other, the enzyme neuraminidase, which destroys these receptors and allows the new virus particles to be released from the cells (Fig. 9.21).

The head of the neuraminidase spike, which contains all the enzymic activity of the molecule, can be released from some influenza virus strains by digesting the virus particles with a protease. The released heads can then be purified and in some cases crystallised. Similarly, hemagglutinin can be separated and purified for detailed study. In 1978 Laver grew the first neuraminidase crystals from influenza virus and, 5 years later, Peter Colman and his colleagues at the CSIRO Division of Protein Chemistry, Melbourne, solved the three dimensional structure using X-ray diffraction. This structure revealed that the neuraminidase was a tetramer composed of four identical monomers, each containing a single polypeptide chain. In the centre of each monomer was a deep cleft, or canyon able to bind sialic acid. This was the active, catalytic site of the enzyme. Different virus strains have different molecular structures. Sixteen different hemagglutin sub-types (H1 to H16) and nine different neuraminidase (N1-N9) have been found and flu strains can be distinguished by the combinations of these molecules e.g. H5N1 is the deadly birdflu. The main discovery was that neuraminidase did not influence the entry of virus into the cell, but was critical for virus release after infection. By removing sialic acid from the cell wall, the neuraminidase allowed detachment.[45]

The drug zanamivir was a neuramidase blocker or inhibitor which was designed from the knowledge of the crystal structure of neuraminidase. Von Itzstein and Wu

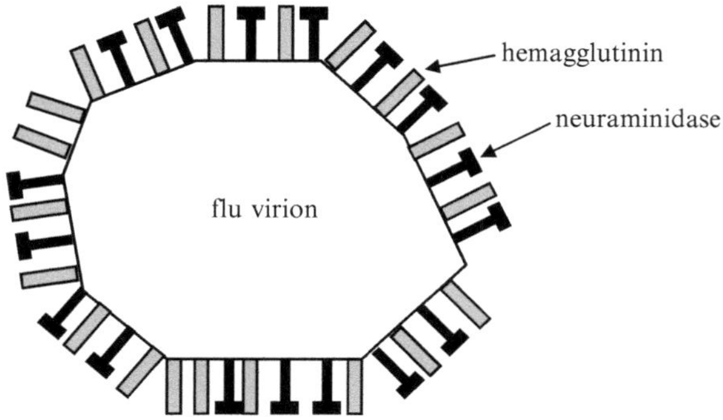

Fig. 9.21 Schematic of virus surface molecules on flu virion

found several molecules which blocked the activity, and selected the most potent for trials. Sialic acid itself, N acetyl neuraminic acid, (NANA) could block the action but the dehydrated version DANA was much better. Four guanadinoDANA was 1,000 times better still and this was the compound which was selected as zanamivir.

Although this is the best example of an adhesion molecule being targeted in drug treatment, a number of other adhesion mechanisms could also be addressed, and these are described in a recent review.[47]

9.12 Conclusions

Virus particle adhesion is a fascinating phenomenon which has been studied by crystallisation, by X-ray analysis, by electron microscopy and more recently by AFM, fluorescence and light scattering. The interparticle forces are very small, comparable to thermal diffusion forces, and are influenced by the presence of surrounding molecules. Typically, viruses can be made to crystallise by adding inert polymers such as PEG which cause a depletion effect between the virions. Too much PEG addition causes too large an adhesion and then the virions aggregate into amorphous clumps, as shown by phase diagrams. Virions can also adhere to each other just like polystyrene particles and the forces measured by a new light scattering laser tracking method. The new method was shown to be good for measuring virus concentrations and also for measuring the doublets caused by van der Waals forces. The self-adhesion was very similar to that observed for polystyrene particles.

The process by which a virus particle infects a cell is complex and involves diffusion towards the cell, periodic contact events, followed by entry and transport to the nucleus. Release of the progeny virions is also complex and depends on surface molecules release. Drugs can be targeted at these adhesion processes by addressing the blocking the surface molecules. Much progress has been made since the 1980s.[48, 49]

References

1. Dimmock, N.J., Easton, A.J., and Leppard, K.N., Introduction to modern virology, 6th ed. 2007.
2. Fields, B.N. (ed., Virology, 5th ed.) Principles of Virology, Wolters Kluwer, New York, 2007.
3. http://phene.cpmc.columbia.edu/index.htm.
4. www.virology.net, The big picture book of viruses,
5. Condit, R.C. (in Fields, B.N. ed., Virology, 5th ed.) Principles of Virology, Wolters Kluwer, New York, 2007.
6. Morgan, C., Ellison, S.A., Rose, H.M. and Moore, D.H., Internal structure in virus particles, Nature 173 (1954) 208–9; Proc Soc Exp Biol & Med 82 (1953) 454–7.
7. Harris, K. and Yang, C., www.stanford.edu/group/virus/herpes/2005/index
8. Kausche, G.A., Pfankuch, E. And Ruska, H., Die Sichtbarmachung von pflanzlichem virus in uebermikroskop, Naturwissenschaften 27 (1939) 292–9.

9. Obenauer-Kutner, L.J., Ihnat, P.M., Yang, T-Y., Dovey-Hartman, B.J., Balu, A., Cullen, C., Bordens, R.W., and Grace, M.J., The Use of Field Emission Scanning Electron Microscopy to Assess Recombinant Adenovirus Stability, Human gene therapy, 13 (2002) 1687–1696.
10. Stanley, W.M., Isolation of a crystalline protein possessing the properties of tobacco mosaic virus, Science 81 (1935) 644–5.
11. Malkin, A.J., Land, T.A., Kuznetsov, Y.G., Mc Pherson, A., De Yoreo, J.J., Phys Rev Lett 75 (1995) 2778–81.
12. Casselyn, M., Tardieu, A., Delacroix, H. and Finet, S., Birth and growth kinetics of Brome mosaic virus microcrystals, Biophys J 87 (2004) 2737–48.
13. Lorber, B. and Witz, J., An Investigation of the crystallogenesis of an icosahedral RNA plant virus with solubility phase diagrams, Cryst growth des 8 (2008) 1522–29.
14. Bawden, F.C. and Pirie, N.W., Crystalline preparations of tomato bushy stunt virus, Br J Exp Path 19 (1938) 251–63.
15. Finch, J.T., Klug, A. and Leberman, R., The structure of turnip crinkle and tomato bushy stunt viruses II, J Mol Biol 50 (1970) 215–22.
16. Bhella, D., Gatherer, D., Chaudhry, Y., Pink, R. and Goodfellow, I.G., Structural insights into Calicivirus attachment and uncoating, J Virology 82 (2008) 8051–8.
17. Makino, A., Shimojima, M., Miyazawa, T., Kato, K., Tohya, Y. and Akashi, H., Junctional adhesion molecule 1 is a functional receptor for feline calicivirus, J Virol. 80 (2006) 4482–90.
18. Ossiboff, R.J. and Parker, J.S., Identification of regions and residues in feline junctional adhesion molecule required for feline calicivirus binding and infection, J Virol. 81 (2007) 13608–621.
19. Liashkovich, I., Hafezi, W., Kuhn, J.E., Oberleithner, H., Kramer, A. and Shahin, V., Exceptional mechanical and structural stability of HSV-1 unveiled with fluid atomic force microscopy, J Cell Sci. 121 (2008) 2287–92.
20. Zhou, Z.H., Dougherty, M., Jakana, J., He, J., Rixon, F.J. and Chin, W., Seeing the Herpes virus capsid at 8.5 Angstrom, Science 288 (2000) 877–80.
21. Binnig, G., Quate, C.F. and Gerber, C., Atomic Force Microscope, Phys Rev Lett 56 (1986) 930–933.
22. Kuznetsov, Y.G., Victoria, J.G., Robinson, W.E. and McPherson, A., Atomic force microscopy investigation of human immunodeficiency virus HIV and HIV infected lymphocytes, J Virol. 77 (2003) 11896–909.
23. Tan, S.S., Sherman, R.L. and Ford, W.T., Nanoscale compression of polymer microspheres by atomic force microscopy, Langmuir 20 (2004) 7015–7020.
24. Roos, W.H., Ivanovska, I.L., Evilevitch, A. and Wuite, G.J.L., Viral capsids: mechanical characteristics, genome packaging and delivery mechanisms, Cell Mol Life Sci. 64 (2007) 1484–97.
25 Hirst, E.R., Yuan, Y.J., Xu, W.L. and Bronlund, J.E., Bond-rupture immunosensors- a review, Biosensors and bioelectronics 23 (2008) 1759–68.
26. Cooper, M.A. and Singleton, V.T., J Molecular recognition, 20 (2007) 154–84.
27. Hoa, X.D., Kirk, A.G. and Tabrizian, M., Biosens. Bioelectron. 23 (2007) 151–160.
28. Berggren, C., Bjarnason, B. and Johansson, G., Electroanalysis 13 (2001) 173–80.
29. Yuan, Y.J., van der Werff, M.J., Chen, H., Hirst, E.R., Xu, W.L. and Bronlund, J.E., Anal Cem 79 (2007) 9039–44.
30. Cooper, M.A., Meas Sci Technol 14 (2003) 1888–93.
31. Chalfie, M., Tu, Y., Euskirchen, G., Ward, W.W. and Prasher, D.C., Green fluorescent protein as a marker for gene expression, Science 263 (1994) 802–5.
32. Desai, P. and Person, S., Incorporation of the green fluorescent protein into the Herpes Simplex virus type 1 capsid, J Virology 72 (1998) 7563–8.
33. Seisenberger, G., Ried, M.U., Endress, T., Buning, H., Hallek, M. and Brauchle, C., Real-time single molecule imaging of the infection pathway of an adeno-associated virus, Science 294 (2001) 1929–1932.
34. Suomalainen, M., Nakano, M.Y., Keller, S., Boucke, K., Stidwell, R.P. and Greber, U.F., Microtubule dependent plus and minus end directed motilities are competing processes for nuclear targeting of adenovirus, J Cell Biol. 144 (1999) 657–672.

35. Lampe, M., Briggs, J.A.G., Endress, T., Glass, B., Riegelsberger, S., Krausslich, H.G., Lamb, D.C., Brauchle, C. and Muller, B., Double labelled HIV particles for study of virus-cell interaction, Virology 360 (2006) 92–104.
36. Cheshenko, N., Liu, W., Satlin, L.M. and Herold, B.C., Multiple Receptor Interactions Trigger Release of Membrane and Intracellular Calcium Stores Critical for Herpes Simplex Virus Entry, MBC 18 (2007) 3119–30.
37. www.Nanosightuk.co.uk
38. Malloy, A. & Carr, R. Nanoparticle tracking analysis - The Halo (TM) system. Part. Syst. Charact. 23 (2006) 197–204.
39. Kendall, K., Dhir, A. and Du. S., A new measure of molecular attractions between nanoparticles near kT adhesion energy, Nanotechnology **20** (2009) (4pp) NANO/305138/PAP
40. Kendall,K., Du, S., Morris, S., Sweet, S., Virus concentration and adhesion measured by laser tracking, J Adhesion (2010) accepted.
41. Kendall, K., Du, S., Sweet, C. and Morris, S., Virus adhesion to polystyrene nanoparticles. In preparation.
42. Romero, J.R., Expert opinion on investigational drugs, 10 (2001) 369–79.
43. Desmond, R.A., Accortt, N.A., Talley, L., Villano, S.A., Soong, S.J. and Whitley, R.J., Entero viral meningitis: Natural history and outcome of Pleconaril therapy, Antimicrobial agents and chemotherapy, 50 (2006) 2409–14.
44. Chen, T, Weng, K, Chang, S, Lin, J, Huang, P and Shih, S., Development of antiviral agents for enteroviruses, J Antimicrobial Chemotherapy 62 (2008) 1169–73.
45. von Itzstein, M, Wu, W-Y, Kok, GB, Pegg, MS, Dyason, JC, Jin, B, et al. Rational design of potent sialidase-based inhibitors of influenza virus replication. Nature 363 (1993) 418–423.
46. Laver, G., Flu drugs- Pathway to discovery, Education in Chemistry Mar 2007.
47. de Clercq, E., Advances in antiviral drug design, vol 5, Elsevier 2007, ch 1.
48. Crowell, R.L., Virus Attachment and Entry into Cells: Proceedings (1986)
49. Compans, R.W., Helenius, A., Oldstone, M.B.A., Compans: Cell Biology of Virus Entry Replication & Pathogenesis, John Wiley 1989.

Chapter 10
Adhesion of Cells

Each of the three kinds of solid support used in the experiments influences the cell movement in its own way

Harrison 1914

This chapter deals with the adhesion of cells, especially the dependence on elasticity. First, we will discuss cell adhesion phenomena starting from the biological foundations. Then, we study artificial model systems to show how the elastic modulus and geometry affect cell adhesion and cause cell differentiation. Finally, at the end of the chapter we describe recent research to demonstrate that complex systems such as living cells within an extracellular matrix can be described by a classical elastic inclusion model.

With respect to adhesion, cells fall into two classes – adherent and non-adherent cells. All of our tissue and organ cells typically are adherent cells, keeping us and the organs together. This demonstrates the paramount importance of adhesion for multi-cellular organisms especially for the evolution of complex living systems.[1]

In fact, adherent cells die if they cannot find a surface to stick to. Merely having the correct adhesion molecules in the solution is not sufficient for survival, so chemistry itself is not enough. Active physical contact with the environment and resulting tensions seem to be necessary for this class of cells to maintain their metabolism.

Non-adherent cells are different. In particular, erythrocytes (red blood cells), responsible for the oxygen and carbon dioxide transport, must be dispersed in the blood and not adhere to the blood vessel if they are to maintain their vital function and prevent life threatening conditions such as a stroke or infarct.

10.1 Cell Adhesion Is a Vital Phenomenon

It is obvious that cell adhesion is essential for all multi-cellular organisms because otherwise the cells, and therefore we, would just fall apart. On the other hand adhesion needs to be potentially well controlled and tuned to enable dynamic processes such as cell division, adaptation, and differentiation i.e. morphogenesis.

K. Kendall et al., *Adhesion of Cells, Viruses and Nanoparticles*,
DOI 10.1007/978-90-481-2585-2_10,

One example, showing that switching between adhesion and a free unbound state is vital, occurs in our blood vessels every now and then. Leukocytes (white blood cells), that are usually floating in the blood stream can leave the vessel and enter tissue after damage or inflammation. This process is called 'extravasation' and demonstrates the ability of cells to switch from a non-adherent state in the blood stream to an adherent state in the tissue. Extravasation is a multi-step process where leukocytes first touch the blood vessel wall, then lightly attach while rolling along the vessel lining, then firmly adhere and finally migrate through the endothelium layer to reach the region of inflammation.[2]

Another interesting example can be seen when adult stem cells from the bone marrow (human mesenchymal stem cells (hMSCs) and human hematopoeitic stem cells (hHSCs) exit from their niche and enter the blood stream as shown in Fig. 10.1. Here, cells that are adherent in the bone marrow matrix switch to a non-adherent state in the blood. From there, they can reach various destinations in different tissues that exhibit drastic differences not only in their biochemical composition but also in mechanical properties such as the Young's elastic modulus E (see Fig. 10.1) which ranges from 1 kPa for brain to 100 kPa in bone precursor and 10 GPa in mineralised bone. Human mesenchymal stem cells (hMSCs) can adapt to this wide range of different tissue types and elasticity and we will discuss the influence of the Young's elastic modulus E of the surrounding on adhesion and differentiation at the end of this chapter. Interestingly, these cells can also go in the reverse direction and go back

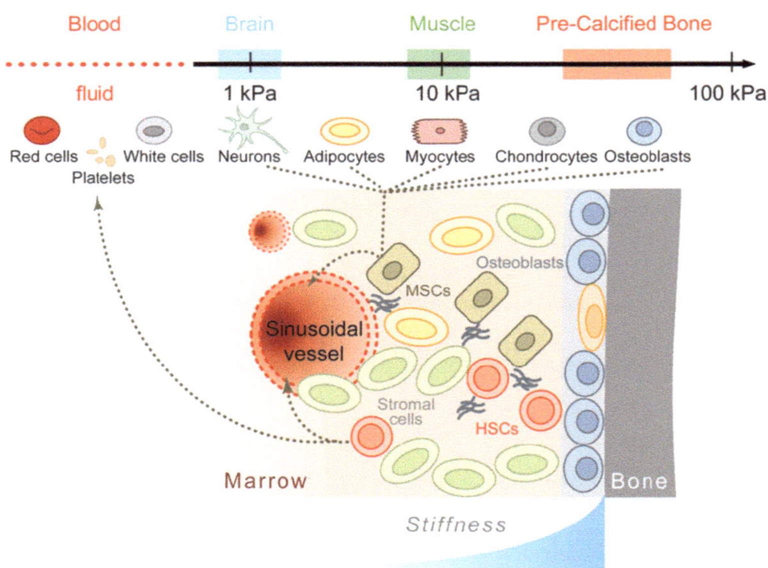

Fig. 10.1 Physiological range of elasticity E ranges from zero (blood as a fluid) to MPa and GPa for fully mineralised bone. HSCs and MSCs are residing in their bone marrow niche but can egress and enter the blood stream. Their differentiation potential ranges from blood cells through to various types of tissue cells[7] (Reproduced with permission)

to their niche, which is called 'homing', a process which neatly demonstrates the reversible aspect of these weak adhesion phenomena.[3]

This reversible mobile, adhesional and adaptive behaviour is happening constantly in our body and helps to maintain homeostasis (i.e. stable dynamical equilibrium of a living organism). Such complex processes are only possible if the cells can finely control and tune the adhesion forces to their surroundings. These and the further upcoming examples show that cell adhesion is vital but also a very complex process, not just dependent on molecules but also strongly influenced by geometry and elasticity.

10.2 The Emergence of Cell Culture as Investigative Tool

In this section we will briefly review cell adhesion history which is tightly connected with the development and progress of cell biology. As already mentioned, most mammalian tissue cells are adherent and so it is necessary to provide culture conditions that enable adhesion to study these cells outside the body (ex vivo). The first reports of isolating and then culturing tissue cells go back to the end of the nineteenth century. As early as 1885, Wilhelm Roux removed a portion of the medullary plate of a chick embryo and maintained it in a warm saline solution contained in a glass vessel (in vitro) for a few days. This first demonstration of maintaining living tissue in a 'synthetic solution' was the precursor of tissue culture. Slightly later in 1907, Ross Granville Harrison demonstrated the in vitro growth of living animal tissue by explanting a fragment of nerve cord from a frog tadpole and placing it in a drop of frog lymph on a glass plate. Within hours nerve fibres begin to sprout from the cells in the cord to wander along the plate, successfully proving that cells could be cultured outside the body. With time, techniques and protocols became more and more sophisticated, allowing growth and maintenance of a wide variety of different cells.[4] Very importantly, cells cultured in vitro also showed various properties that were specific for their type in vivo, opening up possibilities to study these cells in ways that would not have been possible inside an organism.

One of the next important landmarks was the establishment of 'immortalised' cell lines when it was discovered in 1951 that HeLa cells (derived from the cervical cancer cells of Henrietta Lacks) would continue to divide indefinitely. Many cell types do not divide infinitely due to shortening of the telomeres, the 'safety caps' at the end of the chromosomes. The discovery of the telomeres was recognised with the Nobel Prize in Physiology in 2009. In contrast, cancer cells and immortalised cells have an enzyme called telomerase that maintains the telomeres so that the cells can essentially divide forever.[5] These well defined cell lines made it possible to study cell science including adhesion in a reproducible manner. Then the introduction of serum free media with well defined soluble supplements in the 1970s opened up the possibility to investigate the effects of growth factors and adhesion molecules on the immortal cultured cells.

Nowadays, a vast variety of cell lines and primary cells is cultured routinely in sterile polystyrene (PS) cell culture flasks or on glass coverslips that allow for high resolution microscopy. Cell culture has become a very well standardised procedure that can be performed with synthetic and therefore highly reproducible compounds. But a key question already posed by Ross Granville Harrison when he tried different surfaces on which to grow the cells has not been fully answered: 'What are the essential properties of the solid substrate which promote cell growth?'

Despite the great achievements of cell culture to standardise procedures and ingredients one essential aspect was not taken into account for a long time – that is the mechanical properties (e.g. elasticity) of the substrates. The Young's elastic moduli E of glass or PS are in the range of GPa[6] while the elasticity of the native environment of cells is several orders of magnitude smaller, namely in the range of 0.1–100 kPa[7] as illustrated on the top row of Fig. 10.1. This discrepancy might be one reason for distinct differences of cells cultured in vitro compared to their appearance in vivo. For example, cells grown on tissue culture plastic or glass differ significantly in the composition of their cytoskeleton compared to their counterparts in the body[8] reflecting the rigidity of the substrates. So it is just as important to have a range of elastic substrates available for cell culture as it is to have different growth media formulations. We will discuss the effect of substrate stiffness in detail in the following sections.

10.3 Factors in Cell Adhesion – More than Lock and Key

In order to understand the complex processes that are responsible for cell adhesion we first need to know the structures and components that make up a cell and we will put special emphasis on the parts that are most important for adhesion, since plenty of information already exists in cell biology text books.[9] Here, we emphasise animal eukaryotic cells and focus on the building blocks that are most relevant for adhesion (Fig. 10.2).

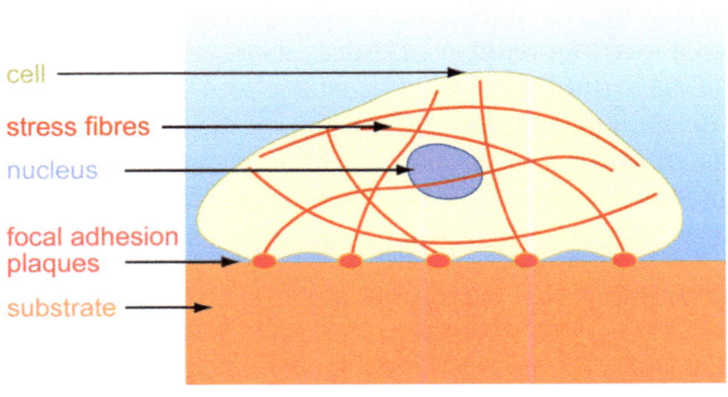

Fig. 10.2 Sketch of a cell adhering to a substrate. Inside the cell there is the nucleus and acto-myosin fibres span a network connecting to the outside via focal adhesion plaques that bind to adhesion molecules at the surface of the substrate

Spotting the parts that are obviously directly influencing adhesion one starts from the outside of the cell. Clearly, a cell is defined by an envelope that separates the inside (cytoplasm) from the extracellular surroundings. This shell is made up by a bilayer membrane of phospholipid molecules which have surfactant characteristics. Within this membrane there is a plethora of proteins, for example ion channels that are necessary to transport ions but also integrins[10] that function as adhesion molecules. Originally discovered by Hynes in 1987,[11] these proteins became the prime suspect of cell adhesion studies. Integrins are located in the cell membrane and can bind selectively to various extracellular matrix proteins covering a substrate. Tissue cells are anchorage dependent i.e. they need to adhere to a solid surface. This surface needs to be covered by cell adhesion molecules to foster cell adhesion and growth but these molecules alone in solution are not sufficient to keep the cells alive. In solution, they can even trigger cell death, although they still bind perfectly well to the cell surface.[12] This shows that cell adhesion is more complex than a simple 'lock and key' principle where the cell just has to find the appropriate molecules.

Inside the cell are the cell organelles (e.g. mitochondria that are critical in generating energy, ribosomes that produce the proteins, the endoplasmic reticulum, the Golgi apparatus, and also the nucleus that contains the DNA). While all these are essential for the cell's life there seems little or no direct relevance for adhesion, although certain proteins are required to produce force and firm adhesion and we will see in a later example that adhesion mechanics can even influence cell differentiation that is directly connected to the cell's nucleus.

The most important structure in the cytoplasm that maintains the structural integrity of a cell is the cytoskeleton.[13,14] The cytoskeleton acts similar to the skeleton in our body including the bones, tendons, and muscles. It is made up of three different types of filaments: the acto-myosin network, microtubules, and intermediate filaments. Microtubules are relatively rigid filaments (persistence length is much larger than a typical cell) that are highly involved in cell division (to separate the chromosomes in a coordinated manner) and generally in intra-cellular transport[15] but not so much in cell adhesion. The intermediate filaments have many diverse functions, especially in cellular mechanics,[16] but do not seem to play much part in adhesion.

The main structure influencing adhesion is the acto-myosin network, composed of actin filaments which are small proteins that can assemble (polymerise) to long semiflexible filaments and also disassemble again. In addition the network contains many variants of actin cross-linking proteins and mini-filaments of myosin motors (in particular non-muscle myosin IIs (NMM II)). These motor proteins can bind to actin and produce mechanical forces due to ATP hydrolysis. Altogether, these components form active networks as the motors can rearrange the filaments and produce contractile forces when coupled in the right geometry. The in vitro analysis of the rheological properties of these active networks has become a very large and active field in biophysics especially the study of their non-equilibrium character.[17,18] As these acto-myosin filaments bundle and team up to larger and more forceful structures, one speaks of 'stress fibres' in the cell.[14] These bundles are spanning the interior of the cell and were already described in the late 1970s and early 1980s. Interestingly they share many similarities with muscle fibres in terms of composition and that they are always contractile,[19] so naively one could call them the muscles of a cell.

The stress fibres contribute most of the cell's mechanical properties and can exert significant forces, but such forces can only be effectively transmitted if there is an anchor point. The cell therefore needs to be able to connect its contractile cytoskeleton to the outside world and this is done via the integrin adhesion molecules. On the cytoplasmic side of the membrane there are multiple linker proteins (i.e. talin, vinculin, paxillin, etc.) that further connect the integrins to the cytoskeleton and on the outside the integrins connect to the extracellular matrix. When a cell attaches, integrins bind and aggregate so that more integrins will bind at the same region. These clusters of integrins and many other linker proteins are called focal adhesions (FAs) and can grow up to several square microns.[20,21] Figure 10.3 shows a sketch of such a focal adhesion plaque with the cytoplasmic side on the upper and the extracellular region on the lower part of the image.

This is a picture based on molecular connections but it is quite clear that there is a mechanical linkage between the contractile stress fibres, the integrin clusters and the substrate matrix. Based on this model, the experiment with soluble adhesion molecules must lead to the conclusion that forces due to adhesion interaction itself are vital for the cell rather than simply the chemical bond of adhesion molecule to the cell. Now it is important to address the substrate properties and how they influence cell adhesion.

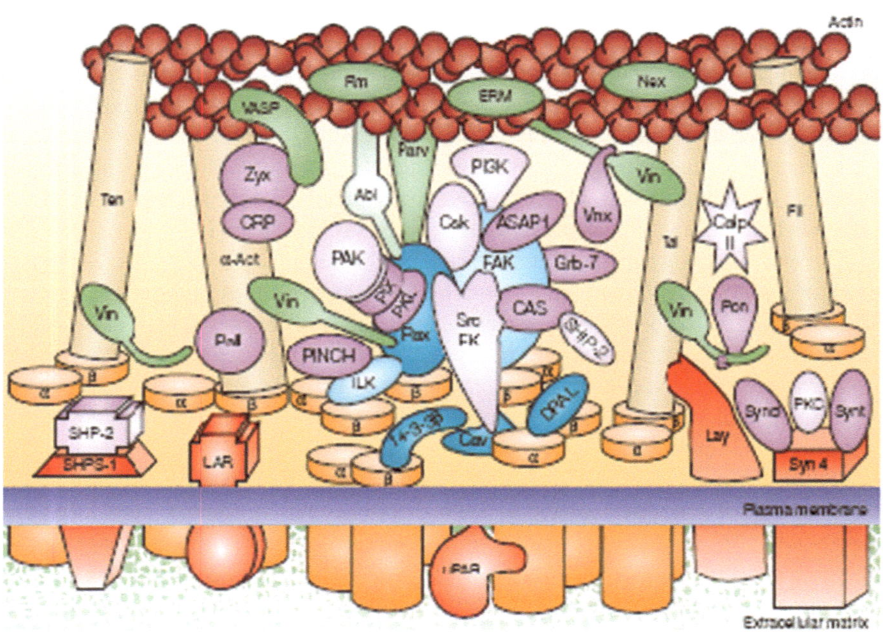

Fig. 10.3 Sketch of focal adhesion from Geiger et al.[20] (with permission)

10.4 Substrate Elasticity and Cell Health

Let us have a look now at the interactions that take place at the cell-matrix interface, especially when the substrate is made softer and more compliant. There is a wide physiological range of elasticity that is much softer than the usual cell culture dishes made of plastic or glass (Fig. 10.1) so it seems obvious to investigate cell adhesion on compliant matrices. Nevertheless it was some time before this was done by Harris in 1980, growing non-muscle cells (fibroblasts) on thin silicone membranes. He found that the cells are able to deform the flexible substrates causing wrinkles (Fig. 10.4).[22,23] So the stress fibres can also exert forces on the substrate even though the cells are not muscle cells. And if the link works in one direction it should also work in the reverse; the cell should be able to respond to the mechanics of the support material.

Following up on these findings, one of the first scientists to study the impact of substrate mechanics on cell behaviour was Opas in 1989.[24] He cultured chick retinal pigmented epithelial (RPE) cells on rigid and elastic substrates that both had a similar surface composition and the cells showed a significantly different response. On the soft thick Matrigel™ (secreted proteins from a sarcoma cell line) substrate the cells did not spread out and maintained their differentiated phenotype being heavily pigmented. On the rigid glass substrate that was covered with Matrigel™ solution and then covalently coupled, the cells began to spread out, formed stress fibres and FA complexes and started to grow in a different structure expressing the dedifferentiated phenotype. Although Opas used only two different conditions – rigid and

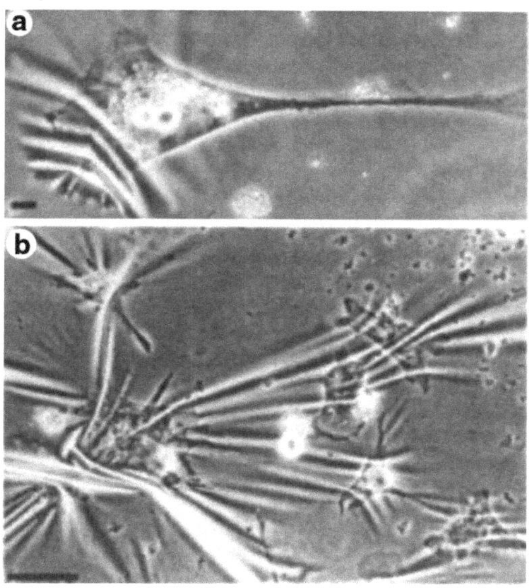

Fig. 10.4 Fibroblast cells on thin silicone films cause wrinkles due to their tensile forces[23] (Reprinted with permission)

elastic – with no quantitative determination of the Young's modulus, this was the first qualitative proof for the mechano-sensitive response to the extracellular support.

Pelham and Wang then significantly advanced the field introducing polyacrylamide (PA) gels as flexible substrates to grow cells on. PA gels are routinely used in gel electrophoresis because of chemical inertness (i.e. proteins do not bond) and the ease of tunability in pore size that allows for a precise control of the mobility of proteins. Using the same chemistry one can control the mechanical properties (i.e. the Young's elastic modulus E) of these PA gels. Soft and rigid gels were tested and distinct differences in cell spread area and formation of the focal adhesion plaques were found.[25] Pelham and Wang also identified myosin as a key player responsible for the elasticity dependent adhesion behaviour. Adding 2,3-butanedione monoxime or KT5926, small molecules that biochemically inhibit the myosin motors, they could demonstrate that cells were no longer able to distinguish a rigid from a soft substrate. This seminal paper finally opened up the field of cell-matrix mechanics and since then a rapidly growing number of studies appeared with astonishingly novel findings, as discussed later.

But first we will take a look at some models of cells adhering to a substrate.

10.5 Model Systems for Cell Adhesion Measurements

To study and understand these complex cell adhesion phenomena it has been very helpful to introduce simplified model systems. One of the simplest models for a cell is a giant unilamellar vesicle (GUV) consisting of a lipid bilayer resembling the outer cell membrane. Such a system is of course passive compared to a living cell but allows investigation of the basic adhesion mechanisms by adding component molecules one by one. The first essential ingredient is a polymer brush that can mimic the function of the glycocalix the pericellular coat of the cell, to prevent non-specific adhesion.

Marx and coworkers performed adhesion experiments with GUVs consisting of the phospholipid DEPC and cholesterol to model the cell[26] and a coated glass surface to act as substrate. The vesicle substrate interaction is composed of repulsive interactions on the subnanometer distances h due to hydration interaction of the phospholipids headgroups and attractive van der Waals forces that further compete with electrostatic repulsion. Using biologically relevant salt concentrations (300 mOsm L^{-1}) the potential has a single minimum, the 'van der Waals minimum' with a depth $W = -V_{(h)}$ of about 10^{-4}–10^{-5} Jm^{-2} at an optimal spacing h^* of about 2–3 nm. Furthermore, membrane undulations due to thermal energy lead to a repulsive interaction – the Helfrich repulsion. In this study, van der Waals interactions were tuned by coating the glass substrates with a thin, nonadhesive, protein passivation film (Blotting Grade Blocker Non-Fat Dry Milk, BioRad, CA). The film thickness was adjusted until the bilayer was in an unbound state corresponding to a value of W of about $1.3 \cdot 10^{-6}$ Jm^{-2}. Otherwise, the vesicle would just adhere tightly and even spread on a clean glass substrate.

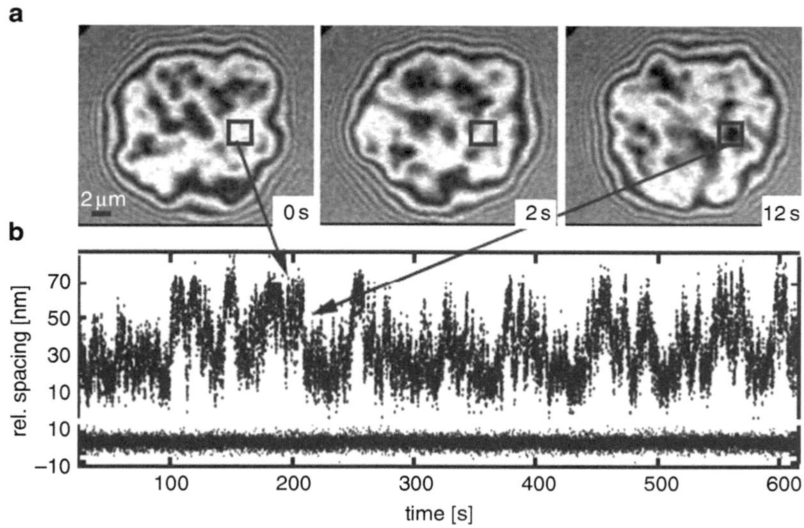

Fig. 10.5 (**a**) RICM images of a giant DEPC (containing 50 mol% cholesterol) vesicle at three consecutive time points. (**b**) Relative spacing between substrate and membrane obtained by an inverse cosine transformation of the RICM intensity image in area of interest depicted by the *boxes* in the *upper panel*. Lower trace shows the signal noise level[26] (Reproduced with permission)

They analysed the adhesion behaviour of these model cells by Reflectance Interference Contrast Microscopy (RICM, see Chapter 7 for details on the technique). Figure 10.5 (top row) shows typical images of an adhering model cell on the coated glass substrate at different times, nicely depicting the fluctuating gap between vesicle and surface as monitored by intensity changes. From the interference pattern one can calculate the distance between the membrane and the surface and analysing a distinct area of interest (black box in the images) one sees the fluctuations in the relative distance vs. time (lower graph (b)). The gap can be quite thick, up to 70 nm, or very thin, down to about 1 nm.

Analysing the relative fluctuations, it is clear that the model cell is not strongly adhering and varying the cholesterol content affects the flexibility of the membrane. One can now refine the model experiment by adding to the vesicle a PEG-lipid that serves as a model for the glycocalix of a cell and is counteracting the *vdW* adhesion. The Flory radius R_g of the PEG-lipid (3.8 nm) is comparable to the position of the minimum h^* of $V(h)$.

Varying the content of PEG-lipid and cholesterol in the vesicles leads to different adhesion profiles as shown in Fig. 10.6 that cannot be fitted with a single Gaussian anymore as it is the case for a pure DEPC vesicle (top left graph). The polymer brushes (PEG-lipids) favour a state with a larger spacing and the increasing stiffness due to the cholesterol concentration makes thermal fluctuations energetically more costly and hence favours a state with a smaller spacing. It is important to note that in all these experiments no 'adhesion molecules' such as integrins are

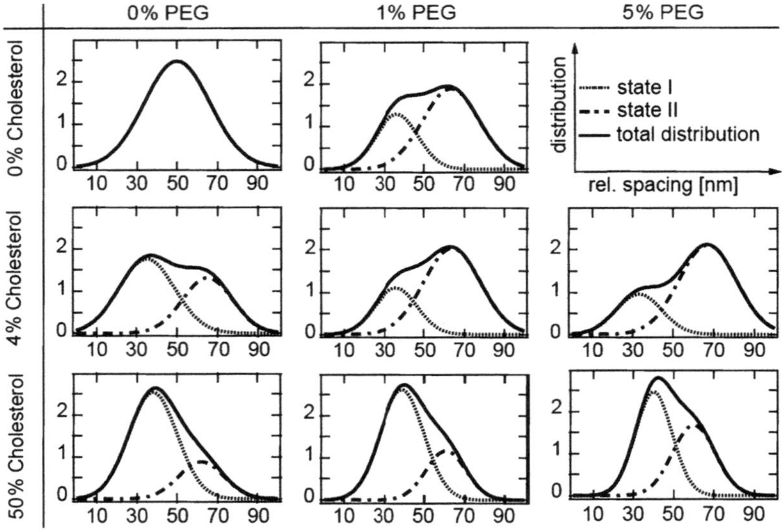

Fig. 10.6 Fitted probability distributions of substrate-membrane spacing for various values of PEG-lipid and cholesterol concentrations in a DEPC vesicle[26] (Reproduced with permission)

used and that the equilibrium state is only influenced by the composition of the model cell membrane.

Using this basic contact model, it was then possible to insert adhesion molecules to study their influence. Sackmann and his colleagues incorporated the lipid DMPE (1,2-dimyristoyl-sn-glycero-3-phosphoethanolamine) with a cyclic RGD peptide (specific amino acid sequence that serves as extracellular ligand for the integrins) headgroup in the vesicles as an adhesion molecule in the model cell and coated the substrate with a lipid bilayer bearing integrins ($\alpha_{IIb}\beta_3$)[27] as substrate anchor.

Again, using RICM to determine cell adhesion they could calculate the free energy of adhesion and found that integrins on a glass surface led to a lower adhesion energy than integrins on a soft polymer cushion (made of cellulose) as shown in Fig. 10.7. Increasing the concentration of PEG-lipid on the outside of the vesicles reduced the adhesion interaction. This shows, that despite the surfaces having the same amount of integrins, the geometry and activity of the integrins was determined by the underlying polymer layer demonstrating that it is essential not only to have the molecules on the surface but also in a well-organised geometry and functional structure.

Of course these are just two examples of model systems of cell adhesion and many more are published. However they serve as plain demonstrations that adhesion itself does not need any 'adhesion molecules' but happens all the time due to attractive van der Waals interactions. Furthermore, molecules (PEG-lipids) that act to reduce adhesion by increasing the gap between the cell and the substrate were needed to keep the model systems in an active and dynamic state that is essential for living systems.

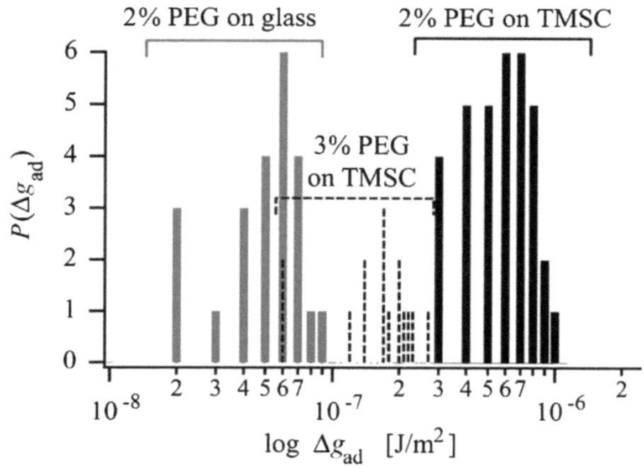

Fig. 10.7 Histograms of the measured free energies of adhesion of vesicles containing 2 and 3 mol% of lipids (PEG lipids) on cellulose (*black* and *dotted lines*, respectively) and of vesicles containing 2 mol% PEG lipids on bare glass substrates (*grey lines*)[27] (Reproduced with permission)

10.6 Substrate Geometry Effects on Cell Health

In the last section we saw how simple model systems of artificial cells help to understand the adhesion process. In the same way we can create well defined substrate geometries to selectively tune parameters and study the effects on adhesion of real cells.

With the advent of soft lithography,[28] it became straightforward and relatively easy to create patterned surfaces on the micrometer scale (Fig. 10.8). Chen et al. used soft lithography to produce substrates of different geometries that were coated with extracellular-matrix molecules.[29] The potential cell spread area governed by the printed patterns was varied while maintaining the total cell-matrix contact area constant by changing the spacing between multiple "*adhesion islands*". There could be one small contact spot, or several subdivided small spots of the same area or one large contact spot of the same diameter. Cell shape was found to vary with the contact geometry and this governed whether individual cells would grow or die, regardless of the type of matrix protein used to promote adhesion.

Although the local adhesion forces were very similar, the global cell behaviour depended on the geometry and size of the available adhesion area. Cell death occurred for too small of an adhesion area, which caused the cell to round up. A large single spot allowed the cell to flatten onto the substrate and thrive. The question then is what happens when a number of smaller focal adhesions are designated (Fig. 10.8)?

In conclusion, there is a minimal overall cell area that is needed for survival but that area might be distributed over several small spots to ensure cell spreading.

Another approach to investigate the influence of geometry was done by Spatz and coworkers.[30,31] They created nanopatterned substrates by deposition of 6 nm gold beads in a well-defined 2D structure using block copolymer micelle nanolithography. The gold beads were functionalised with an (RGD peptide) and were small enough so that only one integrin at the cell surface will bind each bead

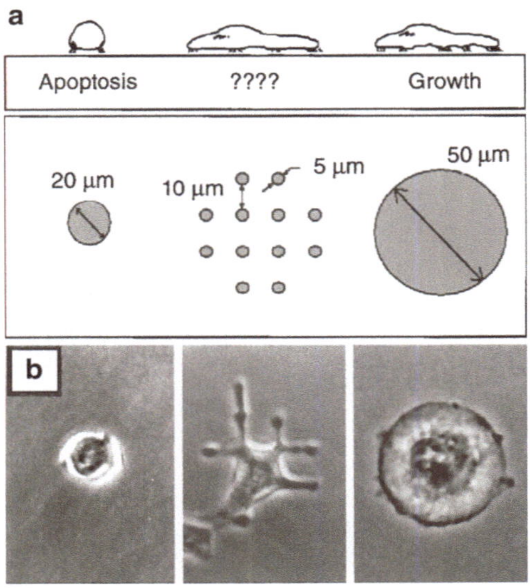

Fig. 10.8 Cell contact area versus cell spreading on micropatterned substrates as a regulator of cell fate. (**a**) Diagram of substrates used to vary cell shape independently of the cell substrate contact area. Substrates were patterned with small, closely spaced circular islands (*centre*) so that cell spreading could be promoted as in cells on larger, single round islands, but the contact area would be low as in cells on the small islands. (**b**) Phase-contrast micrographs of cells spread on single 20 or 50 μm diameter *circles* or multiple 5 μm *circles* patterned as shown in (a)[29] (with permission)

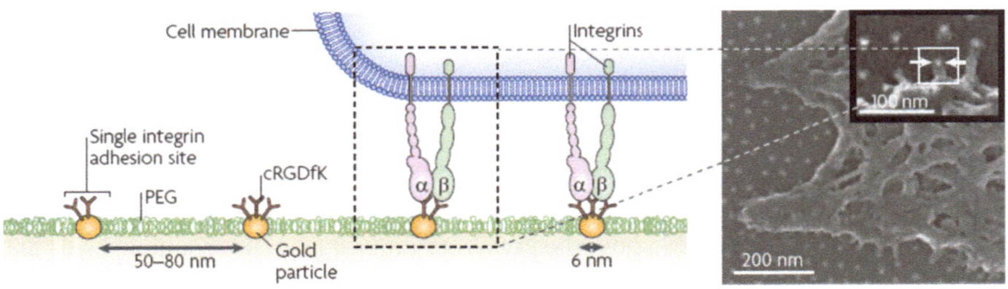

Fig. 10.9 A schematic of a biofunctionalised gold particle substrate in contact with a cell membrane (*left panel*) and a scanning electron micrograph of a cell that is adhering to a gold particle (*right panel*). To enable the specific interaction of gold nanoparticles with integrins, the nanoparticles were functionalised with RGD-peptide. A functionalised gold particle with a diameter of ~6 nm on a PEG passivated background is small enough to allow the binding of only a single integrin protein[32] (Reproduced with permission)

(Fig. 10.9). The trick is now to vary the spacing between the gold beads to control cell adhesion. The experiment shows that when the beads are separated by more than 73 nm, cell spreading and the formation of focal adhesions were abnormal and the cell eventually underwent cell death, but separations smaller than 58 nm allowed for normal adhesion and cell growth.

This second example demonstrates that nanoscale geometry affects cell adhesion by affecting integrin clustering and therefore focal adhesion plaque formation that is essential for normal cell growth.

10.7 Elasticity Drives Stem Cell Differentiation

Having understood the varied contact area effects, it is now important to consider the effect of elasticity of the substrate. A phenomenal landmark study was published in 2006 by Engler et al. from the lab of Dennis E. Discher at the University of Pennsylvania in Philadelphia. They grew hMSCs on flexible substrates and found that these naïve adult stem cells differentiated into different cell lineages depending on the substrate elasticity.[33] Three different lineages of cells could be observed; neurogenic, myogenic and osteogenic and these were recognised by specific fluorescence tests as shown in Fig. 10.10.

The hMSCs plated on elastic PA gels with an elastic Young's modulus of 1, 11, and 34 kPa were immunostained with fluorescent markers for lineage specific molecules: β3 tubulin for the neurogenic, MyoD for the myogenic and CBFα1 for the osteogenic differentiation route. After 1 week, a selective expression of these markers by the cells was observed respectively on 1, 11, and 34 kPa substrates as shown in panel A of Fig. 10.9. These results were supported with array transcription profiling and comparing the results to committed muscle (C2C12) and bone cells (hFOB) as presented in panel B. Panel C shows a graph of fluorescent intensity of the differentiation markers versus substrate elasticity revealing maximal lineage specification at the elasticity E typical for the respective tissue type. The fascinating results are not only that adult stem cell differentiation is driven by matrix elasticity, but that these naïve hMSCs move towards the lineage that is typical for that elasticity (see also Fig. 10.1). Cells therefore adapt their phenotype according to the mechanical properties they feel. It is noteworthy that the biochemical surface composition, using collagen-I as extracellular matrix, was identical for the gels of different stiffness indicating that only the elasticity of the microenvironment was driving the cells toward different lineages, not the biochemistry. It is an unpredictable, but highly repeatable result.

10.8 Physical Models for Cell-Substrate Interactions

These striking experiments by Engler et al. raise several questions about how a relatively simple mechanical property, the Young's elastic modulus E of the substrate, adhering on the outside of the cell can induce such huge effects on the inside of the cell causing changes in cell differentiation. The first question relates to the mechano-transduction of the matrix elasticity information into the cell. Secondly, how does the cell machinery integrates these mechanical cues to produce biochemical signalling that drives differentiation?

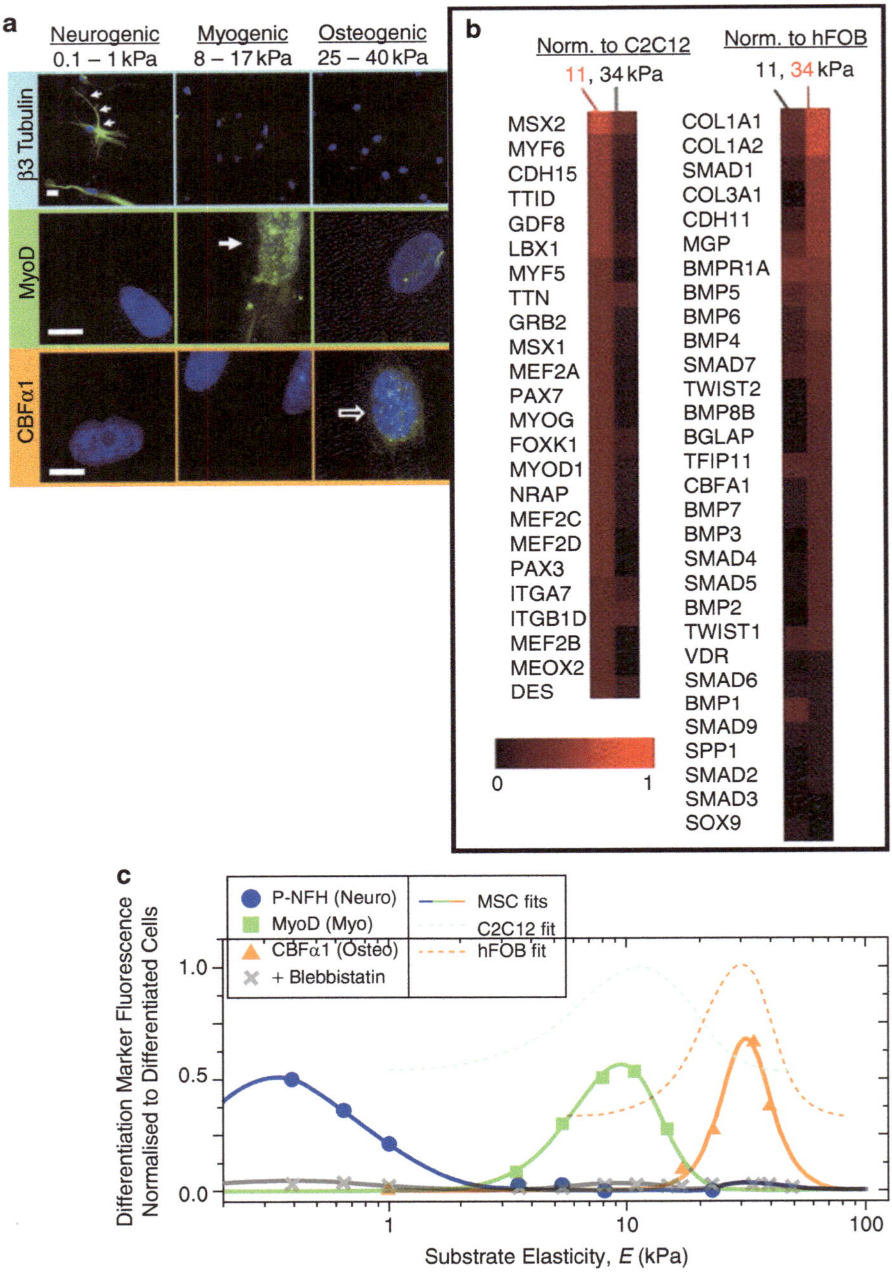

Fig. 10.10 Stem cell differentiation driven by matrix elasticity. (**a**) The neurogenic marker β3 tubulin is expressed in branches (*arrows*) only on the soft matrices. The muscle transcription factor MyoD is produced and nuclear localised (*arrow*) only in hMSCs on matrices of intermediate stiffness. The osteoblast transcription factor CBFa1 (*arrow*) is likewise expressed only on stiff matrices. Scale bar is 5 μm. (**b**) Microarray profiles of hMSCs cultured on 11 or 34 kPa matrices, with expression normalised committed myoblasts (C2C12) and osteoblasts (hFOB) show selective differentiation. (**c**) Fluorescent intensity of differentiation markers versus substrate elasticity reveals maximal lineage specification at the elasticity *E* typical for the respective tissue type[33] (Reproduced with permission)

As described in Section 10.3, the acto-myosin cytoskeleton is the major scaffold in the cell, connects the exterior with the interior of the cell and is also linked mechanically to the nucleus. It is therefore conceivable to speak about a direct mechanical connection of the extracellular matrix and the nucleus. But this is clearly not the only pathway that can lead to differentiation. As the cells were grown for several days the mechanism must be a combination of physical force transduction of biochemical regulation.

To understand the initial impact of cell adhesion on substrates with different elasticity we closely examined the early changes in cytoskeletal arrangement.[34,35] Focusing on the first 24 h ensures that we mainly observe the physical impact of matrix elasticity on the cell that further leads to stem cell differentiation. We plated hMSCs on elastic PA substrates, as described in the last section, fixed the cells and fluorescently stained actin and non-muscle myosin IIa to analyse the organisation of the cytoskeleton. hMSCs fixed after 1 h exhibited a pretty circular spread area as shown in Fig. 10.11, resulting in aspect ratios r close to unity. These cells did not show any significant order or pattern of the acto-myosin filaments as quantified by the order parameter S.[34]

Twenty-four hours after putting the cells on the substrate, the situation looks significantly different. Cells vary not only in shape (aspect ratio r) but also their cytoskeletal organisation is distinct (order parameter S) as seen in Fig. 10.12.

Cells on the soft 1 kPa matrices maintain a small spread area that has a low aspect ratio and very low order parameter indicating isotropic distribution of stress fibres. In contrast, cells on 11 kPa substrates are highly elongated as denoted by an average aspect ratio of 3.3 and exhibit parallel aligned stress fibres (magnified in the inset) resulting in a high order parameter of $S = 0.63$. On even stiffer substrates of 34 kPa, the cells are larger in total area but show reduced aspect ratio and order parameter indicating a more isotropic shape and distribution. This non-monotonic behaviour after 24 h is very similar to elasticity dependence of the myogenic marker proteins showed in panel C of Fig. 10.10 1 week after plating. It seems that the early stage cytoskeletal response is directly linked to the differentiation process occurring over the next several days. Together with the theoretical physicists Assaf Zemel and

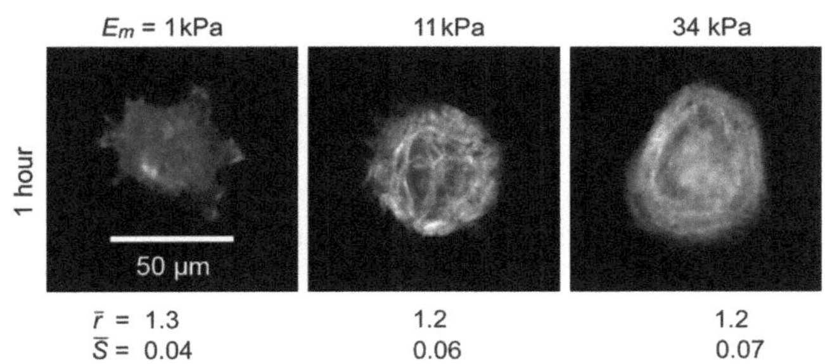

Fig. 10.11 Immunofluorescence images of NMM IIa in hMSCs on elastic substrates of 1, 11, and 34 kPa 1 h after plating. Cells are nearly circular but have an aspect ratio slightly larger than unity, indicating slight deviations from perfect symmetry[31] (Reprinted with permission)

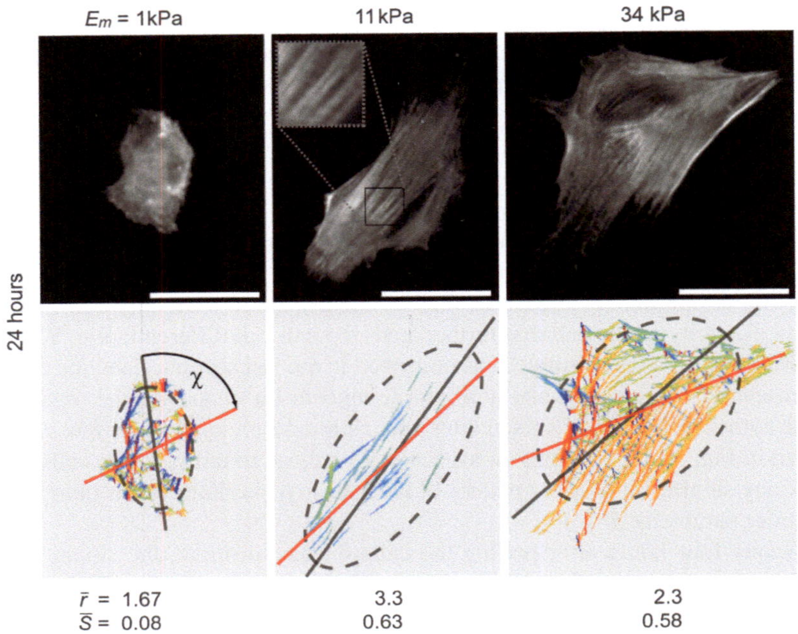

Fig. 10.12 Acto-myosin stress-fibre alignment in hMSCs sparsely plated on substrates of different elasticity. The *top row* shows hMSCs immuno-stained for (NMMIIa) 24 h after plating on elastic substrates with a Young's modulus E_m of 1, 11, and 34 kPa. The *bottom row* shows the respective orientational plots, where the different orientations of myosin filaments are depicted with different colours. The *dark grey dashed ellipses* are calculated from the moments up to the second order and represent cell shape in terms of area and long and short axes, and the *red line* indicates the mean orientation of the stress fibres as determined by the anisotropic filter algorithm. χ is the angle between the mean stress-fibre orientation and the principal axis of the ellipse. Values given for r and S are the mean values of at least 60 cells per condition. All scale bars represent 50 μm[34,35] (Reprinted with permission)

Samuel Safran a model was established based on classical mechanics to explain the non-monotonic cytoskeletal ordering phenomenon observed in the experiments. Extending the theory of Eshelby from the 1950s,[36] the cell is modelled as an active elastic inclusion that has the ability to create a contractile force as shown on the right side of Fig. 10.13. Initially, the cell and matrix are separated and characterised by springs with constants k_c and k_m, respectively. When the cell attaches to the surface after deposition, both springs will be deformed due to the adhesion process. Then, the cell creates contractile forces due to the forming of stress fibres and a new equilibrium state will be established. This one dimensional spring model simply visualises the process that in reality happens in two and three dimensions depending on the experiment. In a stratified system like our elastic PA gels, the cell is modelled as a 2D inclusion in the elastic matrix.

Due to the active contribution of the cell in reaction to the matrix elasticity one finds a resonance effect when both elasticities are of the same order of magnitude

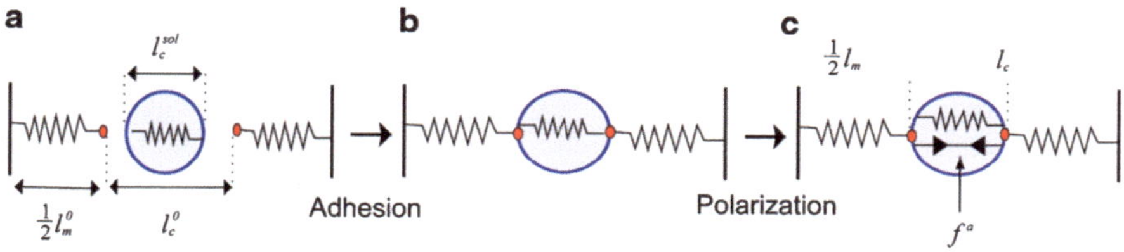

Fig. 10.13 Cell adhesion and polarisation represented by a 1D spring model. The elasticity of the cell and the matrix is represented by springs with constants k_c and k_m, respectively. Elastic morphological changes upon cell adhesion are represented here by a change of cellular spring length. The active cellular forces restore then the length to an equilibrium state[35] (Reprinted with permission)

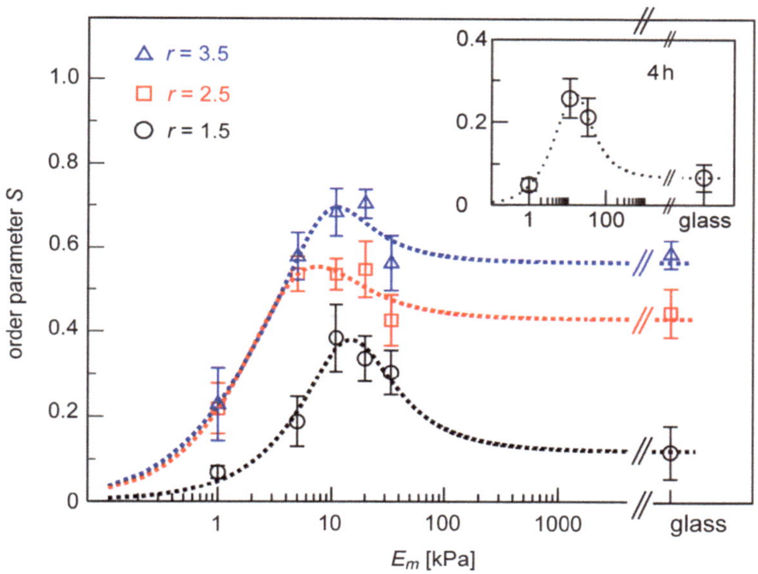

Fig. 10.14 The graph shows experimental values of the stress-fibre order parameter S, 24 h after plating the cells, for the three groups of cells (of aspect ratios $r = 1.5, 2.5, 3.5$) as a function of the Young's modulus of the matrix, E_m. Within each of the different groups, S is maximal for $E_m = 11$ kPa and generally increases with aspect ratio r, in agreement with the theoretical predictions. *Error bars* denote the standard error of the mean and theory curves (*dotted lines*) calculated from the simplified expansion of S are shown to guide the eye. *Inset* shows cells already start to exhibit the non-monotonic behaviour after 4 h[35] (Reprinted with permission)

that leads to a peak in the order parameter. The exact form of the graph also depends on a susceptibility parameter that defines how sensitive the cell is to the matrix elasticity and to the aspect ratio of the cell r.

Therefore we grouped our observed cells according to aspect ratio and plotted the experimental values together with the theoretical predictions in Fig. 10.14. For each of the three groups the maximum value for the order parameter occurs on gels of modulus 11 kPa, approximately equal to the elasticity of the cells. The cytoskeletal ordering also increases with higher aspect ratio *r*.

This study shows that the elasticity of the substrate dictates the organisation of the cytoskeleton and that the cell-matrix interactions can be reasonably well described with a simple classical mechanics model.

This early time regime (24 hours) is just the onset of differentiation and the second step is the biochemical regulation pathway which will kick in to lead to transcriptional changes. However, it is important to combine those with the mechanical interactions and use physical models to refine existing biochemical signalling theories to elucidate the complex interplay of cell adhesion with mechanics and molecules.

10.9 Conclusions

This chapter provides only a brief overview of the nanoscale aspects of cell adhesion phenomena but we hope to convince the reader that inclusion of the physical aspects (e.g. mechanics) are essential for understanding the entire cell development process. For too long, the simplistic biochemical models of ‘lock and key adhesion’ have dominated, whereas we show that there is an interesting interplay between the chemistry, the geometry and the elasticity of both cell and substrate. We understand that a biological cell is a fairly complex structure and that its interactions with the environment that govern adhesion are very versatile. It is obvious that one needs to control the experimental conditions extremely well in order to explore the basic underlying mechanisms. The simple picture of cell adhesion is that the actin cytoskeleton of the cell meets with integrin molecules which then interact with molecules on the substrate surface.

Vesicle models of cells show how the black contact spot is fluctuating between close and distant contact which depends on the presence of polymers and other molecules. These molecules tend to reduce the strong van der Waals bonding which would fix the vesicle strongly to its substrate and rigidify the system. The geometry of the contact is vital to cell viability. Too small a contact spot and the cell dies. A larger contact and subdivision of the focal adhesion points allow spreading and survival of the cell.

The elastic modulus of the substrate plays an important role because the cell is sensing the elastic support using its actin filaments, then responding by altering the structure. A classical elastic inclusion model of the cell in a matrix can provide a neat description of the results.

In the future it will be necessary to dissect the complex interplay of physical and biochemical factors governing cell adhesion and influencing cell behaviour. One last example of the complex interwoven mechanisms is the elasticity dependent efficacy of an anti-cancer drug[37] that might affect the way these drugs are developed and tested. We hope that the mechanics of cell adhesion will contribute to a better understanding of cellular processes and potentially will lead to new strategies in therapeutic applications.[7]

References

1. Szathmary E. and Smith J.M., The major evolutionary transitions, Nature 374 (1995) 227–232.
2. Ley K., Laudanna C., Cybulsky M.I. and Nourshargh S., Getting to the site of inflammation: The leukocyte adhesion cascade updated, Nature Reviews Immunology 7 (2007) 678–689.
3. Papayannopoulou T. and Scadden D.T., Stem-cell ecology and stem cells in motion, Blood 111 (2008) 3923–3930.
4. Alberts B., Johnson A., Lewis J., Raff M., Roberts K. and Walter P., *Molecular biology of the cell*. 4th ed. 2002, New York: Garland Science.
5. Blackburn E.H., Greider C.W. and Szostak J.W., Telomeres and telomerase: The path from maize, tetrahymena and yeast to human cancer and aging, Nat. Med. 12 (2006) 1133–1138.
6. Radmacher M., Measuring the elastic properties of biological samples with the afm, Ieee Engineering in Medicine and Biology Magazine 16 (1997) 47–57.
7. Rehfeldt F., Engler A.J. and Discher D.E., *Stem cells and nanomedicine: Nanomechanics of the microenvironment*, in *Nanotechnology*, V. Vogel, Editor. 2009, Wiley VCH.
8. Elger M., Drenckhahn D., Nobiling R., Mundel P. and Kriz W., Cultured rat mesangial cells contain smooth-muscle alpha-actin not found invivo, Am. J. Pathol. 142 (1993) 497–509.
9. Alberts B., Johnson A., Lewis J., Raff M., Roberts K. and Walter P., *Molecular biology of the cell*. 5th Edition ed. 2007: Garland Science.
10. Hynes R.O., Integrins: Bidirectional, allosteric signaling machines, Cell 110 (2002) 673–687.
11. Hynes R.O., Integrins - a family of cell-surface receptors, Cell 48 (1987) 549–554.
12. Madden H.L. and Henke C.A., Induction of lung fibroblast apoptosis by soluble fibronectin peptides, Am. J. Respir. Crit. Care Med. 162 (2000) 1553–1560.
13. Frixione E., Recurring views on the structure and function of the cytoskeleton: A 300-year epic, Cell Motil. Cytoskeleton 46 (2000) 73–94.
14. Pellegrin S. and Mellor H., Actin stress fibres, Journal of Cell Science 120 (2007) 3491–3499.
15. Olmsted J.B. and Borisy G.G., Microtubules, Annu. Rev. Biochem. 42 (1973) 507–540.
16. Herrmann H., Strelkov S.V., Burkhard P. and Aebi U., Intermediate filaments: Primary determinants of cell architecture and plasticity, J. Clin. Invest. 119 (2009) 1772–1783.
17. Mizuno D., Tardin C., Schmidt C.F. and MacKintosh F.C., Nonequilibrium mechanics of active cytoskeletal networks, Science 315 (2007) 370–373.
18. Koenderink G.H., Dogic Z., Nakamura F., Bendix P.M., MacKintosh F.C., Hartwig J.H., Stossel T.P. and Weitz D.A., An active biopolymer network controlled by molecular motors, Proc. Natl. Acad. Sci. U. S. A. 106 (2009) 15192–15197.
19. Burridge K., Are stress fibers contractile, Nature 294 (1981) 691–692.
20. Geiger B., Bershadsky A., Pankov R. and Yamada K.M., Transmembrane extracellular matrix-cytoskeleton crosstalk, Nat. Rev. Mol. Cell Biol. 2 (2001) 793–805.
21. Goffin J.M., Pittet P., Csucs G., Lussi J.W., Meister J.J. and Hinz B., Focal adhesion size controls tension-dependent recruitment of alpha-smooth muscle actin to stress fibers, J Cell Biol 172 (2006) 259–268.
22. Harris A.K., Wild P. and Stopak D., Silicone-rubber substrata - new wrinkle in the study of cell locomotion, Science 208 (1980) 177–179.
23. Harris A.K., Stopak D. and Wild P., Fibroblast traction as a mechanism for collagen morphogenesis, Nature 290 (1981) 249–251.

24. Opas M., Expression of the differentiated phenotype by epithelial-cells invitro is regulated by both biochemistry and mechanics of the substratum, Developmental Biology 131 (1989) 281–293.
25. Pelham R.J., and Wang Y.L., Cell locomotion and focal adhesions are regulated by substrate flexibility, Proc. Natl. Acad. Sci. U. S. A. 94 (1997) 13661–13665.
26. Marx S., Schilling J., Sackmann E., and Bruinsma R., Helfrich repulsion and dynamical phase separation of multicomponent lipid bilayers, Physical Review Letters 88 (2002).
27. Goennenwein S., Tanaka M., Hu B., Moroder L., and Sackmann E., Functional incorporation of integrins into solid supported membranes on ultrathin films of cellulose: Impact on adhesion, Biophysical Journal 85 (2003) 646–655.
28. Xia Y.N. and Whitesides G.M., Soft lithography, Angewandte Chemie-International Edition 37 (1998) 551–575.
29. Chen C.S., Mrksich M., Huang S., Whitesides G.M., and Ingber D.E., Geometric control of cell life and death, Science 276 (1997) 1425–1428.
30. Arnold M., Cavalcanti-Adam E.A., Glass R., Blummel J., Eck W., Kantlehner M., Kessler H., and Spatz J.P., Activation of integrin function by nanopatterned adhesive interfaces, ChemPhysChem 5 (2004) 383–388.
31. Cavalcanti-Adam E.A., Volberg T., Micoulet A., Kessler H., Geiger B., and Spatz J.P., Cell spreading and focal adhesion dynamics are regulated by spacing of integrin ligands, Biophysical Journal 92 (2007) 2964–2974.
32. Geiger B., Spatz J.P., and Bershadsky A.D., Environmental sensing through focal adhesions, Nat. Rev. Mol. Cell Biol. 10 (2009) 21–33.
33. Engler A.J., Sen S., Sweeney H.L., and Discher D.E., Matrix elasticity directs stem cell lineage specification, Cell 126 (2006) 677–689.
34. Zemel A., Rehfeldt F., Brown A.E.X., Discher D.E., and Safran S., Optimal matrix rigidity for stress-fiber polarization in stem cells, Nature Physics **(doi:10.1038/nphys1613)**, (2010).
35. Zemel A., Rehfeldt F., Brown A.E.X., Discher D.E., and Safran S., Cell shape, spreading symmetry and the polarization of stress-fibers in cells, Journal of Physics: Condensed Matter 22 (2010) 194110.
36. Eshelby J.D., The determination of the elastic field of an ellipsoidal inclusion, and related problems, Proceedings of the Royal Society of London Series a-Mathematical and Physical Sciences 241 (1957) 376–396.
37. Rehfeldt F., Engler A.J., Eckhardt A., Ahmed F., and Discher D.E., Cell responses to the mechanochemical microenvironment - implications for regenerative medicine and drug delivery, Advanced Drug Delivery Reviews 59 (2007) 1329–1339.

Chapter 11
Nanoparticles Adhering to Cells; Toxicity Effects

adaptation of single cells to such minute structures...

Harrison 1914

Just as viruses are known to adhere to cells, then penetrate and kill them as described in Chapter 9, so nanoparticles (NPs) may attach to cells and cause damage. The mechanisms dominating such toxicity are not fully established but it seems clear that the molecules at the NP surface, i.e. the adhesion molecules, must be important. Moreover, the size of the particles is crucial; larger particles of the same material are not as toxic as smaller ones. Many different types of particles, from low modulus polystyrene to intermediate glass, to high modulus carbon have been studied and they all seem to cause problems when ingested in NP form, suggesting that elastic properties are not crucial. In this chapter, the general phenomena observed as NPs contact cells and organisms are described, and damage processes enumerated.

It is shown that NPs can cause inflammatory, respiratory and cardiovascular responses, sometimes leading to death in cells and whole organisms. On the other hand, NPs such as TiO_2 sunscreens could be used positively to protect the skin, deliver drugs to organs, or transfer DNA into cell nuclei. Here, differences between these problematic and beneficial effects are analysed. We also seek to unify the scientific language of diverse disciplines to identify common principles of NP toxicity effects.

11.1 Pathways into the Body

There is evidence that cells have always been exposed to nanoparticles.[1] Cells and organisms like the human body have adapted to natural NP exposures, evolving significant defence mechanisms to limit both dose through physiological defence (e.g. hairs in the nose to filter out nanoparticles) and chemical damage (e.g. blanketing the particles with surfactant proteins, or antioxidants). The problem is that our exposures to diverse insoluble material found in man-made NPs – from car exhaust particles in air, or engineered nanoparticles (ENPs) such as pigments in food – have risen above natural background levels and large exposures have been introduced. As these have

K. Kendall et al., *Adhesion of Cells, Viruses and Nanoparticles*,

DOI 10.1007/978-90-481-2585-2_11,

been studied, more knowledge has been gathered about the effects of large NP doses such that our understanding of the health implications of high exposures, especially in air pollution, has rapidly increased in the last decade.

The four key human exposure routes to fine particles are intravenous injection, inhalation, ingestion, and dermal penetration.[2] Once inside the body, NPs have the potential to be distributed by the circulatory systems and incorporated into cells because of their small size. Indeed, ENPs for drug delivery have been engineered directly for that purpose. The key question is: 'How do these NPs interact with biomolecules, adhere to cells then cause damage to organisms?' Human ENP exposures are relatively new, so only limited data to link exposures to health risks have emerged.[3] However, at extreme high exposure levels, the effects of elevated NP exposures are clear: A striking example is the permanent, light-activated syndrome *argyria* affecting people who drink nanosilver to "protect" their health[4] (Fig 11.1). The silver ENPs deposit in the skin cells, turning the consumer blue when exposed to light.

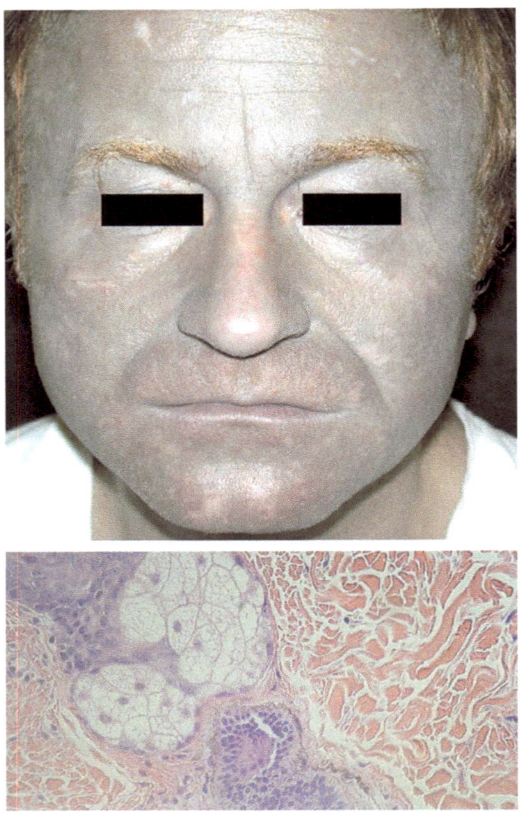

Fig. 11.1 A case of argyria[4]– the permanent blue discolouration of skin caused by the ingestion of silver nanoparticles which deposit in the skin (*dark areas* in lower photograph, with permission)

Consider the pathway these ENPs have traversed to appear in the blue skin, first reaching the gut, then accessing the circulatory systems and finally depositing in sub-cutaneous cells, turning them blue. If the same silver nanoparticles had been painted on the skin, they could not have penetrated so easily. The skin prevents particle uptake from the external environment by forming a continuous protective layer if not disrupted by injury.[5] In contrast, the lung and gut allow nanoscale material (particles, molecules or ions) across cellular barriers and into circulatory systems. While there are removal mechanisms at work, it is known from radiotracer studies in humans that only about 10% of nanoparticles depositing to the human lung are removed within 48 h, and around 1% of nanoparticles are transmitted into the blood circulation.[6] So, although some are cleared following lung deposition, a mechanism exists for a proportion of the NP dose to cross lung membranes, allowing them to reach other areas of the body and potentially cause systemic nanotoxicity. Intravenous injection is the most direct route for NPs into the circulatory system, and is utilised in medical imaging, as described later.

Our purpose in this chapter is to introduce the steps of NP transport, adhesion and toxicity to cells in the body. We focus finally on the most studied translocation processes by taking lessons from the lung, the current major exposure route to NPs. Much cell toxicity work has been completed recently as our ability to grow cells in petri dishes has improved, and we can access stable, cultured cells to dose them with media containing NPs. Animal exposure experiments – where NPs can be washed into the lungs or breathed in using an aerosol – are used to validate these cell studies in whole organisms, to give a complete picture. Epidemiological studies of industrially exposed populations of coal miners, metal welders etc. have provided vital data on human diseases that correlate with high particle exposures. Similarly, environmental epidemiology of air pollution exposure of whole populations informs us of the effects of NPs from car exhausts, for example. The fields of toxicology and epidemiology are therefore working in parallel to establish a biologically plausible mechanism for how sub-micron particles cause cardiovascular[7] and respiratory[8] disease. From these studies, the molecular basis of cell function, human health and disease is emerging, and the role of NP adhesion has become clear. Strangely, the chemical composition of insoluble, nominally inert NP surfaces has little influence, since polymer, carbon, silica and metal nanoparticles all cause toxicity. Here we discuss 'inert nanoparticles', not known toxic materials such as arsenic, lead, thorium and so on, although their effects may also be related to adhesion.

11.2 Variety of Nanoparticles

Nanoparticles are conventionally defined as those below 100 nm diameter, yet particles up to 200 nm are capable of entering cells, and larger grains > 1 μm are capable of entering and translocating within the body. Fibres must be dealt with separately since they exhibit toxicity related to their particular length-diameter ratio and related aerodynamics.

Therefore we define nanoparticles here as particles with any dimension below 1 μm. Many such natural nanoparticles exist in air as a result of normal processes such as forest fires, volcanic eruptions and entrainment of dust and seawater by wind, as described in Chapter 8. However, there is the recent problem of man-made insoluble nanoparticles from vehicle exhausts, power station emissions and synthetic nanoparticles in other media such as pigments and drugs manufactured by the chemical industries, so–called engineered nanoparticles (ENPs).

ENPs can present some new challenges to biological systems precisely because they have been designed for increased reactivity, mobility and persistence.[2] Key ENP exposure routes for humans are direct through medicinal (antibacterial, drug delivery and imaging tools), leisure (sports products), personal care (toothpastes, cosmetics and sun-screens), and food packaging products. Indirect routes exist through degradation and disposal of nanomaterials, plus occupational exposures in manufacturing industries. Hundreds of ENP-containing products are now on the market,[9] as illustrated in Fig. 11.2. In medicine, which employs direct injection or inhalation of nanoparticles into the body, adhesion is undesirable so NP drugs may be coated to evade normal immune defences such as protein attachment.[10] On the other hand, adhesion of nanosilver on surfaces is desirable to confer antimicrobial properties – in socks, plasters and washing machines – without understanding the consequences for health. We need to understand these various effects of adhesion to control nanotoxicity effectively.

The concern is that novel materials may cause novel toxicity. But with such a variety of ENPs now available, it is difficult to pin down the problem areas, especially when disease stems from brief exposures. As with asbestosis or silicosis, years or decades may elapse before the health consequences of exposure become apparent. Therefore it is vital to address the generic toxicity issues involved such as

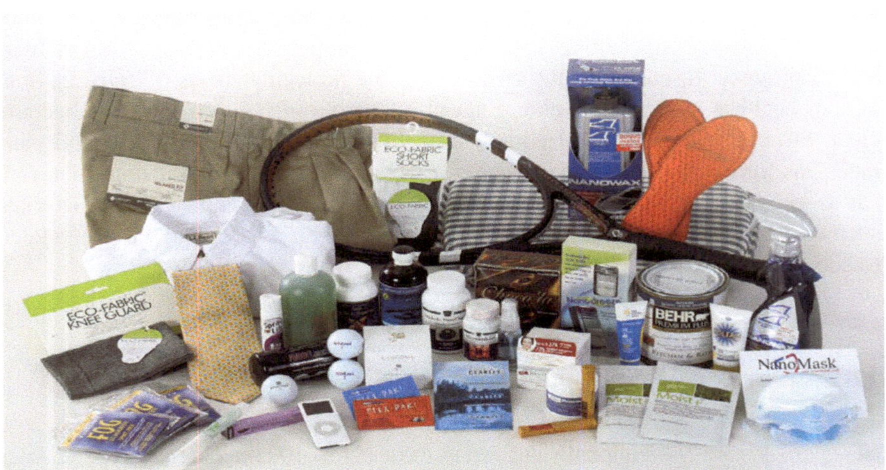

Fig. 11.2 Many products containing nanoparticles are now available on the market (Photo by David Hawxhurst-Woodrow Wilson International Center for Scholars, with permission)

particle dose, mobility across physical barriers, resistance to aggregation, cell penetrability, concentration in particular cells and toxic mechanisms. In the next sections, we consider these factors.

11.3 Measures of Nanoparticle Toxicity

In understanding nanoparticle-cell interactions, the three types of exposure test data – cell tests, animal tests and human epidemiology – must be analysed to assess toxicity.

Cell tests, where specific cell-lines are cultured in petri dishes and exposed to NPs mixed with suspending media, are important in principle because the molecular mechanisms, such as the inflammatory response, can be measured. However, there are criticisms from the medical fraternity because the cells are usually specialised and cannot represent the whole organism. Also the culture medium may be distracting because it modifies particle surfaces and changes particle aggregation state.[11,12] Time is also significant, varying from minutes to hours and days, probably not long enough to reveal long-term effects. But these cellular models have been useful in identifying key mechanistic pathways in cells. The cell culture system as a test bed of particle toxicity is especially useful because of the array of different cell types available, the modes of delivery of particles, the possibility of detailed particle characterisation, the precise range of doses and the adjustability of media conditions. In extreme toxicity cases, the cells die, as in the virus plaque test defined in Chapter 9.

In cell studies, a range of NP characteristics has been linked to toxicity effects.[13] The difficulty is providing evidence of toxicity mechanisms and then linking these to whole animal and epidemiological studies which are both difficult and expensive.

In animal exposure (in vivo) studies, the whole body fate of NPs can be established as they pass through various stages of processing, cellular uptake and or removal. For example, Oberdorster et al.[14] studied TiO_2 and Teflon nanoparticles inhaled by rats showing that the smaller diameters caused more death. But doses in these tests were high (mg kg^{-1} of rat) and interspecies differences were substantial.

Epidemiology is the only method where real world environmental exposures can be linked to incidence of disease, but where cross-correlation and lack of evidence for causality are recurring problems. In epidemiological studies of particle and nanoparticle air pollution exposures, many confounding parameters are now routinely controlled so that we can report findings with considerable confidence. For example, studies of particle exposures[15] of children in California showed that lung function was diminished by living in high atmospheric particle concentrations up to 30 $\mu g\ m^{-3}$ $PM_{2.5}$, as an annual average. These exposures are typical of urbanised areas in the developed world where millions of people live, exposed to urban aerosol materials such as fossil fuel emissions, road dust and natural background material. Measurable effects are reported at particle concentrations hundreds of times lower than in developing countries or in typical animal tests.

The major problem emerging from all three toxicity measures is that particle characteristics and dose measurements (mg kg^{-1} or mg m^{-3}) are inconsistent. In the past, cell studies tended to dose by mass, reasoning that the aim was to link NP mass dose to the physiological response. However, cells do not respond to insoluble particle mass concentration, but to particle number or surface area. Therefore it is much more rational to dose by surface area,[16] especially because extracellular molecules adsorb onto nanoparticle surfaces.

11.4 Nanoparticle Surface Interactions with Extracellular Molecules

Nanoparticles do not land in cells or next to cells, rather they move through a richly complex fluid system laced with components that interact or adhere to surfaces.[8] Studies of particle behaviour in biological fluids have therefore recently emerged as important.[11,17] In contrast to studying NP effects in cells, animals or humans, these studies observe the changes in particles resulting from adsorption of extracellular molecules surrounding the cells. Some of these processes do not involve cells directly, but are mediated by extracellular material alone. They can result in alteration of the physico-chemical properties of the particles such as surface charge or agglomeration. In Fig. 11.3 we show how different functionalised 200 nm polystyrene NPs agglomerated with time in the presence of fibrinogen.[18] We can see that amine groups grafted onto polystyrene surfaces slowed agglomeration of NPs in

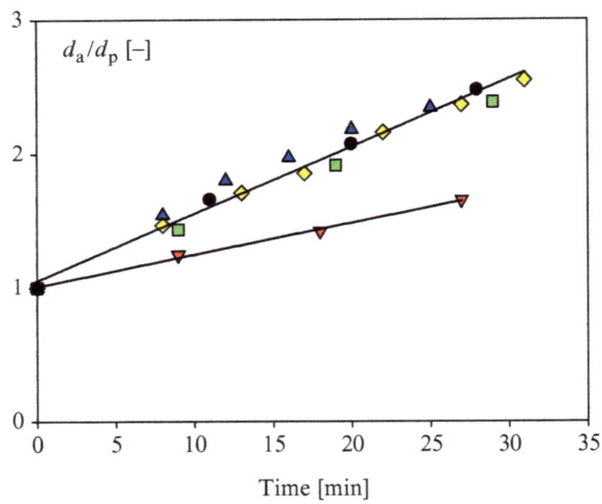

Fig. 11.3 Aggregation kinetics as a function of time for five different polystyrene particle types: particle size ~200 nm and surface area concentration: 1.25 cm^2 ml^{-1}. (○) Polystyrene, PS; (∇) amine functionalised polystyrene; (◊) carboxylate functionalised polystyrene; (□) hydroxylate functionalised polystyrene and (Δ) sulphate functionalised polystyrene. d_a = agglomerate size, d_p = original particle size

fibrinogen, while the other polystyrene particles behaved equally when dosed on a surface area basis. Clearly, small differences at the surface can produce significant agglomeration effects.

Polymer adsorption is fundamental in at least four stages of NP processing in biological systems: (1) initial adsorption to modify particle surfaces, (2) promotion of particle agglomeration or disagglomeration, (3) promotion of NP clearance and (4) downstream cellular interactions and toxicity. There is significant evidence particularly from pharmacology that polymer interaction with NP surfaces dictates cell uptake.[19] Surface adsorbed polymers have long been understood to determine the success of surgical implant materials.[21] Another effect of polymer is the resulting particle agglomeration. Agglomerated NPs are cleared better by cells and this process has a crucial biological function, for example in infection and inflammation control.[20] Think of the way grit or an eye lash is shepherded to the inner corner of your eye by sticky mucus which attaches to foreign surfaces and ushers debris together for removal by tears or the tip of your finger. In humans, globular proteins are especially effective in aggregating bacteria and viruses. Lectin family proteins in particular attach to particles, aggregate them, and shape the immune response.[22]

It is simple to show by experiment that bio-polymers agglomerate NPs long before they reach the cell surfaces.[11,16] A petri dish containing only serum free cell growth media (colourless RPMI) was dosed with 300 cm^2 mL^{-1} of 100 nm polystyrene particles which remained in stable suspension. However, when a petri dish containing the same media was used to grow cells (lung carcinoma cells, A549) for 1 h, the media decanted and centrifuged, then polystyrene nanoparticles added, it was found that the NP size distribution changed immediately in most cases and agglomerated fully over a period of hours. The cells had clearly excreted molecules that caused agglomeration of NPs. NPs formed large agglomerates clearly visible by optical microscopy, whereas the original NPs remained invisible in monodisperse suspension (Fig. 11.4). It seems that this process acts as a natural cell defence mechanism because such agglomeration prevents further translocation and aids collection and clearance by phagocytes.[11]

The same effect was observed when atmospheric NPs were added to human lung lavage liquid, that is the mucus-surfactant lining washed out of patients' lungs.[11] NPs were rapidly agglomerated, consistent with the idea that amino acids, protein fragments and surfactant lipids were attaching to the NP surfaces.[16] It is therefore clear that NPs are influenced by adsorbing surface molecules long before the adhesion and interference with cells occurs.

To show that protein molecules can attach to nanoparticles to promote aggregation, polystyrene particles, 200 nm diameter, were added to phosphate buffered saline solution containing 0.25 mg L^{-1} fibrinogen and agglomeration was observed within 30 min.[18] The agglomerates were dried and viewed in the scanning electron microscope to show polymer bridging between NPs (Fig. 11.5). It was clear that the protein was attached to the surface and was binding the particles together, limiting their mobility. The kinetics of this aggregation were measured and shown to depend on the surface area and functionality of the polystyrene in a predictable way as shown previously in Fig. 11.3.

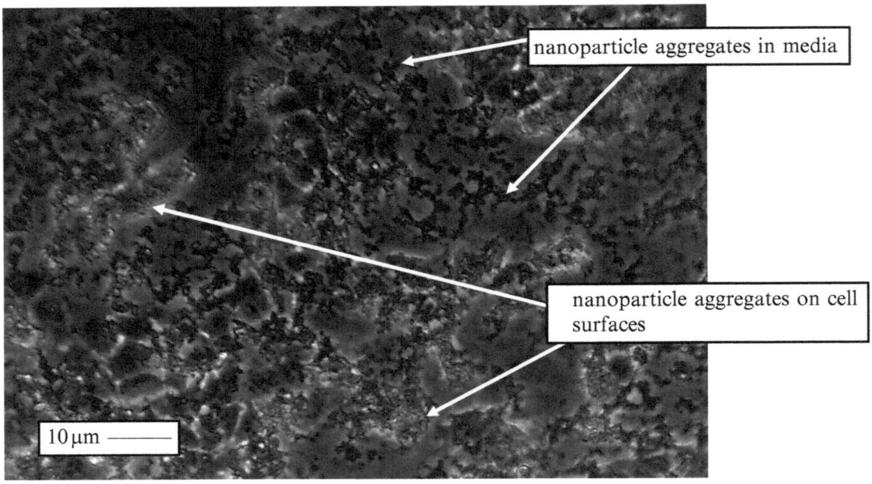

Fig. 11.4 Polystyrene particles (100 nm) agglomerated (*dark grey*) in conditioned serum free media overlying A549 cells, at 24 h. The individual NPs are joined to each other and are no longer free and mobile to damage the cells, which are only visible within this plane of vision when nanoparticles attach. Cells shrank on exposure to these nanoparticles, at a particle concentration = 300 cm^2 mL^{-1}

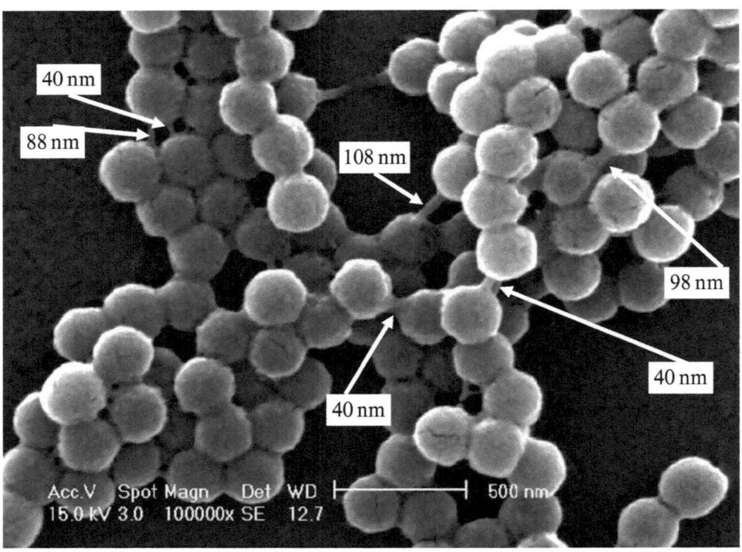

Fig. 11.5 Morphology of 200 nm polystyrene particle agglomerates formed in a fibrinogen PBS suspension. Particle concentration 1.25 cm^2 mL^{-1}; fibrinogen concentration 0.25 mg mL^{-1}

11.5 Nanoparticles Approaching the Cell Surface

After encountering the extracellular polymers, the next step is for the nanoparticle and cell surface to approach one another. This precedes attachment to the cell and subsequent engulfment. Particle-cell interactions begin with the particle approaching a zone of influence around the cell surface, and interactions are mediated by intervening polymer, salt and water layers. Typically, the adhesion process starts as the NPs get within hundreds of nm of the cell membrane, as electrostatic and van der Waals forces cause attraction as described in Chapter 3. Even at this distance, nanoscale features of the cell such as microvilli or pseudopodia may be seen interacting with NPs, changing the cell topography.

A typical experiment is illustrated in Fig. 11.6. Cell culture was incubated with an iron oxide dispersion and the resulting contacts were investigated by transmission electron microscopy[24] to show microvilli projections from the cell membrane addressing the pre-agglomerated nanoparticles. Since many authors report this effect, irrespective of particle type, we infer that polymer attached to the surface influences this process, causing the cell microvilli to approach the NPs.

NP interaction with the cell surface occurs within 1 h or less in cell culture,[25] so that stationary cells attach approximately 50% of particles after 24 h. The theoretical argument is that polymers are excreted routinely by the cells, then attach to available nanoparticle surfaces, agglomerating them into clumps. The agglomerates then attract, attach to, and are swept together by, microvilli at the cell surfaces to form the patterns seen in Fig. 11.4, before being engulfed by the cells. Such agglomeration is dependent on many factors; the concentration of the particles, the particle surface type, and the nature of the extracellular liquid. The polymer attachment

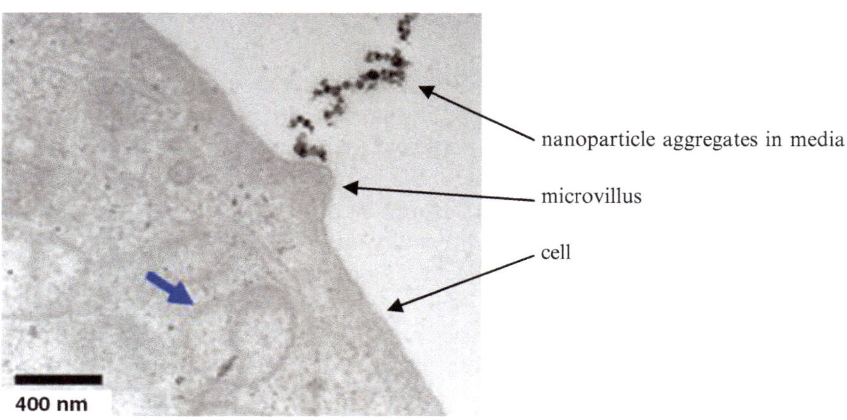

Fig. 11.6 Transmission electron micrograph of a MCF10A cell incubated with iron oxide nanoparticles for 30 min at 37°C. Slices were treated with uranyl acetate to stain membranes. The *thick arrow* points to mitochondria[24] (with permission). Microvilli at the cell surfaces were observed to cluster and extend towards nanoparticle agglomerates in this and other studies

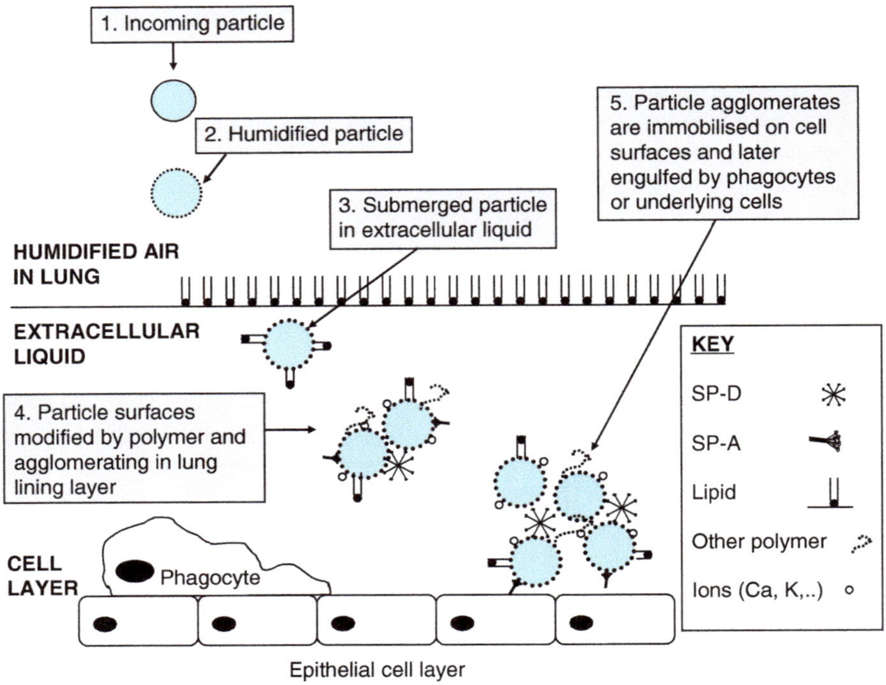

Fig. 11.7 Schematic showing the processes by which NPs in the lung enter extracellular liquid, adsorb molecules, agglomerate, then attach to cell layer. First the NP is covered by water molecules in the humid air, then it picks up lipid, SPA, SPD and other polymers, which cause agglomeration followed by attachment to cells

influences both the behaviour of the NPs and the response of the cells[26], especially aiding clearance by cells which engulf and remove the particles from the system. Pre-agglomerated, cell-surface fixed particles would be cleared better, and therefore be able to do less damage. A scheme of the overall process is shown in Fig. 11.7, depicting wetting of the NP surfaces, adsorption of ions, lipids, polymers, agglomeration, deposition on cells and engulfment by macrophages.

The conclusion is that the biomolecules adsorbed at the surface of the NPs are very important in determining first their agglomeration state, then their approach and adhesion to cell surfaces,[11] clearance rate and ultimately the dose to underlying cells. The process of polymer attachment therefore performs three key protective functions:

- Enhances agglomeration to aid clearance
- Reduces unwanted adhesive interactions at particle surfaces to minimise interference with the cell innards
- Tags particles for downstream cell recognition and stimulation of cell responses

11.6 Nanoparticles Entering Cells

One of the best images of nanoparticle entry into a cell is shown in Fig. 11.8. To engulf the particles, an invagination of the cell membrane forms, encloses the particle agglomerate in an internal space filled with liquid, engulfs it completely, and the vesicle moves rapidly into the cell. These are termed endocytotic vesicles, which exhibit different mechanisms depending on the material being internalised, and different sizes, from 100–500 nm in diameter. Various complex chemistries have been proposed for this engulfment, but it is also described as non-specific.

These observations fit the usual descriptions of vesicle formation in animal cells[19] (phagocytosis, pinocytosis, receptor mediated endocytosis). But the main observation in NP treated cells is that nanoparticle agglomerates amplify vesicle formation as shown in the micrographs of Fig. 11.9, in which the addition of silica NPs (116 nm) has multiplied both the number and the size of vesicles.[29]

Cellular uptake of NPs through the membrane can be rapid depending on particle size and cell type: Rejman[31] reported 50% of 50 nm polystyrene particles were internalised, within 30 min, much more than 500 nm NPs. Slower rates were reported for 200 nm particles, although similar incorporation was observed after 2–3 h. Kemp[25] reported similar rates for 50 nm NPs – 40% within 30 min – with significant differences in uptakes between different cell types and different particle surfaces (although behaviour in the media was not well accounted for). Alveolar type I (ATI) cells took up much greater quantities of NPs compared to smaller alveolar type II (ATII) cells, preferentially taking up negatively charged particles of identical size.[25] The effects of media-NP interactions on agglomeration and particle size were not considered.

The main cells involved in NP clearance process in the lungs are macrophages which are mobile to crawl along the epithelial cells, engulfing agglomerated foreign

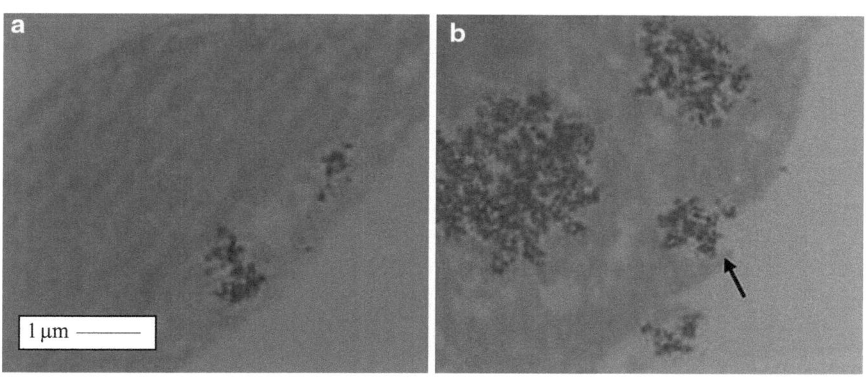

Fig. 11.8 TEM micrographs of a single cell (mouse keratinocyte) containing agglomerated titanium dioxide (TiO_2) nanoparticles. (**a**) The cell took up the TiO_2 NPs and localised agglomerates in membrane-bound vacuoles. (**b**) The cell engulfed the agglomerate (as shown by the *black arrow*), localising it into a membrane-bound vacuole, the process of endocytosis[28] (with permission)

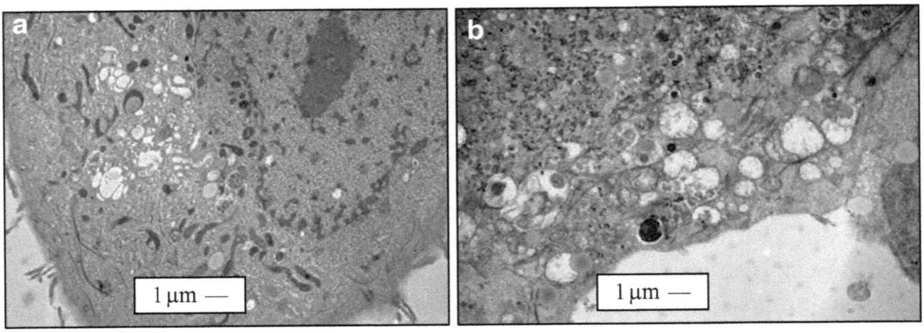

Fig. 11.9 TEM images of the A459 lung cell-line dosed with NPs. (**a**) An untreated healthy A459 cell, with white vacuoles. (**b**) Similar cell treated with 2 cm^2 mL^{-1} of 116 nm diameter Aerosil Si-NP agglomerates, showing many more larger, irregularly shaped white vacuoles and dark grey silica depositions in the cytoplasm

material as they progress (see Fig. 11.7). The phagocytes are the first cellular line of defence, to engulf particles in extracellular spaces.[8] The lung has the largest population of macrophages which remove depositing particulate and cellular debris, so that particles are partially cleared before encountering underlying cells. Phagocytic cells thereby modulate the effects of epithelial cells. They also excrete molecules which excite the inflammatory response system, drawing in more phagocytes and more serum. Together, these processes initiate the lung inflammatory response, in which cells release particular molecules which enhance the immune reaction to elevated concentrations of inhaled NPs, but this takes around 60 min in humans.[23] Prior to this cellular response, acellular polymer and lipid attachment dominate as the primary defence process.

Nanoparticle entry into cells has been reviewed extensively for drug delivery and cell imaging because particle coatings encourage entry of the NPs.[30] In these papers, cellular uptake is clearly dependent on the size of the particles, the type of cell and the prior surface coating which dictates extra-cellular-polymer attachment. Rejman et al.[31] reported that polystyrene particles up to 500 nm were attached to cells following cell exposures, but that 500 nm particles were only ever detected at the periphery of cells and not fully internalised. In contrast, <200 nm sized particles were distributed throughout the cells in the perinuclear region. Accumulation of smaller particles at the cell membrane occurred in a size dependent way. In drug delivery, nanoparticles are coated by a range of cationic peptides to gain entry to the cell[10,19] as well as providing stability in the NP dispersion as prepared. Xia et al.[10] reported the use of polyethyleneimine coatings on silica NPs and showed several interesting influences on toxicity including molecular weight of the polymer.

A typical patent[38] published in 2009 describes how nanoparticles can be dispersed in an organic solvent with polymer in solution, then blown up with supercritical carbon dioxide to produce very fine coated material which can be used in medical and other applications.

Magnetic nanoparticles as shown in Fig. 11.10 have also been studied since the 1970s because they offer possibilities for external control and separation.

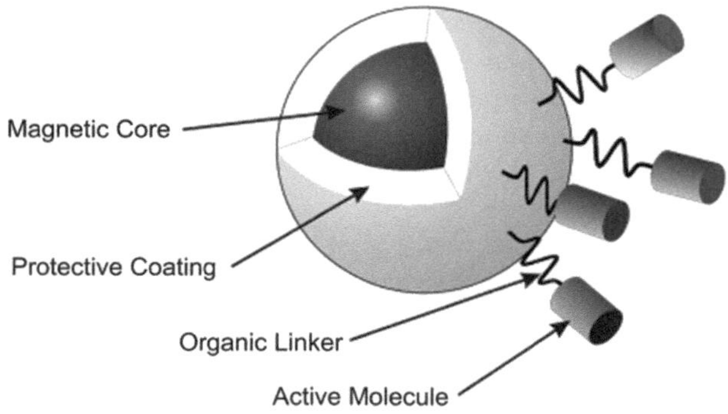

Fig. 11.10 Theoretical picture of magnetic nanoparticle prepared for injection into the bloodstream[39](with permission)

For example, a magnetic field can be used to direct the therapeutic drug to the target organ. Several different polymer protective coatings have been used including dextran, starch, albumin, polylactic acid, polyethylene glycol and polyvinyl acetate[39] Linking the active drug molecule to the magnetic nanoparticle requires organic linkers which can be amines, thiols, acids or aldehydes. Several clinical trials have shown effects of these injected NPs on cancers, but there remains concern that long-term NP toxicity may be associated with short-term use.

The main summary of observations obtained from drug delivery experiments shows that:

- Smaller insoluble particles tend to translocate further via circulatory systems than larger particles
- Smaller insoluble particles get into cells easier
- The surface molecules of NPs and particles are important for cell entry and subsequent toxicity

11.7 Toxicity Mechanics

Not only can nanoparticles be used to identify and treat illness; there is no doubt that some nanoparticles cause disease.[1–3] As nanoparticles are transported into the body according to Fig. 11.7, they can cause toxicity in several different ways. For example, even before the NPs make contact with cells, they can adsorb the polymer molecules in the extracellular medium. If some essential molecules are depleted, such as lung surfactant, then the organism will suffer. Subsequently, the NPs may agglomerate in the extracellular space and cause physical blockages. Once the NPs contact the cells or interact with specific polymers then there can be an immune response. Later inside the cell, the smallest NPs (<20 nm) may interfere with the nucleus or the cell apparatus such as mitochondria by traversing the semi-permeable membranes such as the nuclear pore complex. NP overload may also occur if a cell engulfs too many NPs, causing cell death.

An example of critical molecules is described by Wright[34] who showed that surfactant protein D (SPD) was essential to regulate the immune response in animals. Since the main concentration of SPD is in the lungs, where nanoparticles from smoking and air pollution deposit, one possible toxic mechanism[12] is the adsorption of SP-D on the extensive NP surface area, leading to increased infections and emphysema.[35]

A recent significant advance was to show how mice drinking titania nanoparticles caused significant genetic change together with gut inflammation.[41] Figure 11.11 shows a clear dose-response relationship between the dose of TiO_2 nanoparticles and genetically altered cells.

It had been known that titanium dioxide, produced in several million tons per year worldwide as white pigment, can cause respiratory cancers in rats at high doses. This latest study showed that drinking 160 nm TiO_2 (Degussa, now Evonik) suspensions in water caused DNA changes in mice. The crystal structure was a mixture of 75% anatase and 25% rutile TiO_2, purity was at least 99.5% TiO_2, and primary particle size was 21 nm with a specific surface area of 50 ± 15 m^2 g^{-1}. Mice drank solutions of dispersed TiO_2 nanoparticles in drinking water at 60, 120, 300, and 600 μg mL^{-1} concentrations. Daily intake ranged from 3 to 7 mL $mouse^{-1}$, consistent with normal daily water intake, over 5 days. NP exposure studies in mice demonstrate potential toxicity, but interspecies differences make extrapolation to human effects difficult.

Another significant toxicity effect is revealed once nanoparticles are engulfed by macrophages, the critical cells in lung particle clearance. Mouse macrophages

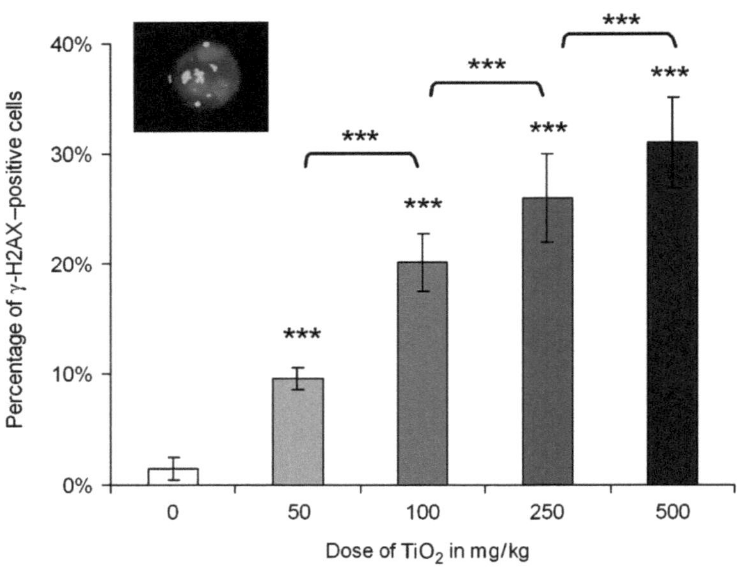

Fig. 11.11 Percentage of {gamma}-H2AX-positive cells in bone marrow in untreated and TiO2 nanoparticles-treated mice and a picture of a {gamma}-H2AX-positive cell with more than four foci. Columns are means for five mice[41] (with permission)

exposed to nanoparticles may experience overload where they have a distinctive foamy appearance, swollen with fluid filled vesicles containing particles.[42] Under these conditions whole macrophage populations may die, and the nucleus swells and disintegrates. Particles appear as agglomerates within vesicles spread through the cytoplasm, similar to non-phagocytic cells[29] (Figs. 11.8 and 11.9).

Size is the major parameter critically linked to NP toxicity in organisms, presumably because such small particles are less well cleared, may circulate more widely and may penetrate protective cell membranes more easily, thereby presenting a foreign surface capable of interference with normal cell processes and organelles. After size, surface chemistry plays its part via control of polymer attachment and agglomeration.[43] The effect of particle elasticity does not seem to have been tested.

In whole animals, but not necessarily in cell culture, NPs stimulate an inflammatory response. The role of NPs and foreign surfaces in inflammatory responses has been examined for decades because of prostheses wear or leaking silicone breast implants.[44–46] Polyethylene, metal, carbon fibre and silicone particles from prostheses have migrated significant distances within the body via the circulatory systems, at times creating localised and systemic inflammation. All such NP effects have been attributed to oxidative stress where an imbalance between oxidants and anti-oxidants develops.[44] The role of polymer attachment in this process is yet to be explained, although hydrogen ion release on protein attachment is well understood.

An interesting experiment[61] has shown that some NPs are able to stimulate aggregation of platelets (Fig. 11.12, left). The experiment measured the aggregation rates of washed platelets in Tyrode's solution at 37°C, using a light-based aggregometer. First, nanoparticles were added at a specific mass concentration range. It was observed that 500 and 100 nm polystyrene particles agglomerated

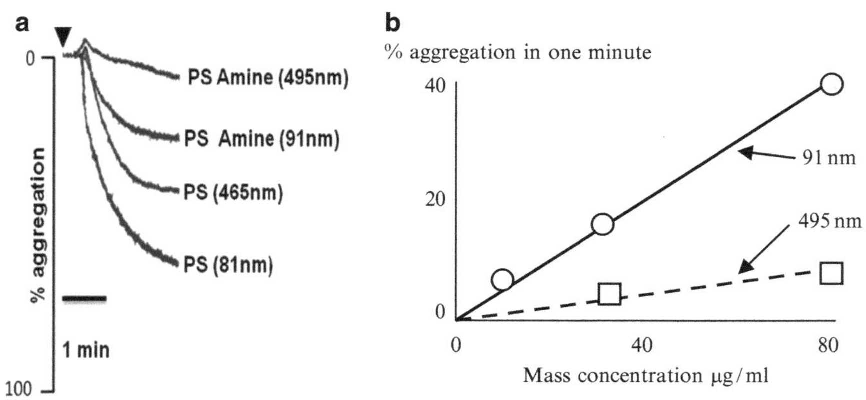

Fig. 11.12 (**a**) *Left*, Platelet aggregation can be stimulated by the presence of polystyrene particles at equal mass concentration, but amine coated particles slow that aggregation; (**b**) *Right*, aggregation rate for amine coated polystyrene showing linear response and increase for smaller particles

platelets faster than identical particles with an amine coat. Obviously the amine surface molecules inhibited the platelet response.

The rate of aggregation increased with the mass concentration of the particles for both surface types. For the amine coated particles, the response was linear as shown in Fig. 11.12b (right). More significantly, aggregation increased for the fine particles compared to the coarse ones at the same mass or surface area concentrations. The five times smaller amine coated particles aggregated the platelets five times faster than the 495 nm diameter particles at the same mass concentration.

An interpretation of this result is that platelet aggregation is triggered by a nanoparticle adhesion event, so that the number of particles is the important parameter. But the amine coated particles give reduced adhesion and so the number of particles is not the only important factor; the surface molecules also play a significant part in the process.

11.8 Translocation of Nanoparticles Within Organisms

The cardiovascular effects of nanoparticle exposures (including air pollution and smoking) have led to a growing interest in how these particles affect cells dispersed around the body, well beyond the first point of contact. The key question is: Do NPs just act locally or do they travel to other parts of the body? First let us consider the destination of injected nanoparticles, then address inhalation and ingestion routes.

Ink NPs injected into the skin for tattoos may remain indefinitely if injected at the right layer of skin, and only rarely cause toxicity, such as allergy, reaction to sunlight, interference with MRI scans, granulomas and accumulation in lymph nodes. Nanoparticles injected directly into the bloodstream are potentially more damaging since they may travel around the body and deposit in various tissues over a period of hours[6]. Given the same surface type, smaller NPs have longer half lives in the circulatory systems than larger ones.[19] Particles which evade coverage by biopolymers appear to remain in circulation for longer. Lymphatic uptake following intravenous injection is dependent on particle size, with larger particles (>20 nm) being retained in lymph nodes. Surface coating with protein can effectively redirect NPs to different target organs at low concentrations of 0.1–0.5% by mass.[47] Two of the most widely reported translocation routes with toxic potential for injected particles[49] are crossing the blood-brain barrier and entering the lymph system (lymph nodes, spleen and liver).

Inhaled NPs behave differently because particles first interact with lung surfactant, a mixture of phospholipids and protein polymers. The majority of particles remain trapped in the lung, around 25% are gradually removed by macrophages within 24 h, and only a small proportion of NPs circulate in the bloodstream. Particle size dictates ability of NPs to translocate from the lung and the residence time in the circulatory system, but surface type is also influential. This in turn affects the cells where NPs can cause damage. In studies of ultrafine iridium particles

(15 and 80 nm median diameter) inhaled by healthy rats, deposited particles were partially cleared from the airways, including the alveolar region, into the gastrointestinal tract and faeces. Less than 1% of the lung deposited particles translocated into secondary organs such as liver, spleen, heart, and brain.[6] Eighty nanometre particles translocated ten times less than 15 nm particles. Studies following administration of identical particles by either intratracheal instillation or intravenous injection confirmed that (1) these particles were neither dissolved nor absorbed from the gut, (2) circulating particles were rapidly accumulated in the liver and spleen and retained there, and (3) soluble 192 Ir instilled in the lungs was rapidly excreted via urine with little retention in the lungs and other organs.[6] Geiser et al.[48] showed that inhaled TiO_2 nanoparticles crossed cellular membranes by nonphagocytic mechanisms in the lungs and were then found in capillaries. NPs (approximately 30 nm) suspended in the air can also deposit in the nasopharynx of rats, from which they can migrate through the olfactory nerves into the brain.[49]

Drinking or eating nanoparticles poses different barriers to translocation than inhalation and injection. Most ingested nanoparticles end up in the faeces while only about 1% gets through the gut barrier and into circulation.[41] However, as the silver man picture (Fig. 11.1) shows, once a few of the ingested nanoparticles have entered the blood, then they can become embedded in the skin and show colourful effects. Experiments on European drinking water have shown that 1 L contains typically 7.10^{11} natural nanoparticles, averaging 15 nm diameter, which have passed through the normal filtration and treatment processes.[57] These particles were mainly humic acid and polysaccharides. The fate of these NPs in the body is not known.

11.9 Nanoparticle Toxicity: Lessons from the Lung

A new theory of nanoparticle damage to the lungs does not depend on direct adhesion of NPs to lung cells. Instead it suggests that the normally protective lung polymer surfactants (proteins and phospholipids) are mopped up by the nanoparticles, removing their protective action.[11,16] The rate and extent of polymer adhesion to surfaces therefore significantly influences the extent of NP toxicity.

In the lungs, certain polymers like surfactant proteins SP -A and SP-D (Fig. 11.13) have specific defensive roles to protect against biological particle invasion and consequent infection.[34] For example, SP-A and SP-D act as surfactants (sometimes called opsonins), coating invading micro-organisms in the lung, modulating phagocytosis by binding and capture of bacterial toxins, and suppressing or stimulating inflammatory responses.[50] SP-D is present at highest concentrations in the lung, but is found in all compartments where foreign material needs to be cleared; in the circulatory system,[51] in amniotic fluid and the female reproductive tract,[52] the ear, intestine and stomach, even in tears.[53] Clearly, this is a surfactant which has wide-ranging influence on the whole body, and specifically on immunity.

SP-A and -D defend the lungs by aggregating bacteria and other pathogens.[54] By taking these proteins and mixing them with nanoparticles, it was shown that the

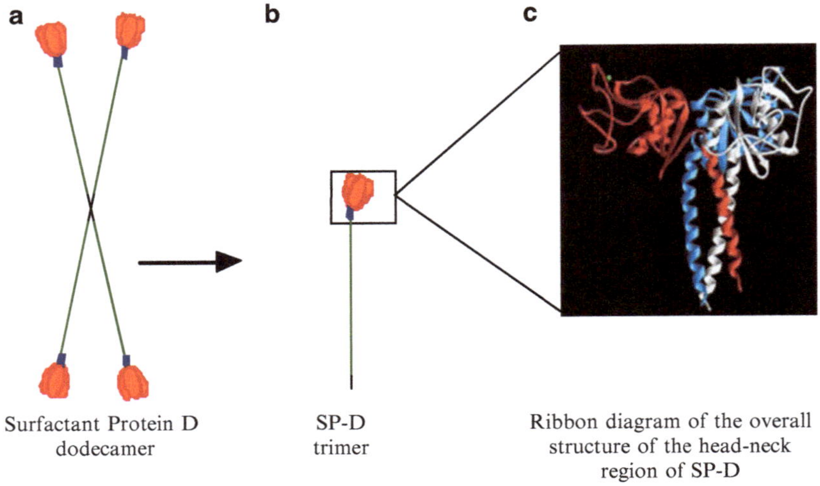

Fig. 11.13 Surfactant protein D (SP-D)[34] (with permission)

SP-A and SP-D coated the nanoparticle surfaces.[16] Therefore the proteins were not available to carry out their normal defence duties. Sequestration of these molecules from the lung in this way could cause the onset of inflammation, and trigger the inflammatory cascade. For example, *P. aeruginosa* is a major cause of pneumonia in patients with impaired host defence, and is one of the most common causes of pneumonia in intensive care units. It is also associated with air pollution nanoparticle exposure deaths. Similarly, childhood exposure to airborne pollution is linked to the development of acute respiratory infection in infants, making it one of the biggest killers of children in the world.[51] SP-A and SP-D enhance pulmonary clearance of many infectious particles including *P. aeruginosa* in the lungs,[55] but nanoparticles disrupt these polymers, allowing the pathogens to evade removal. In this way, NPs can promote infection and inflammation.[56,58]

Since the first papers linking air quality to disease in the early 1990s, an internationally derived dose-response curve for human disease through breathing airborne NPs has emerged as shown in Fig. 11.14. This was obtained largely from epidemiology conducted in and by North America, but has been upheld by studies around the world.

The exposure-response relationship between cardiovascular disease mortality and NPs is relatively steep around 100 μg m^{-3} of exposure and flattens out at higher exposures. This agrees with the aggregation effect theory whereby agglomeration increases at higher particle concentrations, making the fine particles less damaging. First, the aerodynamic size of the particles influences toxicity, by determining the site of deposition in the respiratory tract. Particles breathed into the lung are partially filtered by the structure of the respiratory tract so that only the finer fractions reach the lower

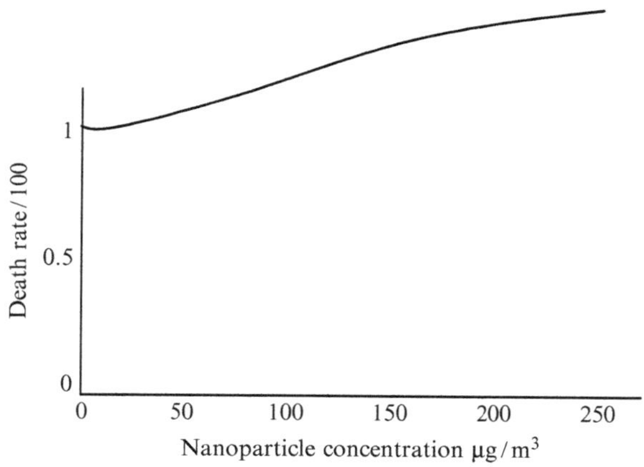

Fig. 11.14 Plot of death rate versus NP concentration in air, where NP is given by PM2.5 i.e. mass of particles

alveolar compartments where oxygen exchange takes place. Second, the aggregated nanoparticles are less mobile and so cannot diffuse into damaging locations.

To explain the large number of cardiovascular deaths, the inflammatory effect of nanoparticles may be invoked. This effect starts after the NP agglomerates have deposited on the alveolar cells and the macrophages have begun to mop them up. The macrophages release cytokines into the lung fluid, stimulating an inflammatory reaction which affects the whole body via the circulatory system. These cytokines have been observed when nanoparticles are added to lung cell cultures in petri dishes.[63]

It is important to note that nanoparticle exposures tend to affect those with existing cardio-vascular or respiratory illnesses most significantly, where lung surfactant performance is already compromised,[59] or those at the age extremes, both children and the elderly.[60] The timescales are of the order of minutes, hours or days for acute effects such as heart failure, and years for chronic diseases such as cancer. Failure to agglomerate and clear particles in the lung immediately may therefore have long-term consequence for disease development. Although the exact mechanisms underlying these PM effects remain unknown, more recent work has shown that the cardiovascular system (acutely) and the developing respiratory system (chronic effects) are significantly affected. Crucially, air pollution alters the developing lung irreversibly, leading to a lifetime of reduced function even after removal from the pollution. Other effects are also measureable – following a coal ban in Dublin and a 70% reduction in black smoke concentrations, death rates in the city declined by 6%;[62] during traffic intervention schemes at the Atlanta and Beijing Olympics, air pollution dropped and city-wide health improved. Evidence points to NPs causing localised inflammation in the lung, which may contribute to a systemic inflammatory state, impaired growth and cardiovascular damage. For example, the collapse of the Twin Towers in New York released mixed dust and fibres (Fig. 11.15) inducing pulmonary and systemic pathologies in rescue workers and NYC residents.

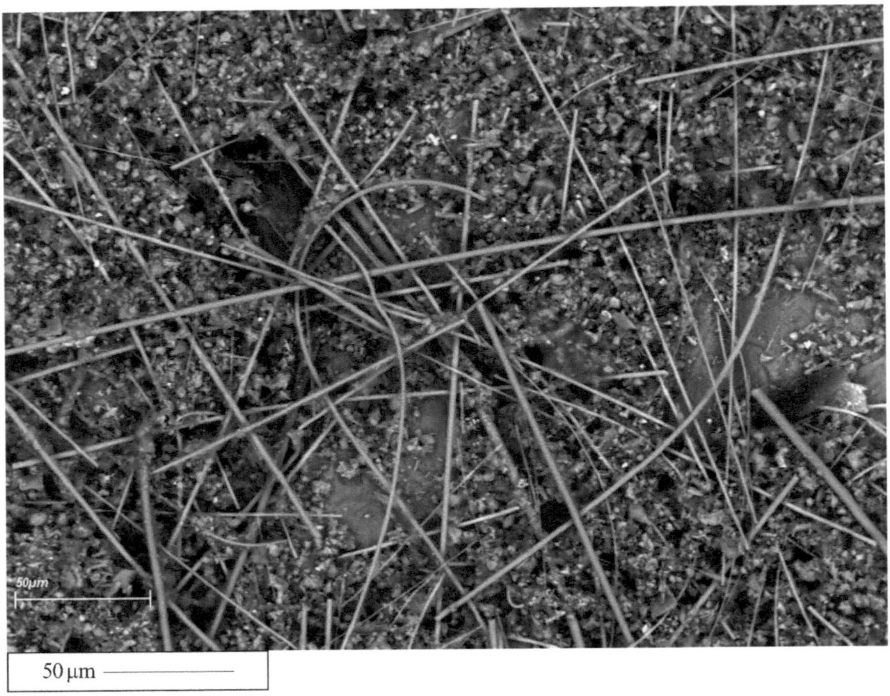

Fig. 11.15 Particles and fibres released in the World Trade Center (WTC) collapse

In NP and ENP exposure related diseases, there are still large gaps in our knowledge regarding the passage of particles through the lung barrier. While we can clearly link particle exposure and effect, the precise molecular biology underlying the associated lung and heart diseases remains unclear. This may explain why current NP toxicology fails to link convincingly to the epidemiology to date, and presents an opportunity for immediate advance.

11.10 Conclusions

There is a rapid growth of understanding in the effects that nanoparticles have on cells, organisms and humans. The effects are specifically related to their small size which allows significant mobility and penetration. The negative effects produced by NPs are modest, of the order of 1% increase in deaths with every 10 $\mu g\ m^{-3}$ PM_{10}, but consistently detectable in populations around the world. They include cardiovascular, respiratory and inflammatory responses, which may result in increased infections, impaired growth, altered immune responses (e.g. allergy) and premature death. When considering the potential benefits of ENPs including protection from sunlight by NP sunscreens, delivery of organ specific drugs via NP drug delivery vehicles, and delivery of DNA by NPs, these potential negative effects should be considered.

Adhesion plays a role in these phenomena. In the first place, adhesion of molecules like proteins to the nanoparticle surfaces is important because this both modifies the fate of the particles and also can deplete the body's protein reserves which have defensive roles. Second, the control of NP aggregation is vital. Agglomeration of the nanoparticles through adhesion immobilises the discrete and agglomerated particles, allowing them to be removed faster by biological clearance mechanisms. Third, once coated, the nanoparticles can adhere to cells and become immobilised at the cell surfaces where they may initiate inflammation processes which damage the body. If a nanoparticle gets past these immobilising steps, it may get into the central machinery of the cell or organism and disrupt reactions, for example causing damage to cellular polymers such as DNA.

The first line of defence in the body is interaction with polymer molecules such as albumin and fibrinogen which influence agglomeration; agglomeration in turn controls cell penetration, clearance, immune responses and tissue targeting. Such adsorption at particle surfaces therefore plays a large part in particle motion and adhesion. There are many examples, where nanoparticles are used in medical imaging and drug delivery for instance, in which various molecular coatings have been used to control NP surface properties in order to adjust NP fate in organisms.

The processes of NP toxicity after exposures of cell cultures, animals and humans have been described in terms of five steps:

- Coating of NPs by protein molecules
- Agglomeration of NPs
- Adhesion to cell walls
- Engulfment by cells
- Interference of NPs with cell components

The effect of air pollution by NPs on the lung has illustrated these features and provoked new arguments about smoking, car emissions and other environmental NP and ENP releases. The effects are greater in scale than obesity in the developed world. New investigations into the ecological and human health implications of ENP release have provided more detailed evidence of effect.

To avoid future toxicity problems a key principle is that new nanoparticles be evaluated before widespread dissemination. Exposure of cell cultures, animals and humans to airborne particles (e.g. tobacco smoke, air pollution, NPs and ENPs) has been consistently linked with subsequent development of disease and premature death. We now understand that adhesion of nanoparticles to proteins, to aggregates, to cells and to cell machinery is important in disease development.

References

1. Russell, A. G. & Brunekreef, B., A focus on particulate matter and health. Environmental Science & Technology 43 (2009) 4620.
2. Royal Commission on Environmental Pollution, Novel materials in the environment: The case of nanotechnology. 27th Report (2008).
3. Song, Y., Li, X. & Du, X., Exposure to nanoparticles is related to pleural effusion, pulmonary fibrosis and granuloma. Eur Respir J 34 (2009) 559–567.

4. Tomi, N. S., Kränke, B. & Aberer, P. W., A silver man. The Lancet 363 (2004) 532.
5. Alvarez-Román, R., Naik, A., Kalia, Y. N., Guy, R. H. & Fessi, H., Skin penetration and distribution of polymeric nanoparticles, Journal of Controlled Release 99 (2004) 53.
6. Kreyling, W. G. et al., Translocation of ultrafine insoluble iridium particles from lung epithelium to extrapulmonary organs is size dependent but very low, Journal of Toxicology and Environmental Health-Part A 65 (2002) 1513–1530.
7. Araujo, J. & Nel, A., Particulate matter and atherosclerosis: role of particle size, composition and oxidative stress, Particle and Fibre Toxicology 6 (2009) 24.
8. Gehr, P., Blank, F. & Rothen-Rutishauser, B. M., Fate of inhaled particles after interaction with the lung surface, Paediatric Respiratory Reviews 7 (2006) S73–S75.
9. Woodrow Wilson Center for International Scholars, Project on Emerging Nanotechnologies: Nanotechnology Consumer Products Inventory, (2009).
10. Xia, T. et al., Polyethyleneimine coating enhances the cellular uptake of mesoporous silica nanoparticles and allows safe delivery of siRNA and DNA constructs, ACS Nano 3 (2009) 3273.
11. Kendall, M. et al., Lung lining liquid modifies $PM_{2.5}$ in favor of particle aggregation: a protective mechanism, American Journal of Physiology-Lung Cellular and Molecular Physiology 282 (2002) L109–L114.
12. Kendall, M.,Fine airborne urban particles ($PM_{2.5}$) sequester lung surfactant and amino acids from human lung lavage, American Journal of Physiology-Lung Cellular and Molecular Physiology 293 (2007) L1053–L1058.
13. Nel, A., Xia, T., Madler, L. & Li, N., Toxic potential of materials at the nanolevel, Science 311 (2006) 622–627.
14. Oberdorster, G., Oberdorster, E. & Oberdorster, J., Nanotoxicology: an emerging discipline evolving from studies of ultrafine particles, Environ Health Perspect 113 (2005) 823–839.
15. Gauderman, W. J. et al., Effect of exposure to traffic on lung development from 10 to 18 years of age: a cohort study, Lancet 369 (2007) 571–577.
16. Kendall, M., Brown, L. & Trought, K., Molecular adsorption at particle surfaces: A PM toxicity mediation mechanism, Inhalation Toxicology 16 (2004) 99–105.
17. Lundqvist, M. et al., Nanoparticle size and surface properties determine the protein corona with possible implications for biological impacts, Proceedings of the National Academy of Sciences 105 (2008) 14265–14270.
18. Kendall, M., Ding, P. and Kendall, K., Submicron and nanoparticle interactions with fibrinogen: The importance of aggregation state (*Nanotoxicology; submitted October 2009, accepted April 2010*).
19. Hillaireau, H. & Couvreur, P., Nanocarriers' entry into the cell: relevance to drug delivery, Cellular and Molecular Life Sciences 66 (2009) 2873–2896.
20. Esmon, C. T., Basic mechanisms and pathogenesis of venous thrombosis, Blood Reviews 23 (2009) 225.
21. Nilsson, B., Ekdahl, K. N., Mollnes, T. E. & Lambris, J. D., The role of complement in biomaterial-induced inflammation, Molecular Immunology 44 (2007) 82.
22. Geijtenbeek, T. B. H. & Gringhuis, S. I., Signalling through C-type lectin receptors: shaping immune responses, Nat Rev Immunol 9 (2009) 465.
23. Alexis, N. E. et al., In vivo particle uptake by airway macrophages in healthy volunteers, Am. J. Respir. Cell Mol. Biol., 2005–0373OC (2005).
24. Zhang, C., Tang, T., Ren, W., Zhang, X. & Dai, K., Influence of mouse genetic background on wear particle-induced in vivo inflammatory osteolysis, Inflammation Research 57 (2008) 211–215.
25. Kemp, S. J. et al., Immortalization of human alveolar epithelial cells to investigate nanoparticle uptake, Am. J. Respir. Cell Mol. Biol. 39 (2008) 591–597.
26. Dutta, D. et al., Adsorbed proteins influence the biological activity and molecular targeting of nanomaterials, Toxicological Sciences 100 (2007) 303–315.
27. Braydich-Stolle, L. K. et al., Crystal structure mediates mode of cell death in TiO_2 nanotoxicity, Journal of Nanoparticle Research 11 (2008) 361–1374.

28. Bouclier, C., Moine, L., Hillaireau, H., et al Physicochemical characteristics and preliminary in vivo biological evaluation of nanocapsules, Biomacromolecules 9 (2008) 2881–2890.
29. Parveen, N. Masters Thesis. Supervisors: Dr M Kendall and Dr N Hodges. University of Birmingham, 2009.
30. Thurn, K. et al., Nanoparticles for applications in cellular imaging, Nanoscale Research Letters 2 (2007) 430.
31. Rejman, J., Oberle, V., Zuhorn, I. S. & Hoekstra, D., Size-dependent internalization of particles via the pathways of clathrin- and caveolae-mediated endocytosis, Biochem. J. 377 (2004) 159–169.
32. Stearns, R. C., Paulauskis, J. D. & Godleski, J. J., Endocytosis of ultrafine particles by A549 Cells, Am. J. Respir. Cell Mol. Biol. 24 (2001) 108–115.
33. Han, Y., Wang, X. & Li, S., A simple route to prepare stable hydroxyapatite nanoparticles suspension, Journal of Nanoparticle Research 11 (2009) 1235–1240.
34. Wright, J. R., Immunoregulatory functions of surfactant proteins, Nature Reviews Immunology 5 (2005) 58–68.
35. Wert, S., Jones, T., Korfhagen, T., Fisher, J. & Whitsett, J., Spontaneous emphysema in surfactant protein D gene-targeted mice, Chest 117 (2000) 248S–248S.
36. deGennes, P.G., Passive entry of a DNA molecule into a small pore, PNAS 96 (1999) 7262–64.
37. Guenoune-Gelbart, D., Elbaum, M., Sagi, G., Levy, A., Epel, B.L., Mol Plant Microbe Interact 21 (2008) 335–345.
38. Wang, Y., Pfeffer,R., Dave, R., US Patent 7537803 (2009), 'Polymer coating/encapsulation of nanoparticles…'.
39. McBain, S.C., Yiu, H.P., Dobson, J., Magnetic nanoparticles for gene and drug delivery, Int J Nanomedicine 3 (2008) 169–180.
40. Ma, Y.-J. & Gu, H.-C., Study on the endocytosis and the internalization mechanism of aminosilane-coated Fe3O4 nanoparticles in vitro, Journal of Materials Science: Materials in Medicine 18 (2007) 2145–2149.
41. Trouiller, B., Reliene, R., Westbrook, A., Solaimani, P., Schiestl, R.H., Titanium dioxide nanoparticles induce DNA damage and genetic instability in vivo in mice, Cancer Research 69 (2009) 8784–9; Westbrook AM, Wei B, Braun JB Schiestl RH. Intestinal mucosal inflammation leads to systemic genotoxicity in mice. Cancer Res 69 (2009) 4827–34.
42. Hirama, N. et al., Increased surfactant protein-D and foamy macrophages in smoking-induced mouse emphysema, Respirology 12 (2007) 191–201.
43. Warheit, D. B., Webb, T. R., Colvin, V. L., Reed, K. L. & Sayes, C. R., Pulmonary bioassay studies with nanoscale and fine-quartz particles in rats: Toxicity is not dependent upon particle size but on surface characteristics, Toxicological Sciences 95 (2007) 270–280.
44. Tang, L. P. & Eaton, J. W., Adsorbed fibrinogen triggers acute inflammatory responses to biomaterials, Clinical Research 41 (1993) A164–a164.
45. Uchida, M. et al., Reduced platelet adhesion to titanium metal coated with apatite, albumin-apatite composite or laminin-apatite composite, Biomaterials 26 (2005) 6924.
46. Reddy, S. T., Swartz, M. A. & Hubbell, J. A., Targeting dendritic cells with biomaterials: developing the next generation of vaccines, Trends in Immunology 27 (2006) 573.
47. Araujo, L., LÃbenberg, R. & Kreuter, J., Influence of the surfactant concentration on the body distribution of nanoparticles, Journal of Drug Targeting 6 (1999) 373–385.
48. Geiser, M. et al., Ultrafine particles cross cellular membranes by nonphagocytic mechanisms in lungs and in cultured cells. Environmental Health Perspectives 113 (2005) 1555–1560.
49. Oberdorster, G. et al. Translocation of inhaled ultrafine particles to the brain. Inhalation Toxicology 16 (2004) 437–445.
50. Mason, R. J., Greene, K. & Voelker, D. R. Surfactant protein A and surfactant protein D in health and disease. American Journal of Physiology-Lung Cellular and Molecular Physiology 19 (1998) L1–L13.
51. Stahlman, M. T., Gray, M. E., Hull, W. M. & Whitsett, J. A. Immunolocalization of surfactant protein-D (SP-D) in human fetal, newborn, and adult tissues. Journal of Histochemistry & Cytochemistry 50 (2002) 651–660.

52. Oberley, R. E., Goss, K. L., Ault, K. A., Crouch, E. C. & Snyder, J. M. Surfactant protein D is present in the human female reproductive tract and inhibits Chlamydia trachomatis infection. Molecular Human Reproduction 10 (2004) 861–870.
53. Kishore, U. et al., Surfactant proteins SP-A and SP-D: Structure, function and receptors. Molecular Immunology 43 (2006) 1293–1315.
54. Kingma, P. S. & Whitsett, J. A. In defense of the lung: surfactant protein A and surfactant protein D. Current Opinion in Pharmacology 6 (2006) 277–283.
55. Giannoni, E., Sawa, T., Allen, L., Wiener-Kronish, J. & Hawgood, S. Surfactant proteins A and D enhance pulmonary clearance of Pseudomonas aeruginoso. American Journal of Respiratory Cell and Molecular Biology 34 (2006) 704–710.
56. Hartl, D. & Griese, M. Surfactant protein D in human lung diseases. European Journal of Clinical Investigation 36 (2006) 423–435.
57. Kaegi, R., Nanoparticles in drinking water, EAWAG News 66 (2009) 7–9.
58. Muhlfeld, C. et al., Interactions of nanoparticles with pulmonary structures and cellular responses. American Journal of Physiology-Lung Cellular and Molecular Physiology 294 (2008) L817–L829.
59. Seaton, A., Macnee, W., Donaldson, K. & Godden, D. Particulate air pollution and acute health effects. Lancet 345 (1995) 176–178.
60. Peel, J. L. et al., Ambient air pollution and respiratory emergency department visits. Epidemiology 16 (2005) 164–174.
61. Olushola, D., MSc Thesis, Public and Environmental Health Sciences. Supervisor: Dr M Kendall. University of Birmingham 2009.
62. Clancy, L., Goodman, P, Sinclair, H and Dockery DW Effect of air pollution control on death rates in Dublin, Ireland: an intervention study. Lancet, 360 (9341) 1210–4 (2002).
63. Kendall, M., in preparation

Chapter 12
Cell, Virus and Nanoparticle Adhesion: Significance and Future

solid support...no movement takes place in its absence

Harrison 1914

Harrison[1] originally showed 100 years ago that animal cells required adhesion to a solid surface if they were to move, grow and reproduce. That observation was the basis for tissue culturing which has been universally applied over the past century for growing and studying human cells. Following that invention of cultured cells, viruses could then be grown controllably within the cells so that the processes of virus attack and spread began to be understood. In a similar way, cultured cells have been exposed to nanoparticles in order to define how cell damage occurs, causing nanoparticle toxicity.

The purpose of this chapter is to consider how this knowledge will grow during the next century. First we review the questions about adhesion which have caused argument. Then follows a summary of the solutions to some of those controversies. Finally it is possible to speculate about potential advances and breakthroughs to come in future.

12.1 Key Questions of Adhesion

There are several glib explanations of nanoparticle, virus and cell adhesion which recur constantly and which need refuting vigilantly.[2,3]

Perhaps the most common is the idea that cells have projections and use a lock and key mechanism to bind to each other or to surrounding surfaces (Fig. 12.1a). Yet it has been shown that cells and viruses can stick to surfaces which are smooth at the atomic scale, following Newton's[4] original idea that smooth surfaces stick best. As Newton showed, roughnesses or intervening dust particles inhibit adhesion in a predictable way. Fuller and Tabor[5] confirmed that concept in 1975. In fact, cells which need to disperse, such as puff-ball fungal spores, are seen to have sharp projections and spikes which prevent their adhesion and allow the spores to spread without aggregating. Geckos[6] have feet covered with millions of hairs but it has been shown that these do not interlock with the surface. Instead, each hair has a

K. Kendall et al., *Adhesion of Cells, Viruses and Nanoparticles*,
DOI 10.1007/978-90-481-2585-2_12, © Springer Science+Business Media B.V. 2011

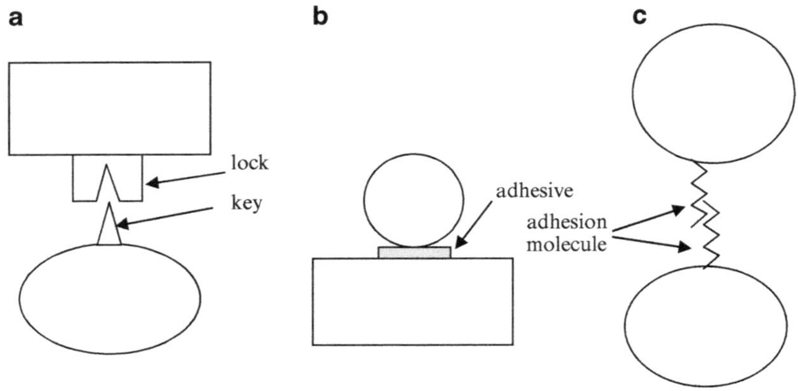

Fig. 12.1 (**a**) Lock and key fallacy; (**b**) adhesive argument; (**c**) adhesion molecule concept

smooth spatula at its end which makes smooth contact with the substrate. One must conclude that the idea of mechanical lock-and-key mechanisms operating at the molecular level to bind cells together is fatuous.

A second common argument is that cells secrete some adhesive material which acts as a glue to bind the surfaces together (Fig. 12.1b). This explanation arises because we are very familiar with using adhesives to glue broken pots or to stick bricks together with mortar. However, it is easy to show that smooth clean surfaces in atomic contact stick best and that contamination with glue actually reduces the adhesion force at the atomic level. Geckos do not produce glue at their feet, yet they can hang from the ceiling and their smooth setai stick best to silicon wafers which are atomically perfect. The conclusion is that clean smooth surfaces produce ideal adhesion and that contaminants like adhesives weaken the atomic bonds between molecules, as demonstrated in 1971.[7] While adhesives are certainly excreted by many organisms, it is evident that such adhesives fill gaps and spread the contact area. In other words the material is more like a sealant which extends contact considerably while reducing the adhesion force for each molecule.

Related to this is the adhesion molecule argument which suggests that cells will stick to a surface if is coated with a special protein like fibronectin.[8] There is a grain of truth in this theory because there is no doubt that surface active molecules exist which are attracted to surfaces and which interfere with wetting and adhesion processes. Indeed, all surfaces on earth are covered with contaminating molecules such as oxygen, water or grease so the problem is actually finding clean surfaces to start with. Generally these contaminants reduce adhesion and perfect contact can only be made by squeezing these molecules out of the way. Experiments show that when contact is first made, adhesion can be weak, but after a time, the adhesion increases because the surfaces pull each other together and expel the contaminating molecules. For example the adhesion of rubber to glass doubled in 10,000 s of contact[9] as gas and water molecules were squeezed out. So adhesion molecules should really be viewed as special surfactants which play a part in the process of contact between cells and surfaces, first keeping them apart, then allowing them to touch. An associated problem is the diverse nomenclature describing adhesion molecules (e.g. antibodies, ligands, receptors, agonists etc.), words which are based on the false lock and key model. Our suggestion is that adhesion molecule is the phrase which describes these various materials and that it is confusing to ascribe lock and key functions to these polymers.

12.2 Adhesion Fundamentals

The answer to the question 'What causes adhesion between nanoparticles, viruses and cells?' is not given by accounts of 'lock and key', 'adhesives' or 'adhesion molecules'. Adhesion is a natural process of van der Waals attraction.[3] Any particle, virus or cell is naturally attracted to a neighbouring surface because all the atoms within it exert electromagnetic instantaneous dipole adhesion forces on surrounding atoms, acting over nanometre distances. This idea leads to the concept of 'work of adhesion' W which is the reversible energy, due to the van der Waals force, needed to separate 1 m^2 of contact. A particle approaching a surface will increasingly experience an attractive van der Waals force within a few nanometres or less, causing the particle to jump into contact, releasing energy W. Application of a tensile force to break the van der Waals bonds injects the work W back into the surfaces and the particle then jumps out of contact. Consequently, no other mechanism is needed to explain the adhesion phenomenon. Neither hooks, velcro, locks and keys, adhesives nor adhesion molecules are necessary to account for this electromagnetic attraction.[2]

Certain factors modify the force: For example, if the particle is not smooth, then the close approach required for van der Waals attraction may not be achievable. Remember 99% of adhesion is below 1 nm gap. Thus, surface projections and roughness inhibit adhesion in a predictable way.[5] Again, if the surface is covered with contaminant material, then the van der Waals forces are shielded and adhesion is further reduced. For example, rubber adheres ten times less when wet.[7] Therefore adhesives, surfactant molecules or adhesion molecules decrease the natural adhesion between surfaces by keeping them apart. In other words they reduce W the work of adhesion by reducing the van der Waals force of attraction.

Altering the shape of the body also alters the adhesion force. A smaller diameter cell must give proportionally lower adhesion force for the same van der Waals attraction. Sometimes the elastic modulus of the cell is important. For example, wedging a sheet of cells from a surface requires a larger force if the cells are stiffer, the force rising with square root of Young's modulus. Typically, equations for adhesion can be derived for adhesion force in different tests using the energy balance theory of brittle fracture[10]. The simplest is 90° peeling of a sheet of cells; a more complex equation applies to zero degree peeling as shown in Fig. 12.2.

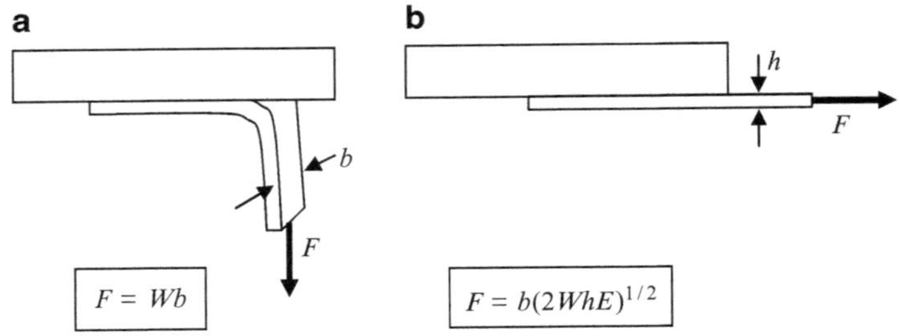

Fig. 12.2 (**a**) Ninety degree peeling of a sheet of cells from a plastic block. (**b**) Zero degree peeling of cell sheet from block

It follows that macroscopic adhesion can be determined using the work of adhesion W as the measure of van der Waals force while taking into account the geometrical and elastic properties of the test pieces.

12.3 Adhesion at the Molecular Level

The problem then arises that adhesion at the molecular level needs more than the single W parameter to describe it. Essentially, we need to know not only the energy but also the range of the adhesion, in other words how far the van der Waals forces extend from the surface. This demands at least two parameters and may require several more if the energy versus separation curve (i.e. the potential) is complicated in shape. The simplest approximation mathematically is the square well potential shown in Fig. 12.3a, but a more realistic potential is the Lennard-Jones curve shown in Fig. 12.3b.[11] Knowing the potential of a virus, for example, allows one to calculate when the virus will solidify into a crystal as the conditions such as concentration and temperature are altered, giving an explanation of the phase diagram which can be defined experimentally.[12]

The Lennard-Jones equation contains two terms, the negative r^{-6} attraction and the positive repulsion which rises more steeply with distance,

$$V(r) = 4\varepsilon\left[\left(\frac{\sigma}{r}\right)^{12} - \left(\frac{\sigma}{r}\right)^{6}\right]$$

where ε is adhesion energy and σ is bond length. For more complex potentials, other terms are needed to account for electrostatic charges or interactions with solvent molecules which alter the adhesion.

The most complex potentials are those in which surface molecules are squeezed out in steps to give oscillations of force as the gap is reduced (Fig. 12.4). These oscillations have been seen experimentally and modelled by molecular dynamics,

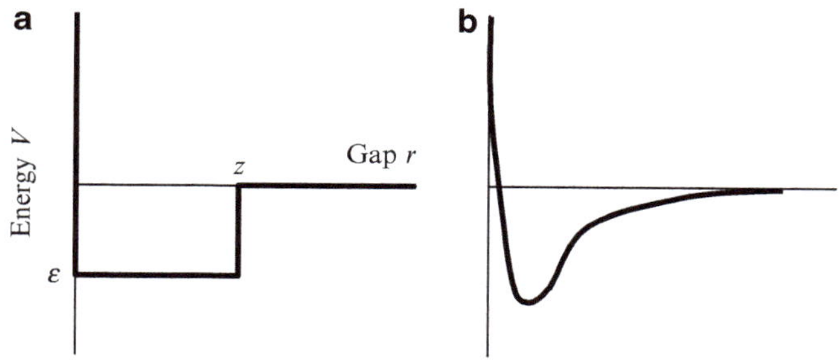

Fig. 12.3 (**a**) Square well potential between two particles. (**b**) Lennard-Jones potential

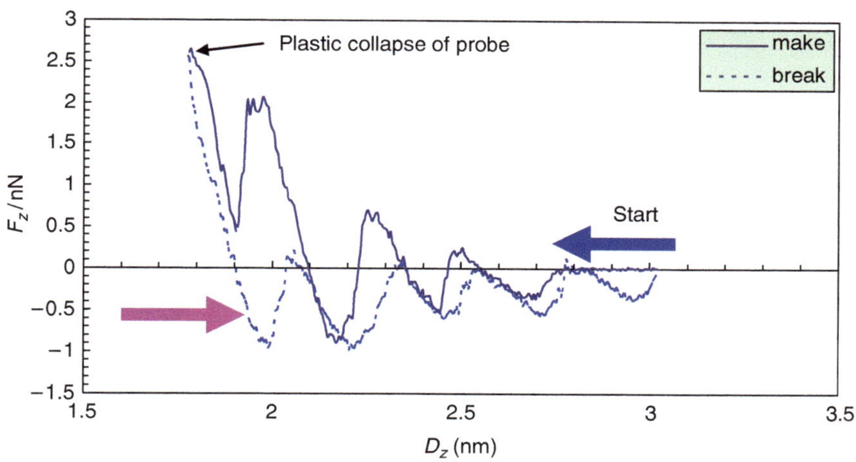

Fig. 12.4 Molecular dynamics model of making and breaking a contact between MgO surfaces in water. Four oscillations per nm gap are seen, each oscillation almost corresponding to the diameter of a water molecule

showing that the process of adhesion can proceed in steps as molecules are expelled from the narrow gap at the contact.[13]

However, when large swollen polymer molecules are trapped between the surfaces, as shown experimentally for polyacrylic acid in Fig. 12.5, there are longer range oscillations, depending on the diameter of the polymer blobs. In this example there was one oscillation in 10 nm, suggesting that the adhesion molecule was 10 nm in diameter in this case.[14]

This complexity of the adhesion process, in which contact can only be made by first stepwise squeezing out large adhesion molecules, gives a clue to the operation of adhesion molecules. There is no longer one jump to contact in this complex system, but a whole series of steps towards contact, each of which can be influenced by small alterations of adhesion molecule structure. A larger adhesion molecule can give a weak minimum of energy, allowing Brownian movement to break apart cells,

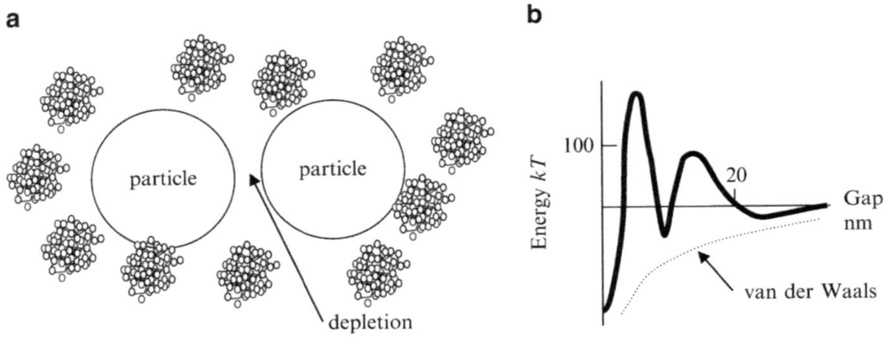

Fig. 12.5 (**a**) Representation of swollen polyacrylic acid molecules in water around two silica spheres; (**b**) Interaction energy between particles in the presence of polymer compared with van der Waals interaction

whereas a slightly smaller adhesion molecule can give a deeper adhesion minimum, much larger than kT, allowing permanent adhesion. Thus there are several quasi-stable adhesion levels as the particles approach stepwise. An adhesion molecule can act both as a dispersing agent and as a coagulant, depending on the gap.

This brings us to the other massive change at the molecular level; the adhesion bond is no longer static but subject to molecular collisions and statistics, consequently making and breaking in a dynamic equilibrium. Therefore, in a suspension of cells or viruses viewed at one moment, some will be adhering and some not. The adhesion energy is the same for all cells, but statistical mechanics tells us that the identical cells can attach for a time but later be detached by a Brownian collision. On average, this results in the population being partly dispersed and partly adhering, with the number of adhering cells giving a measure of the parameters in the potential equation. Contact and adhesion therefore are not predictable uniquely for each cell but can only be viewed as an average over the population.[15]

12.4 Subdivision of Contact Spots

Similarly, the contact spot is often split up into separated ever-changing multiple contact areas. Yeast cells, for example, are almost spherical and behave like rubber latex particles, obeying JKR theory. Whereas animal cells do not normally have a single contact point but in general make several micro-contacts which are in constant motion. Two consequences can follow from this: sharp protuberances can inhibit the growth of contact area and therefore prevent adhesion as with spiky pollen grains; alternatively, increasing the number of contact spots by creating fine hairs with smooth ends can increase adhesion as in the gecko foot. Many insects, spiders and lizards which need to crawl up walls and adhere to ceilings do not have large single contact spots but instead have feet covered with millions of smooth-ended hairs (setai). A general rule is that more hairs and smaller contact spots are needed for larger creatures, with the number of setai varying with mass to the power 0.7. A simple analysis based on JKR[7] contact mechanics suggests that adhesion should rise with $N^{1/2}$, but more detailed investigation shows that force can rise with N for independent contacts or with $N^{3/4}$ for interacting contacts. These ideas have been applied to the design of 'gecko tape' which can adhere to smooth surfaces by a similar mechanism, and to design of nano-patterned surfaces which can give interesting adhesion effects. The same concept may be extended to the projections from cells: pseudopodia and cilia both act to move material in the body as individual or groups of independent extensions. By subdividing these extensions, the ease of making and breaking of contact is achieved and intimate contact of foreign material with cell membranes is prevented.

A great achievement of the last decade has been to measure the nano-forces of adhesion for each hair and to observe the interesting contact versus release mechanics. Quick and controllable release is just as important to the gecko as strong adhesion, because the creature must scuttle rapidly up walls to escape predators.[16]

Such breakthroughs show that the importance of new measurement methods is crucial. The atomic force microscope (AFM), the interference contrast microscope, the fluorescence microscope have all had a major impact on adhesion measurements. This is illustrated in the brilliant micrograph of Fig. 12.6, showing multiple yellow contact spots of a single cell attaching to glass,[17] each contact spot regulated by the red actin fibre of the cell skeleton.

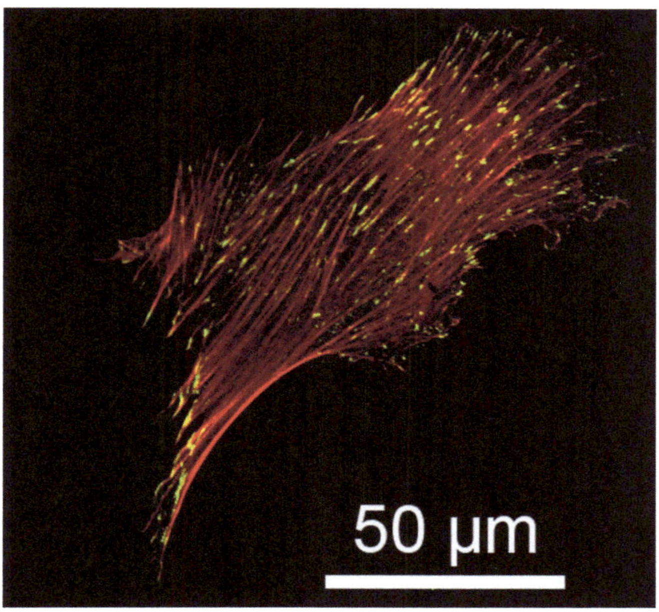

Fig. 12.6 Fluorescence microscope picture of a human mesenchymal stem cell adhering to glass, actin cytoskeleton in red, focal adhesion contacts (vinculin) are stained green and show up colocalized with actin in yellow

In the last few years, the optical tweezer and the laser tracking microscope have become much more prominent in measuring small contact spots, their forces and their fluctuations. Because it is possible with these techniques to identify energy changes around the Brownian energy kT, much smaller than those detectable by AFM, we can now begin to understand how small alterations in a protein molecule can influence the adhesion process.[18]

12.5 Connexion Between Nanoparticles, Viruses and Cells

All these advances in adhesion science; identification of van der Waals forces, understanding the energetic of making and breaking, seeing the fluctuations at the molecular level, observing the stepwise-jump-to-contact, noting the subdivision of the contact spots; have enabled us to view nanoparticles, viruses and cells in a new way. They are all elastic blobs in the size range between the nanometre scale where everything sticks and the macro-world above 10 μm where adhesion is unlikely.

Nanoparticles are simplest and provide clues to the adhesion of viruses and cells. They are often spheres with surfaces which can interact with the proteins and surfactants used by living particles. Therefore they are tools for understanding the pure interface phenomena of cell adhesion. By coating nanoparticles with known molecules as shown in Fig. 12.7a, their toxicity can be moderated and this leads to a theory that adhesion of nanoparticles is crucial to cell and organism health.[19]

Viruses are nanoparticles self-assembled from protein subunits which package their reproductive RNA or DNA (Fig. 12.7b). By observing such viruses as they aggregate, interact with nanoparticles, or disassemble to reveal their surface protein molecules, we can understand the complex adhesion mechanisms which allow viruses to attach to neighbouring bodies and thus invade ordinary cells[20], and detach to release progeny.

Cells are much more complex (Fig. 12.7c) and therefore are at the end of the chain of understanding. The wide variety of cells existing in nature leads us to think

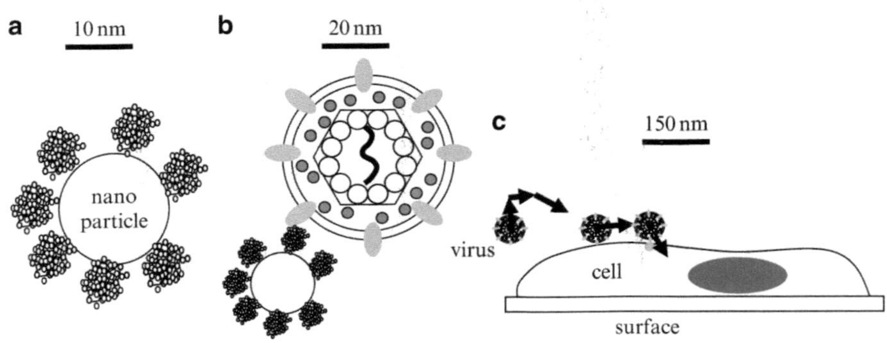

Fig. 12.7 (**a**) Nanoparticle with surface molecules in water. (**b**) Nanoparticle interacting with a virus. (**c**) Virus approaching cell adhering to surface

that there are many different mechanisms operating to make or break adhesive bonds. These fascinating possibilities will occupy us for some time yet.

12.6 Future Problems and Trends

The key questions posed at the start of this book are not fully resolved. Adhesion is clearly important in toxicity of particles, during infection of viruses, for sexual contact of cells, in signalling, embryonic development, movement of foreign material and cells using pseudopodia and cilia, growth of large organisms, inflammation, wound healing and metastasis of tumours. It is oft repeated that these functions can be explained by particular adhesion molecules and specific structures of proteins operating a lock and key mechanism, but observation tells us that the adhesion process is governed by van der Waals attractions, moderated by geometry, elasticity and Brownian motion. We must therefore avoid such explanations.[3]

The van der Waals forces are weakened by water and further weakened by large molecules like proteins which hinder the contact and have to be squeezed out. During this squeezing, the adhesion can take several values, gradually increasing towards the true equilibrium value. Thus there is not one adhesion value, but several adhesion levels depending on how far the squeezing has progressed. Understanding the details of this intricate process is the next stage of discovery, especially to understand how minor changes in a protein molecule can influence the contact mechanism.

One of the simplest questions is 'How do nanoparticles kill people?' Nanoparticles are the simplest structures studied in this book but can have a variety of compositions ranging from silica to carbon to polystyrene. It turns out that the particle chemistry is perhaps less important than the large insoluble, surface area which adsorbs lung surfactant, containing complex protein adhesion molecules that coat the alveolar cells. The coated nanoparticles can aggregate and remove this vital lung fluid, promoting inflammation and even death. Testing this toxicity mechanism is a task which has been accelerating recently.[19]

A second question is 'How does a virus target a cell before penetrating it?' Virus particles have been measured in suspension and were shown to have a slight tendency to aggregate depending on the adhesion potential, the viral concentration and the presence of surrounding polymer molecules. Introducing larger particles into the mixture changes the aggregation and shows that viruses can be attracted to the cells preferentially under certain circumstances. The idea that the viral aggregation onto the cells can be controlled is a research topic gaining ground.[20]

The third question 'How does a cell stick to and move around on a solid substrate?' is more complex. A marvellous experiment observing cells on equal molecular surfaces of differing Young's modulus demonstrated that the cells adhered differently. It seemed that the animal cells could sense the surface elasticity, extrude various polymers onto the substrate and incessantly probe the underlying material. New techniques are being applied to study this problem in many laboratories across the world.[17]

Of course, these major questions lead to more fundamental studies which can fill the gaps in the puzzle. A typical theoretical problem is modelling a protein molecule as it sits on a surface and moderates the adhesion process. Molecular dynamics has made great progress in visualizing the complex polymer structures and functions and further insights are anticipated.

The process by which a protein molecule can accelerate adhesion between two particles, either nanoparticles viruses or cells, has been explained by suggesting that the molecule has to be squeezed out of the gap in steps, allowing not just one adhesive state, but several. This means that the phase diagram of the system is much more complex than formerly thought. Also it suggests that there are significant energy barriers between the states which allow catalytic action to accelerate the kinetic process of contact. The idea that an adhesion molecule catalyses change of the energy barriers near a surface is a new one, with great significance.

Yet another fundamental detail is the elucidation of lipid membrane interactions which can be understood in terms of bilayer structures which can entrain protein molecules.[21] It is evident that interesting structures can form at the nanometre level, based on the tiny adhesion forces described in this book.

12.7 A Vision of the Future

The conclusion from this review is that the key questions of nanoparticle, virus and cell adhesion remain tantalisingly difficult. However, rapid progress is being made to understand the fundamental issues of adhesion, especially focusing on the nanometre processes which govern contact.

In the first place, the theoretical concepts have moved away from the trivial lock and key models and are now based on the correct van der Waals principles.

At the macroscopic level, the description of adhesion in terms of W (the work of adhesion) E the elastic modulus and h the geometrical parameter has been established.

Nanoscopically, the influence of Brownian movement has been emphasised to reveal that adhesion is a statistical process which leads to a dynamic equilibrium depending on the particle concentration, on the attractive potential and the thermal energy kT. Understanding the different phases of such systems is an exciting field.

Improved measurements of particle adhesion continue to be developed and can sense very small energies around the kT level, which will allow more precise detection of adhesion statistics in future.

Simultaneous advances in molecular dynamics modelling allow much improved visualisation of the detailed adhesion processes in which individual atoms in a protein can be seen to be important.

In conclusion, substantial progress is being made to advance the science and arts of adhesion and significant breakthroughs are expected in the future.

References

1. Harrison, R.G., The reaction of embryonic cells to solid structures, J Expt Zool 17 (1914) 521–44.
2. Kendall, K., Adhesion: Molecules and Mechanics, Science 263 (1994) 1720–25
3. Kendall, K., Molecular adhesion and its applications, Kluwer New York 2001.
4. Newton, I., Opticks, Smith and Walford, London 1704, reprinted Dover, New York, 1952, p.201.
5. Fuller, K.N.G., Tabor, D., The effect of surface roughness on the adhesion of elastic solids, Proc. R. Soc. Lond. A 345 (1975) 327–342.
6. Autumn, K., Properties, principles and parameters of the gecko adhesive system, in Biological Adhesives, (AM Smith and JA Callow eds) Springer, Berlin 2006, 225–256.
7. Johnson, K.L., Kendall, K., Roberts, A.D., Surface energy and the contact of elastic solids, Proc R Soc Lond A324 (1971) 301–313.
8. Umemori, H., The sticky synapse: Cell adhesion molecules, Springer Berlin 2009
9. Kendall, K., Rolling friction and adhesion between smooth solids, Wear 33 (1975) 351–8.
10. Kendall, K., The adhesion and surface energy of elastic solids, J Phys D: Appl Phys 4 (1971) 1186–1195.
11. Kendall, K., Liang, W., Stainton, C., New theory and observations of cell adhesion, Proc R Soc Lond 454 (1998) 2529–33.
12. Tsonchev, S., Niece, K.L., Schatz, G.C., Ratner, M.A., Stupp, S.I., Phase diagram for assembly of biologically-active peptide amphiphiles, J Phys Chem B112 (2008) 441–7.
13. Kendall, K., Du, S., Dhir, A., Strength by Atomic Force Microscopy: squeezing water from an MgO suface, Phil Mag 2010 in press.
14. Milling, A.J., and Kendall, K., Depletion, adsorption and structuring of sodium poly(acrylate) at the water-silica interface 1. An Atomic Force Microscope force study, Langmuir 16 (2000) 5106–5115
15. Kendall, K., Liang, W., Stainton, C., Adhesion of Fine Particles in Dispersions, J Adhesion, 67 (1998) 97–109.
16. Autumn, K., Liang, Y.A., Hsieh, S.T., Zesch, W., Chan, W.P., Kenny, T.W., Fearing, R., Full, R.J., Adhesive force of a single gecko foot hair, Nature 405 (2000) 681–5.
17. Zemel, A., Rehfeldt, F., Brown, A.E.X., Discher, D.E., Safran, S. A., Matrix rigidity optimizes the polarization of stem cells, Phys Rev (2010) submitted
18. Kendall, K., Dhir, A., Du, S., A new measure of molecular attractions between nanoparticles near kT adhesion energy, Nanotechnology 20 (2009) 275701.
19. Kendall, M., Toxicity related to nanoparticles and proteins 2010 submitted
20. Kendall, K., and Du, S., Virus adhesion to particles (2010) in preparation.
21. Davies, D.M., Mechanisms and functions for the duration of intercellular contacts made by lymphocytes, Nature Reviews Immunology 9 (2009) 543–55.

Index

K. Kendall et al., *Adhesion of Cells, Viruses and Nanoparticles*,
DOI 10.1007/978-90-481-2585-2, © Springer Science+Business Media B.V. 2011

P

R

S

T

MIX
Papier aus verantwortungsvollen Quellen
Paper from responsible sources
FSC® C105338

If you have any concerns about our products,
you can contact us on
ProductSafety@springernature.com

In case Publisher is established outside the EU,
the EU authorized representative is:
Springer Nature Customer Service Center GmbH
Europaplatz 3, 69115 Heidelberg, Germany

Printed by Libri Plureos GmbH
in Hamburg, Germany